MW01484028

AROUND

ON A BICYCLE

BY

THOMAS STEVENS

FROM SAN FRANCISCO TO TEHERAN

WITH OVER ONE HUNDRED ILLUSTRATIONS

British Library Cataloguing-in-Publication Data
A catalogue record for this book is available from the
British Library

A History of the Bicycle

As means of human transportation go, bicycles (vehicles that have two wheels and require balancing by the rider), have a relatively short history. They date back to the early nineteenth-century, with the first model, the German 'draisine', dating back to 1817. Despite their short history, bicycles have become immensely popular, and there are now over one billion worldwide – twice as many as automobiles.

The first bicycles are more properly termed 'velocipedes'; a word first coined by Frenchman Nicéphore Niépce in 1818, to describe his version of the Laufmaschine, which was invented by the German Karl Drais in 1817. Otherwise known as the 'dandy horse' and the 'draisienne', this contraption was the was the first human means of transport to use only two wheels in tandem. Its rider sat astride a wooden frame supported by two in-line wheels and pushed the vehicle along with his/her feet while steering the front wheel (hence the name 'laufmaschine' - directly translating as 'running machine'). Apart from use on a well-maintained pathway in a park or garden, this early velocipede had extremely limited use however, and riders soon found out that their boots wore out surprisingly rapidly!

Nevertheless, Drais's velocipede provided the basis for further developments. In fact, it was a draisine which inspired a French metalworker around 1863 to add rotary cranks and pedals to the front-wheel hub; to create the first pedal-operated 'bicycle' as we today understand the word.

Though technically not part of two-wheel bicycle-history, the intervening decades of the 1820s –1850s witnessed many developments similar to the draisine, even if the idea of a workable two-wheel design, requiring the rider to balance, had been dismissed. These new machines had three wheels (tricycles) or four (quadracycles), and came in a very wide variety of designs, using pedals, treadles and hand-cranks – but these designs often suffered from high weight and high rolling resistance.

The first mechanically-propelled, two-wheeled vehicle may have been built by Kirkpatrick MacMillan, a Scottish blacksmith, in 1839 – although this claim is often disputed. He is also associated with the first recorded instance of a cycling traffic offense, when a Glasgow newspaper in 1842 reported an accident in which an anonymous 'gentleman from Dumfries-shire... bestride a velocipede... of ingenious design' knocked over a little girl in Glasgow and was fined five shillings. In the early 1860s, Frenchmen Pierre Michaux and Pierre Lallement took bicycle design in a new direction by adding a mechanical crank drive with pedals on an enlarged front wheel. In 1869, bicycle wheels with wire spokes were patented by Eugène Meyer of Paris. The French vélocipède, made of iron and wood, developed into the 'penny-farthing' (historically known as an 'ordinary bicycle', a retronym, since there was then no other kind). It featured a tubular steel frame on which were mounted wire-spoked wheels with solid rubber tyres.

These bicycles were difficult to ride due to their high seat and poor weight distribution however – and extremely dangerous. They were fast, but unsafe. The rider was high up in the air and travelling at a great speed. If they hit a bad spot in the road they could easily be thrown over the front wheel and be seriously injured (two broken wrists were common, in attempts to break a fall) or even killed. 'Taking a header' (also known as 'coming a cropper'), was not at all uncommon. The rider's legs were frequently caught underneath the handlebars, so falling free of the machine was often not possible. The dangerous nature of these bicycles (as well as Victorian mores) made cycling the preserve of adventurous young men. The risk averse, such as elderly gentlemen, preferred the more stable tricycles (most popular in England) or quadracycles.

The 'dwarf ordinary' addressed some of these faults by reducing the front wheel diameter and setting the seat further back. This, in turn, required gearing – effected in a variety of ways – to efficiently use pedal power. Having to both pedal and steer via the front wheel remained a dilemma. J. K. Starley, J. H. Lawson, and Shergold solved this problem by introducing the chain drive (originated by the unsuccessful 'bicyclette' of Englishman Henry Lawson), connecting the frame-mounted cranks to the rear wheel. These models were known as 'safety bicycles', 'dwarf safeties', or 'upright bicycles' for their lower seat height and better weight distribution, although without pneumatic tyres, the ride of the smaller-wheeled bicycle would be much rougher than that of the larger-wheeled variety. Starley's 1885 Rover, manufactured in Coventry is usually described as the first recognizably modern bicycle. Soon the 'seat tube' was added, creating the modern bike's double-triangle diamond frame.

Further innovations increased comfort and ushered in a second bicycle craze; the 1890s 'Golden Age of Bicycles'. Since women could not cycle in the then-current fashions for voluminous and restrictive dress, the bicycle craze fed into a movement for so-called rational dress, which helped liberate women from corsets and ankle-length skirts and other encumbering garments, substituting the then-shocking bloomers. In 1888, Scotsman John Boyd Dunlop introduced the first practical pneumatic tire, which soon became universal. Soon after, the rear freewheel was developed, enabling the rider to coast. This refinement led to the 1890s invention of coaster brakes. Dérailleur gears and hand-operated Bowden cable-pull brakes were also developed during these years, but were only slowly adopted by casual riders. By the turn of the century, cycling clubs flourished on both sides of the Atlantic, and touring and racing became widely popular.

The bicycle's invention has had an enormous effect on society, both in terms of culture and of advancing modern industrial methods. Several components that eventually played a key role in the development of the automobile were initially invented for use in the bicycle, including ball bearings, pneumatic tires, chain-driven sprockets, and tension-spoked wheels. Today's bicycle is used as a mode of transport, entertainment and as a sporting endeavour that has provided countless people with a practical, as well as fun means of travel. Although it has a relatively short history, its development has occurred rapidly, and continues to see technical improvements in the present day. We hope the current reader enjoys this book on the subject.

PREFACE.

SHAKESPEARE says, in *All's Well that Ends Well*, that "a good traveller is something at the latter end of a dinner;" and I never was more struck with the truth of this than when I heard Mr. Thomas Stevens, after the dinner given in his honor by the Massachusetts Bicycle Club, make a brief, off-hand report of his adventures. He seemed like Jules Verne, telling his own wonderful performances, or like a contemporary Sinbad the Sailor. We found that modern mechanical invention, instead of disenchanting the universe, had really afforded the means of exploring its marvels the more surely. Instead of going round the world with a rifle, for the purpose of killing something,—or with a bundle of tracts, in order to convert somebody,—this bold youth simply went round the globe to see the people who were on it; and since he always had something to show them as interesting as anything that they could show him, he made his way among all nations.

What he had to show them was not merely a man perched on a lofty wheel, as if riding on a soap-bubble; but he was also a perpetual object-lesson in what Holmes calls "genuine, solid old Teutonic pluck." When the soldier rides into danger he has comrades by his side, his country's cause to defend, his uniform to vindicate, and the bugle to cheer him on; but this solitary rider had neither military station, nor an oath of allegiance, nor comrades, nor bugle; and he went among men of unknown languages, alien habits and hostile faith with only his own tact and courage to help him through. They proved sufficient, for he returned alive.

I have only read specimen chapters of this book, but find in them the same simple and manly quality which attracted us all when Mr. Stevens told his story in person. It is pleasant to know that while peace reigns in America, a young man can always find an opportunity to take his life in his hand and originate some exploit as good as those of the much-wandering Ulysses. In the German story "Titan," Jean Paul describes a manly youth who "longed for an adventure for his idle bravery;" and it is pleasant to read the narrative of one who has quietly gone to work, in an honest way, to satisfy this longing.

THOMAS WENTWORTH HIGGINSON.

CAMBRIDGE, MASS., April 10, 1887.

CONTENTS.

X CONTENTS.

LIST OF ILLUSTRATIONS.

a

FROM SAN FRANCISCO TO TEHERAN.

CHAPTER I.

OVER THE SIERRAS NEVADAS.

THE beauties of nature are scattered with a more lavish hand across the country lying between the summit of the Sierra Nevada Mountains and the shores where the surf romps and rolls over the auriferous sands of the Pacific, in Golden Gate Park, than in a journey of the same length in any other part of the world.

Such, at least, is the verdict of many whose fortune it has been to traverse that favored stretch of country. Nothing but the limited power of man's eyes prevents him from standing on the top of the mountains and surveying, at a glance, the whole glorious panorama that stretches away for more than two hundred miles to the west, terminating in the gleaming waters of the Pacific Ocean. Could he do this, he would behold, for the first seventy-five or eighty miles, a vast, billowy sea of foot-hills, clothed with forests of sombre pine and bright, evergreen oaks ; and, lower down, dense patches of white-blossomed chaparral, looking in the enchanted distance like irregular banks of snow. Then the world-renowned valley of the Sacramento River, with its level plains of dark, rich soil, its matchless fields of ripening grain, traversed here and there by streams that, emerging from the shadowy depths of the foot-hills, wind their way, like gleaming threads of silver, across the fertile plain and join the Sacramento, which receives them, one and all, in her matronly bosom and hurries with them on to the sea.

Towns and villages, with white church-spires, irregularly sprinkled over hill and vale, as though sown like seeds from the giant

1

hand of a mighty husbandman, would be seen nestling snugly
amid groves of waving shade and semi-tropical fruit trees. Beyond
all this the lower coast-range, where, toward San Francisco, Mount
Diablo and Mount Tamalpais—grim sentinels of the Golden Gate
—rear their shaggy heads skyward, and seem to look down with
a patronizing air upon the less pretentious hills that border the
coast and reflect their shadows in the blue water of San Francisco Bay. Upon the sloping sides of these hills sweet, nutritious
grasses grow, upon which peacefully graze the cows that supply
San Francisco with milk and butter.

Various attempts have been made from time to time, by ambitious cyclers, to wheel across America from ocean to ocean ; but
—"Around the World !"

" The impracticable scheme of a visionary," was the most charitable verdict one could reasonably have expected.

The first essential element of success, however, is to have sufficient confidence in one's self to brave the criticisms—to say nothing of the witticisms—of a sceptical public. So eight o'clock on
the morning of April 22, 1884, finds me and my fifty-inch machine
on the deck of the Alameda, one of the splendid ferry-boats plying
between San Francisco and Oakland, and a ride of four miles
over the sparkling waters of the bay lands us, twenty-eight minutes later, on the Oakland pier, that juts far enough out to allow
the big ferries to enter the slip in deep water. On the beauties
of San Francisco Bay it is, perhaps, needless to dwell, as everybody has heard or read of this magnificent sheet of water, its surface flecked with snowy sails, and surrounded by a beautiful
framework of evergreen hills ; its only outlet to the ocean the famous Golden Gate—a narrow channel through which come and
go the ships of all nations.

With the hearty well-wishing of a small group of Oakland and
'Frisco cyclers who have come, out of curiosity, to see the start, I
mount and ride away to the east, down San Pablo Avenue, toward
the village of the same Spanish name, some sixteen miles distant.
The first seven miles are a sort of half-macadamized road, and I
bowl briskly along.

The past winter has been the rainiest since 1857, and the continuous pelting rains had not beaten down upon the last half of
this imperfect macadam in vain ; for it has left it a surface of
wave-like undulations, from out of which the frequent bowlder

protrudes its unwelcome head, as if ambitiously striving to soar above its lowly surroundings. But this one don't mind, and I am perfectly willing to put up with the bowlders for the sake of the undulations. The sensation of riding a small boat over "the gently-heaving waves of the murmuring sea" is, I think, one of the pleasures of life; and the next thing to it is riding a bicycle over

'The Start.

the last three miles of the San Pablo Avenue macadam as I found it on that April morning.

The wave-like macadam abruptly terminates, and I find myself on a common dirt road. It is a fair road, however, and I have plenty of time to look about and admire whatever bits of scenery happen to come in view. There are few spots in the "Golden State" from which views of more or less beauty are not to be obtained; and ere I am a baker's dozen of miles from Oakland pier I

find myself within an ace of taking an undesirable header into a
ditch of water by the road-side, while looking upon a scene that
for the moment completely wins me from my immediate surroundings. There is nothing particularly grand or imposing in the outlook here ; but the late rains have clothed the whole smiling face
of nature with a bright, refreshing green, that fails not to awaken
a thrill of pleasure in the breast of one fresh from the verdureless
streets of a large sea-port city. Broad fields of pale-green, thrifty-
looking young wheat, and darker-hued meads, stretch away on
either side of the road ; and away beyond to the left, through an
opening in the hills, can be seen, as through a window, the placid
waters of the bay, over whose glittering, sunlit surface white-
winged, aristocratic yachts and the plebeian smacks of Greek and
Italian fishermen swiftly glide, and fairly vie with each other in
giving the finishing touches to a picture.

So far, the road continues level and fairly good ; and, notwithstanding the seductive pleasures of the ride over the bounding billows of the gently heaving —— macadam, the dalliance with the
scenery, and the all too frequent dismounts in deference to the
objections of phantom-eyed roadsters, I pulled up at San Pablo
at ten o'clock, having covered the sixteen miles in one hour and
thirty-two minutes ; though, of course, there is nothing speedy
about this—to which desirable qualification, indeed, I lay no
claim.

Soon after leaving San Pablo the country gets somewhat
" choppy," and the road a succession of short-hills, at the bottom
of which modest-looking mud-holes patiently await an opportunity
to make one's acquaintance, or scraggy-looking, latitudinous wash-
outs are awaiting their chance to commit a murder, or to make the
unwary cycler who should venture to "coast," think he had
wheeled over the tail of an earthquake. One never minds a hilly
road where one can reach the bottom with an impetus that sends
him spinning half-way up the next ; but where mud-holes or wash-
outs resolutely "hold the fort" in every depression, it is different,
and the progress of the cycler is necessarily slow.

I have set upon reaching Suisun, a point fifty miles along the
Central Pacific Railway, to-night ; but the roads after leaving San
Pablo are anything but good, and the day is warm, so six P.M.
finds me trudging along an unridable piece of road through the
low tuile swamps that border Suisun Bay. "Tuile" is the name

given to a species of tall rank grass, or rather rush, that grows to
the height of eight or ten feet, and so thick in places that it is diffi-

The Burning Tuiles.

cult to pass through, in the low, swampy grounds in this part of
California. These tuile swamps are traversed by a net-work of

small, sluggish streams and sloughs, that fairly swarm with wild
ducks and geese, and justly entitle them to their local title of " the
duck-hunters' paradise." Ere I am through this swamp, the shades
of night gather ominously around and settle down like a pall over
the half-flooded flats ; the road is full of mud-holes and pools of
water, through which it is difficult to navigate, and I am in something of a quandary. I am sweeping along at the irresistible velocity of a mile an hour, and wondering how far it is to the other
end of the swampy road, when thrice welcome succor appears from
a strange and altogether unexpected source. I had noticed a small
fire, twinkling through the darkness away off in the swamp ; and
now the wind rises and the flames of the small fire spread to the
thick patches of dead tuile. In a short time the whole country, including my road, is lit up by the fierce glare of the blaze ; so that
I am enabled to proceed with little trouble. These tuiles often catch
on fire in the fall and early winter, when everything is comparatively
dry, and fairly rival the prairie fires of the Western plains in the
fierceness of the flames.

The next morning I start off in a drizzling rain, and, after going
sixteen miles, I have to remain for the day at Elmira. Here,
among other items of interest, I learn that twenty miles farther
ahead the Sacramento River is flooding the country, and the only
way I can hope to get through is to take to the Central Pacific track
and cross over the six miles of open trestle-work that spans the
Sacramento River and its broad bottom-lands, that are subject to
the annual spring overflow. From Elmira my way leads through
a fruit and farming country that is called second to none in the
world. Magnificent farms line the road ; at short intervals appear
large well-kept vineyards, in which gangs of Chinese coolies are
hoeing and pulling weeds, and otherwise keeping trim. A profusion of peach, pear, and almond orchards enlivens the landscape
with a wealth of pink and white blossoms, and fills the balmy
spring air with a subtle, sensuous perfume that savors of a tropical
clime.

Already I realize that there is going to be as much " foot-riding "
as anything for the first part of my journey ; so, while halting for
dinner at the village of Davisville, I deliver my rather slight shoes
over to the tender mercies of an Irish cobbler of the old school,
with *carte blanche* instructions to fit them out for hard service.
While diligently hammering away at the shoes, the old cobbler

grows communicative, and in almost unintelligible brogue tells a complicated tale of Irish life, out of which I can make neither head, tail, nor tale ; though nodding and assenting to it all, to the great satisfaction of the loquacious manipulator of the last, who in an hour hands over the shoes with the proud assertion, "They'll last yez, be jabbers, to Omaha."

Reaching the overflowed country, I have to take to the trestle-work and begin the tedious process of trundling along that aggra-vating roadway, where, to the music of rushing waters, I have to step from tie to tie, and bump, bump, bump, my machine along for six weary miles. The Sacramento River is the outlet for the tremendous volumes of water caused every spring by the melting snows on the Sierra Nevada Mountains, and these long stretches of open trestle have been found necessary to allow the water to pass beneath. Nothing but trains are expected to cross this trestle-work, and of course no provision is made for pedestrians. The en-gineer of an approaching train sets his locomotive to tooting for all she is worth as he sees a " strayed or stolen " cycler, slowly bumping along ahead of his train. But he has no need to slow up, for occasional cross-beams stick out far enough to admit of stand-ing out of reach, and when he comes up alongside, he and the fire-man look out of the window of the cab and see me squatting on the end of one of these handy beams, and letting the bicycle hang over.

That night I stay in Sacramento, the beautiful capital of the Golden State, whose well-shaded streets and blooming, almost tropical gardens combine to form a city of quiet, dignified beauty, of which Californians feel justly proud. Three and a half miles east of Sacramento, the high trestle bridge spanning the main stream of the American River has to be crossed, and from this bridge is obtained a remarkably fine view of the snow-capped Sierras, the great barrier that separates the fertile valleys and glori-ous climate of California, from the bleak and barren sage-brush plains, rugged mountains, and forbidding wastes of sand and alkali, that, from the summit of the Sierras, stretch away to the eastward for over a thousand miles. The view from the American River bridge is grand and imposing, encompassing the whole foot-hill country, which rolls in broken, irregular billows of forest-crowned hill and charming vale, upward and onward to the east, gradually getting more rugged, rocky, and immense, the hills changing to

mountains, the vales to cañons, until they terminate in bald, hoary
peaks whose white rugged pinnacles seem to penetrate the sky, and
stand out in ghostly, shadowy outline against the azure depths of
space beyond.

After crossing the American River the character of the country
changes, and I enjoy a ten-mile ride over a fair road, through one
of those splendid sheep-ranches that are only found in California,
and which have long challenged the admiration of the world.
Sixty thousand acres, I am informed, is the extent of this pasture,
all within one fence. The soft, velvety greensward is half-shaded
by the wide-spreading branches of evergreen oaks that singly and
in small groups are scattered at irregular intervals from one end of
the pasture to the other, giving it the appearance of one of the old
ancestral parks of England. As I bowl pleasantly along I involuntarily look about me, half expecting to see some grand, stately
old mansion peeping from among some one of the splendid oak-
groves ; and when a jack-rabbit hops out and halts at twenty paces
from my road, I half hesitate to fire at him, lest the noise of the
report should bring out the vigilant and lynx-eyed game-keeper,
and get me "summoned" for poaching. I remember the pleasant
ten-mile ride through this park-like pasture as one of the brightest
spots of the whole journey across America. But "every rose conceals a thorn," and pleasant paths often lead astray ; when I emerge
from the pasture I find myself several miles off the right road and
have to make my unhappy way across lots, through numberless
gates and small ranches, to the road again.

There seems to be quite a sprinkling of Spanish or Mexican
rancheros through here, and after partaking of the welcome noontide hospitality of one of the ranches, I find myself, before I realize
it, illustrating the bicycle and its uses, to a group of sombrero-decked
rancheros and darked-eyed señoritas, by riding the machine round
and round on their own ranch-lawn. It is a novel position, to say
the least ; and often afterward, wending my solitary way across
some dreary Nevada desert, with no company but my own uncanny shadow, sharply outlined on the white alkali by the glaring
rays of the sun, my untrammelled thoughts would wander back to
this scene, and I would grow "hot and cold by turns," in my
uncertainty as to whether the bewitching smiles of the señoritas
were smiles of admiration, or whether they were simply "grinning" at the figure I cut. While not conscious of having cut a

sorrier figure than usual on that occasion, somehow I cannot rid myself of an unhappy, harrowing suspicion, that the latter comes nearer the truth than the former.

The ground is gradually getting more broken ; huge rocks intrude themselves upon the landscape. At the town of Rocklin we are supposed to enter the foot-hill country proper. Much of the road in these lower foot-hills is excellent, being of a hard, stony character, and proof against the winter rains.

Everybody who writes anything about the Golden State is expected to say something complimentary—or otherwise, as his experience may seem to dictate—about the " glorious climate of California ; " or else render an account of himself for the slight, should he ever return, which he is very liable to do. For, no matter what he may say about it, the " glorious climate " generally manages to make one, ever after, somewhat dissatisfied with the extremes of heat and cold met with in less genial regions.

This fact of having to pay my measure of tribute to the climate forces itself on my notice prominently here at Rocklin, because, indirectly, the "climate" was instrumental in bringing about a slight accident, which, in turn, brought about the—to me—serious calamity of sending me to bed without any supper. Rocklin is celebrated—and by certain bad people, ridiculed—all over this part of the foot-hills for the superabundance of its juvenile population. If one makes any inquisitive remarks about this fact, the Rocklinite addressed will either blush or grin, according to his temperament, and say, "It's the glorious climate." A bicycle is a decided novelty up here, and, of course, the multitudinous youth turn out in droves to see it. The bewildering swarms of these small mountaineers distract my attention and cause me to take a header that temporarily disables the machine. The result is, that, in order to reach the village where I wish to stay over night, I have to "foot it" over four miles of the best road I have found since leaving San Pablo, and lose my supper into the bargain, by procrastinating at the village smithy, so as to have my machine in trim, ready for an early start next morning. If the " glorious climate of California " is responsible for the exceedingly hopeful prospects of Rocklin's future census reports, and the said lively outlook, materialized, is responsible for my mishap, then plainly the said " G. C. of C." is the responsible element in the case. I hope this compliment to the climate will strike the Californians as about the correct thing ; but, if it should happen to work the other way, I beg of them at once to pour out the vials of their wrath on the heads of the 'Frisco Bicycle Club, in order that their fury may be spent ere I again set foot on their auriferous soil.

"What'll you do when you hit the snow?" is now a frequent question asked by the people hereabouts, who seem to be more conversant with affairs pertaining to the mountains than they are of what is going on in the valleys below. This remark, of course, has reference to the deep snow that, toward the summits of the mountains, covers the ground to the depth of ten feet on the level, and from that to almost any depth where it has drifted and accumulated. I have not started out on this greatest of all bicycle tours without looking into these difficulties, and I remind them that the long snow-sheds of the Central Pacific Railway make it possible for one to cross over, no matter how deep the snow may lie on the ground outside. Some speak cheerfully of the prospects for getting over, but many shake their heads ominously and say, "You'll never be able to make it through."

Rougher and more hilly become the roads as we gradually penetrate farther and farther into the foot-hills. We are now in far-famed Placer County, and the evidences of the hardy gold diggers' work in pioneer days are all about us. In every gulch and ravine are to be seen broken and decaying sluice-boxes. Bare, whitish-looking patches of washed-out gravel show where a "claim" has been worked over and abandoned. In every direction are old water-ditches, heaps of gravel, and abandoned shafts—all telling, in language more eloquent than word or pen, of the palmy days of '49, and succeeding years; when, in these deep gulches, and on these yellow hills, thousands of bronzed, red-shirted miners dug and delved, and "rocked the cradle" for the precious yellow dust and nuggets. But all is now changed, and where were hundreds before, now only a few "old timers" roam the foot-hills, prospecting, and working over the old claims; but "dust," "nuggets," and "pockets" still form the burden of conversation in the village bar-room or the cross-roads saloon. Now and then a "strike" is made by some lucky—or perhaps it turns out, unlucky—prospector. This for a few days kindles anew the slumbering spark of "gold fever" that lingers in the veins of the people here, ever ready to kindle into a flame at every bit of exciting news, in the way of a lucky "find" near home, or new gold-fields in some distant land.

These occasions never fail to have their legitimate effect upon the business of the bar where the "old-timers" congregate to learn the news ; and, between drinks, yarns of the good old days of '49 and '50, of "streaks of luck," of "big nuggets," and "wild times," are spun over and over again. Although the palmy days of the "diggin's" are no more, yet the finder of a "pocket" these days seems not a whit wiser than in the days when "pockets" more frequently rewarded the patient prospector than they do now ; and at Newcastle—a station near the old-time mining camps of Ophir and Gold Hill—I hear of a man who lately struck a "pocket," out of which he dug forty thousand dollars ; and forthwith proceeded to imitate his reckless predecessors by going down to 'Frisco and entering upon a career of protracted sprees and debauchery that cut short his earthly career in less than six months, and wafted his riotous spirit to where there are no more forty thousand dollar pockets, and no more 'Friscos in which to squander it.

In this instance the "find" was clearly an unlucky one. Not quite so bad was the case of two others who, but a few days before my arrival, took out twelve hundred dollars ; they simply, in the language of the gold fields "turned themselves loose," "made things hum," and "whooped 'em up" around the bar-room of their village for exactly three days ; when, "dead broke," they took to the gulches again, to search for more. "Yer oughter hev happened through here with that instrumint of yourn about that time, young fellow ; yer might hev kept as full as a tick till they war busted," remarked a slouchy-looking old fellow whose purple-tinted nose plainly indicated that he had devoted a good part of his existence to the business of getting himself "full as a tick" every time he ran across the chance.

Quite a different picture is presented by an industrious old Mexican, whom I happen to see away down in the bottom of a deep ravine, along which swiftly hurries a tiny stream. He is diligently shovelling dirt into a rude sluice-box which he has constructed in the bed of the stream at a point where the water rushes swiftly down a declivity. Setting my bicycle up against a rock, I clamber down the steep bank to investigate. In tones that savor of anything but satisfaction with the result of his labor, he informs me that he has to work "most infernal hard" to pan out two dollars' worth of "dust" a day. "I have had to work over all that pile of gravel you see yonder to clean up seventeen dollars' worth of dust," further volunteered the old "greaser," as I picked up a spare shovel and helped him remove a couple of bowlders that he was trying to roll out of his way. I condole with him at the low grade of the gravel he is working, hope he may " strike it rich " one of these days, and take my departure.

Up here I find it preferable to keep the railway track, alongside of which there are occasionally ridable side-paths; while on the wagon roads little or no riding can be done on account of the hills, and the sticky nature of the red, clayey soil. From the railway track near Newcastle is obtained a magnificent view of the lower country, traversed during the last three days, with the Sacramento River winding its way through its broad valley to the sea. Deep cuts and high embankments follow each other in succession, as the road-bed is now broken through a hill, now carried across a deep gulch, and anon winds around the next hill and over another ravine. Before reaching Auburn I pass through "Bloomer Cut," where perpendicular walls of bowlders loom up on both sides of the track looking as if the slightest touch or jar would unloose them and send them bounding and crashing on the top of the passing train as it glides along, or drop down on the stray cycler who might venture through. On the way past Auburn, and on up to Clipper Gap, the dry, yellow dirt under the overhanging rocks, and in the crevices, is so suggestive of " dust," that I take a small prospecting glass, which I have in my tool-bag, and do a little prospecting ; without, however, finding sufficient "color" to induce me to abandon my journey and go to digging.

Before reaching Clipper Gap it begins to rain ; while I am taking dinner at that place it quits raining and begins to come down by buckets full, so that I have to lie over for the remainder of the day. The hills around Clipper Gap are gay and white with chaparral blossom, which gives the whole landscape a pleasant, gala-day appearance. It rains all the evening, and at night turns to heavy, damp snow, which clings to the trees and bushes. In the morning the landscape, which a few hours before was white with chaparral bloom, is now even more white with the bloom of the snow.

My hostelry at Clipper Gap is a kind of half ranch, half road-side inn, down in a small valley near the railway ; and mine host, a jovial Irish blade of the good old " Donnybrook Fair " variety, who came here in 1851, during the great rush to the gold fields, and, failing to make his fortune in the " diggings," wisely decided to send for his family and settle down quietly on a piece of land, in preference to returning to the "ould sod." He turns out to be a "bit av a sphort meself," and, after showing me a number of minor pets and favorites, such as game chickens, Brahma geese, and a litter of young bull pups, he proudly leads the way to the barn to show me "Barney," his greatest pet of all, whom he at present keeps securely tied up for safe-keeping. More than one evil-minded person has a hankering after Barney's gore since his last battle for the championship of Placer County, he explains, in which he inflicted severe punishment on his adversary and resolutely refused to give in ; although his opponent on this important occasion was an imported dog, brought into the county by Barney's enemies, who hoped to fill their pockets by betting against the local champion. But Barney, who is a medium-sized, ferocious-looking bull terrier, "scooped" the crowd backing the imported dog, to the extent of their "pile," by "walking all round" his adversary ; and thereby stirring up the enmity of said crowd against himself, who—so says Barney's master—have never yet been able to scare up a dog able to "down" Barney. As we stand in the barn-door Barney eyes me suspiciously, and then looks at his master ; but luckily for me his master fails to give the word. Noticing that the dog is scarred and seamed all over, I inquire the reason, and am told that he has been fighting wild boars in the chaparral, of which gentle pastime he is extremely fond. "Yes, and he'll tackle a cougar too, of which there are plenty of them around here, if that cowardly animal would only keep out of the trees," admiringly continues mine host, as he orders Barney into his empty salt-barrel again.

To day is Sunday, and it rains and snows with little interruption, so that I am compelled to stay over till Monday morning. While it is raining at Clipper Gap, it is snowing higher up in the mountains, and a railway employee volunteers the cheering information that, during the winter, the snow has drifted and accumulated in the sheds, so that a train can barely squeeze through, leaving no room for a person to stand to one side. I have my own ideas of whether this state of affairs is probable or not, however, and determine to pay no heed to any of these rumors, but to push ahead. So I pull out on Monday morning and take to the railway track again, which is the only passable road since the tremendous downpour of the last two days.

The first thing I come across is a tunnel burrowing through a hill. This tunnel was originally built the proper size, but, after

Crossing the Sierra Nevadas.

being walled up, there were indications of a general cave-in ; so the company had to go to work and build another thick rock-wall inside the other, which leaves barely room for the trains to pass through without touching the sides. It is anything but an inviting path around the hill; but it is far the safer of the two. Once my foot slips, and I unceremoniously sit down and slide around in the soft yellow clay, in my frantic endeavors to keep from slipping down the hill. This hardly enhances my personal appearance; but it doesn't matter much, as I am where no one can see, and a clay-besmeared individual is worth a dozen dead ones. Soon I am on the track again, briskly trudging up the steep grade toward the snow-line, which I can plainly see, at no great distance ahead, through the windings around the mountains.

All through here the only riding to be done is along occasional short stretches of difficult path beside the track, where it happens to be a hard surface; and on the plank platforms of the stations, where I generally take a turn or two to satisfy the consuming curiosity of the miners, who can't imagine how anybody can ride a thing that won't stand alone; at the same time arguing among themselves as to whether I ride along on one of the rails, or bump along over the protruding ties.

This morning I follow the railway track around the famous "Cape Horn," a place that never fails to photograph itself permanently upon the memory of all who once see it. For scenery that is magnificently grand and picturesque, the view from where the railroad track curves around Cape Horn is probably without a peer on the American continent.

When the Central Pacific Railway company started to grade their road-bed around here, men were first swung over this precipice from above with ropes, until they made standing room for themselves; and then a narrow ledge was cut on the almost perpendicular side of the rocky mountain, around which the railway now winds.

Standing on this ledge, the rocks tower skyward on one side of the track so close as almost to touch the passing train; and on the other is a sheer precipice of two thousand five hundred feet, where one can stand on the edge and see, far below, the north fork of the American River, which looks like a thread of silver laid along the narrow valley, and sends up a far-away, scarcely perceptible roar, as it rushes and rumbles along over its rocky bed. The railroad track is carefully looked after at this point, and I was able, by turning round and taking the down grade, to experience the novelty of a short ride, the memory of which will be ever welcome should one live to be as old as "the oldest inhabitant." The scenery for the next few miles is glorious; the grand and imposing mountains are partially covered with stately pines down to their bases, around which winds the turbulent American River, receiving on its boisterous march down the mountains tribute from hundreds of smaller streams and rivulets, which come splashing and dashing out of the dark cañons and crevasses of the mighty hills.

The weather is capricious, and by the time I reach Dutch Flat, ten miles east of Cape Horn, the floodgates of heaven are thrown open again, and less than an hour succeeds in impressing Dutch Flat upon my memory as a place where there is literally "water, water, everywhere, but not a drop to— ;" no, I cannot finish the quotation! What is the use of lying? There is plenty to drink at Dutch Flat; plenty of everything.

But there is no joke about the water ; it is pouring in torrents from above ; the streets are shallow streams ; and from scores of ditches and gullies comes the merry music of swiftly rushing waters, while, to crown all, scores of monster streams are rushing with a hissing sound from the mouths of huge pipes or nozzles, and playing against the surrounding hills ; for Dutch Flat and neighboring camps are the great centre of hydraulic mining operations in California at the present day. Streams of water, higher up the mountains, are taken from their channels and conducted hither through miles of wooden flumes and iron piping ; and from the mouths of huge nozzles are thrown with tremendous force against the hills, literally mowing them down.

The rain stops as abruptly as it began. The sun shines out clear and warm, and I push ahead once more.

Gradually I have been getting up into the snow, and ever and anon a muffled roar comes booming and echoing over the mountains like the sound of distant artillery. It is the sullen noise of monster snow-slides among the deep, dark cañons of the mountains, though a wicked person at Gold Run winked at another man and tried to make me believe it was the grizzlies "going about the mountains like roaring lions, seeking whom they might devour." The giant voices of nature, the imposing scenery, the gloomy pine forests which have now taken the place of the gay chaparral, combine to impress one who, all alone, looks and listens with a realizing sense of his own littleness.

What a change has come over the whole face of nature in a few days' travel! But four days ago I was in the semi-tropical Sacramento Valley; now gaunt winter reigns supreme, and the only vegetation is the hardy pine.

This afternoon I pass a small camp of Digger Indians, to whom my bicycle is as much a mystery as was the first locomotive; yet they scarcely turn their uncovered heads to look; and my cheery greeting of "How," scarce elicits a grunt and a stare in reply. Long years of chronic hunger and wretchedness have well-nigh eradicated what little energy these Diggers ever possessed. The discovery of gold among their native mountains has been their bane; the only antidote the rude grave beneath the pine and the happy hunting-grounds beyond.

The next morning finds me briskly trundling through the great, gloomy snow-sheds that extend with but few breaks for the next forty miles. When I emerge from them on the other end I shall be over the summit and well down the eastern slope of the mountains. These huge sheds have been built at great expense to protect the track from the vast quantities of snow that fall every winter on these mountains. They wind around the mountain-sides, their roofs built so slanting that the mighty avalanche of rock and snow that comes thundering down from above glides harmlessly over, and down the chasm on the other side, while the train glides along unharmed beneath them. The section-houses, the water-tanks, stations, and everything along here are all under the gloomy but friendly shelter of the great protecting sheds.

Fortunately I find the difficulties of getting through much less than I had been led by rumors to anticipate; and although no riding can be done in the sheds, I make very good progress, and trudge merrily along, thankful of a chance to get over the mountains without having to wait a month or six weeks for the snow outside to disappear. At intervals short breaks occur in the sheds, where the track runs over deep gulch or ravine, and at one of these openings the sinuous structure can be traced for quite a long distance, winding its tortuous way around the rugged mountain sides, and through the gloomy pine forest, all but buried under the snow. It requires no great effort of the mind to imagine it to be some wonderful relic of a past civilization, when a venturesome race of men thus dared to invade these vast wintry solitudes and burrow their way through the deep snow, like moles burrowing through the

loose earth. Not a living thing is in sight, and the only sounds
the occasional roar of a distant snow-slide, and the mournful sighing of the breeze as it plays a weird, melancholy dirge through the
gently swaying branches of the tall, sombre pines, whose stately
trunks are half buried in the omnipresent snow.

In the Central Pacific Snow-sheds.

To-night I stay at the Summit Hotel, seven thousand and seventeen feet above the level of the sea. The "Summit" is nothing if
not snowy, and I am told that thirty feet on the level is no unusual
thing up here. Indeed, it looks as if snow-balling on the " Glorious Fourth " were no great luxury at the Summit House ; yet not-

withstanding the decidedly wintry aspect of the Sierras, the low temperature of the Rockies farther east is unknown ; and although there is snow to the right, snow to the left, snow all around, and ice under foot, I travel all through the gloomy sheds in my shirt-sleeves, with but a gossamer rubber coat thrown over my shoulders to keep off the snow-water which is constantly melting and dripping through the roof, making it almost like going through a shower of rain. Often, when it is warm and balmy outside, it is cold and frosty under the sheds, and the dripping water, falling among the rocks and timbers, freezes into all manner of fantastic shapes. Whole menageries of ice animals, birds and all imaginable objects, are here reproduced in clear crystal ice, while in many places the ground is covered with an irregular coating of the same, that often has to be chipped away from the rails.

East of the summit is a succession of short tunnels, the space between being covered with snow-shed ; and when I came through, the openings and crevices through which the smoke from the engines is wont to make its escape, and through which a few rays of light penetrate the gloomy interior, are blocked up with snow, so that it is both dark and smoky ; and groping one's way with a bicycle over the rough surface is anything but pleasant going. But there is nothing so bad, it seems, but that it can get a great deal worse ; and before getting far, I hear an approaching train and forthwith proceed to occupy as small an amount of space as possible against the side, while three laboriously puffing engines, tugging a long, heavy freight train up the steep grade, go past. These three puffing, smoke-emitting monsters fill every nook and corner of the tunnel with dense smoke, which creates a darkness by the side of which the natural darkness of the tunnel is daylight in comparison. Here is a darkness that can be felt ; I have to grope my way forward, inch by inch ; afraid to set my foot down until I have felt the place, for fear of blundering into a culvert ; at the same time never knowing whether there is room, just where I am, to get out of the way of a train. A cyclometer wouldn't have to exert itself much through here to keep tally of the revolutions ; for, besides advancing with extreme caution, I pause every few steps to listen ; as in the oppressive darkness and equally oppressive silence the senses are so keenly on the alert that the gentle rattle of the bicycle over the uneven surface seems to make a noise that would prevent me hearing an approaching train.

This finally comes to an end ; and at the opening in the sheds I climb up into a pine-tree to obtain a view of Donner Lake, called the "Gem of the Sierras." It is a lovely little lake, and amid the pines, and on its shores occurred one of the most pathetically tragic events of the old emigrant days. Briefly related : A small party of emigrants became snowed in while camped at the lake, and when, toward spring, a rescuing party reached the spot, the last survivor of the party, crazed with the fearful suffering he had undergone, was sitting on a log, savagely gnawing away at a human arm, the last remnant of his companions in misery, off whose emaciated carcasses he had for some time been living !

My road now follows the course of the Truckee River down the eastern slope of the Sierras, and across the boundary line into Nevada. The Truckee is a rapid, rollicking stream from one end to the other, and affords dam-sites and mill-sites without limit.

There is little ridable road down the Truckee cañon ; but before reaching Verdi, a station a few miles over the Nevada line, I find good road, and ride up and dismount at the door of the little hotel as coolly as if I had rode without a dismount all the way from 'Frisco. Here at Verdi is a camp of Washoe Indians, who at once showed their superiority to the Diggers by clustering around and examining the bicycle with great curiosity. Verdi is less than forty miles from the summit of the Sierras, and from the porch of the hotel I can see the snow-storm still fiercely raging up in the place where I stood a few hours ago ; yet one can feel that he is already in a dryer and altogether different climate. The great masses of clouds, travelling inward from the coast with their burdens of moisture, like messengers of peace with presents to a far country, being unable to surmount the great mountain barrier that towers skyward across their path, unload their precious cargoes on the mountains ; and the parched plains of Nevada open their thirsty mouths in vain. At Verdi I bid good-by to the Golden State and follow the course of the sparkling Truckee toward the Forty-mile Desert.

CHAPTER II.

OVER THE DESERTS OF NEVADA.

GRADUALLY I leave the pine-clad slopes of the Sierras behind, and every revolution of my wheel reveals scenes that constantly remind me that I am in the great "Sage-brush State." How appropriate indeed is the name! Sage-brush is the first thing seen on entering Nevada, almost the only vegetation seen while passing through it, and the last thing seen on leaving it. Clear down to the edge of the rippling waters of the Truckee, on the otherwise barren plain, covering the elevated table-lands, up the hills, even to the mountain-tops—everywhere, everywhere, nothing but sage-brush. In plain view to the right, as I roll on toward Reno, are the mountains on which the world-renowned Comstock lode is situated, and Reno was formerly the point from which this celebrated mining-camp was reached.

Before reaching Reno I meet a lone Washoe Indian; he is riding a diminutive, scraggy-looking mustang. One of his legs is muffled up in a red blanket, and in one hand he carries a rudely-invented crutch. "How will you trade horses?" I banteringly ask as we meet in the road; and I dismount for an interview, to find out what kind of Indians these Washoes are. To my friendly chaff he vouchsafes no reply, but simply sits motionless on his pony, and fixes a regular "Injun stare" on the bicycle. "What's the matter with your leg?" I persist, pointing at the blanket-be-muffled member.

"Heap sick foot" is the reply, given with the characteristic brevity of the savage; and, now that the ice of his aboriginal reserve is broken, he manages to find words enough to ask me for tobacco. I have no tobacco, but the ride through the crisp morning air has been productive of a surplus amount of animal spirits, and I feel like doing something funny; so I volunteer to cure his "sick foot" by sundry dark and mysterious manœuvres, that I unblushingly intimate are "heap good medicine." With owlish solemnity my small monkey-wrench is taken from the tool-bag and waved around the "sick foot" a few times, and the operation is completed by squirting a few drops from my oil-can through a hole in the blanket. Before going I give him to understand that, in order to have the "good medicine" operate to his advantage, he will have to soak his copper-colored hide in a bath every morning for a week, flattering myself that, while my mystic manœuvres will do him no harm, the latter prescription will certainly do him good if he acts on it, which, however, is extremely doubtful.

Rolling into Reno at 10.30 A.M. the characteristic whiskey-straight hospitality of the Far West at once asserts itself, and one individual with sporting proclivities invites me to stop over a day or two and assist him to "paint Reno red" at his expense. Leaving Reno, my route leads through the famous Truckee meadows—a strip of very good agricultural land, where plenty of money used to be made by raising produce for the Virginia City market.

"But there's nothing in it any more, since the Comstock's played out," glumly remarks a ranchman, at whose place I get dinner. "I'll take less for my ranch now than I was offered ten years ago," he continues.

The "meadows" gradually contract, and soon after dinner I find myself again following the Truckee down a narrow space between mountains, whose volcanic-looking rocks are destitute of all vegetation save stunted sage-brush. All down here the road is ridable in patches; but many dismounts have to be made, and the walking to be done aggregates at least one-third of the whole distance travelled during the day. Sneakish coyotes prowl about these mountains, from whence they pay neighborly visits to the chicken-roosts of the ranchers in the Truckee meadows near by. Toward night a pair of these animals are observed following behind at the respectful distance of five hundred yards. One need not be apprehensive of danger from these contemptible animals, however; they are simply following behind in a frame of mind similar to that of a hungry school-boy's when gazing longingly into a confectioner's window. Still, night is gathering around, and it begins to look as though I will have to pillow my head on the soft side of a bowlder, and take lodgings on the footsteps of a bald mountain to-night; and it will scarcely invite sleep to know that two pairs of sharp, wolfish eyes are peering wistfully through the darkness at one's prostrate form, and two red tongues are licking about in hungry anticipation of one's blood. Moreover, these animals have an unpleasant habit of congregating after night to pay their compliments to the pale moon, and to hold concerts that would put to shame a whole regiment of Kilkenny cats ; though there is but little comparison between the two, save that one howls and the other yowls, and either is equally effective in driving away the drowsy Goddess. I try to draw these two animals within range of my revolver by hiding behind rocks ; but they are too chary of their precious carcasses to take any risks, and the moment I disappear from their sight behind a rock they are on the alert, and looking " forty ways at the same time," to make sure that I am not creeping up on them from some other direction.   Fate, however, has decreed that I am not to sleep out to-night—not quite out. A lone shanty looms up through the gathering darkness, and I immediately turn my footsteps thitherwise.   I find it occupied.   I am all right now for the night.   Hold on, though ! not so fast !   " There is many a slip," etc.   The little shanty, with a few acres of rather rocky ground, on the bank of the Truckee, is presided over by a lonely bachelor of German extraction, who eyes me with evident suspicion, as, leaning on my bicycle in front of his rude cabin door I ask to be accommodated for the night.   Were it a man on horseback, or a man with a team, this hermit-like rancher could satisfy himself to some extent as to the character of his visitor, for he sees men on horseback or men in wagons, on an average, perhaps, once a week during the summer, and can see plenty of them any day by going to Reno. But me and the bicycle he cannot " size up " so readily.   He never saw the like of us before, and we are beyond his Teutonic frontier-like comprehension.   He gives us up ; he fails to solve the puzzle ; he knows not how to unravel the mystery ; and, with characteristic Teutonic bluntness, he advises us to push on through fifteen miles of rocks, sand, and darkness, to Wadsworth.   The prospect of worrying my way, hungry and weary, through fifteen miles of rough, unknown country, after dark, looms up as rather a formidable task.   So summoning my reserve stock of persuasive eloquence, backed up by sundry significant movements, such as setting the bicycle up against his cabin-wall, and sitting down on a block of wood under the window, I finally prevail upon him to accommodate me with a blanket on the floor of the shanty.   He has just finished supper, and the remnants of the frugal repast are still on the table ; but he says nothing about any supper for me : he scarcely feels satisfied with himself yet : he feels that I have, in

some mysterious manner, gained an unfair advantage over him, and
obtained a foothold in his shanty against his own wish—jumped
his claim, so to speak.   Not that I think the man really inhospitable
at heart ; but he has been so habitually alone, away from his fellow-
men so much, that the presence of a stranger in his cabin makes
him feel uneasy ; and when that stranger is accompanied by a
queer-looking piece of machinery that cannot stand alone, but
which he nevertheless says he rides on, our lonely rancher is perhaps not so much to be wondered at, after all, for his absent-mind-
edness in regard to my supper.   His mind is occupied with other
thoughts.   " You couldn't accommodate a fellow with a bite to eat,
could you ? " I timidly venture, after devouring what eatables are
in sight, over and over again, with my eyes.   " I have plenty of
money to pay for any accommodation I get," I think it policy to
add, by way of cornering him up and giving him as little chance to
refuse as possible, for I am decidedly hungry, and if money or
diplomacy, or both, will produce supper, I don't propose to go to
bed supperless.   I am not much surprised to see him bear out my
faith in his innate hospitality by apologizing for not thinking of
my supper before, and insisting, against my expressed wishes, on
lighting the fire and getting me a warm meal of fried ham and coffee, for which I beg leave to withdraw any unfavorable impressions
in regard to him which my previous remarks may possibly have
made on the reader's mind.

After supper he thaws out a little, and I wheedle out of him a
part of his history.   He settled on this spot of semi-cultivable
land during the flush times on the Comstock, and used to prosper
very well by raising vegetables, with the aid of Truckee-River
water, and hauling them to the mining-camps ; but the palmy days
of the Comstock have departed and with them our lonely rancher's
prosperity.   Mine host has barely blankets enough for his own
narrow bunk, and it is really an act of generosity on his part when
he takes a blanket off his bed and invites me to extract what comfort I can get out of it for the night.   Snowy mountains are round
about, and curled up on the floor of the shanty, like a kitten under
a stove in mid-winter, I shiver the long hours away, and endeavor
to feel thankful that it is no worse.

For a short distance, next morning, the road is ridable, but
nearing Wadsworth it gets sandy, and " sandy," in Nevada means
deep, loose sand, in which one sinks almost to his ankles at every

step, and where the possession of a bicycle fails to awaken that degree of enthusiasm that it does on a smooth, hard road.　At Wadsworth I have to bid farewell to the Truckee River, and start across the Forty-mile Desert, which lies between the Truckee and Humboldt Rivers.　Standing on a sand-hill and looking eastward across the dreary, desolate waste of sand, rocks, and alkali, it is with positive regret that I think of leaving the cool, sparkling stream that has been my almost constant companion for nearly a hundred miles.　It has always been at hand to quench my thirst or furnish a refreshing bath.　More than once have I beguiled the tedium of some uninteresting part of the journey by racing with some trifling object hurried along on its rippling surface.　I shall miss the murmuring music of its dancing waters as one would miss the conversation of a companion.

This Forty-mile Desert is the place that was so much dreaded by the emigrants *en route* to the gold-fields of California, there being not a blade of grass nor drop of water for the whole forty miles ; nothing but a dreary waste of sand and rocks that reflects the heat of the sun, and renders the desert a veritable furnace in midsummer ; and the stock of the emigrants, worn out by the long journey from the States, would succumb by the score in crossing.　Though much of the trail is totally unfit for cycling, there are occasional alkali flats that are smooth and hard enough to play croquet on ; and this afternoon, while riding with careless ease across one of these places, I am struck with the novelty of the situation.　I am in the midst of the dreariest, deadest-looking country imaginable.　Whirlwinds of sand, looking at a distance like huge columns of smoke, are wandering erratically over the plains in all directions.　The blazing sun casts, with startling vividness on the smooth white alkali, that awful scraggy, straggling shadow that, like a vengeful fate, always accompanies the cycler on a sunny day, and which is the bane of a sensitive wheelman's life !　The only representative of animated nature hereabouts is a species of small gray lizard that scuttles over the bare ground with astonishing rapidity.　Not even a bird is seen in the air.　All living things seem instinctively to avoid this dread spot save the lizard.　A desert forty miles wide is not a particularly large one ; but when one is in the middle of it, it might as well be as extensive as Sahara itself, for anything he can see to the contrary, and away off to the right I behold as perfect a mirage as one could wish to see.

The " Forty-mile Desert."

A person can scarce help believing his own eyes, and did one not have some knowledge of these strange and wondrous phenomena, one's orbs of vision would indeed open with astonishment; for seemingly but a few miles away is a beautiful lake, whose shores are fringed with wavy foliage, and whose cool waters seem to lave the burning desert sands at its edge.

A short distance to the right of Hot Springs Station broken clouds of steam are seen rising from the ground, as though huge caldrons of water were being heated there. Going to the spot I find, indeed, "caldrons of boiling water;" but the caldrons are in the depths. At irregular openings in the rocky ground the bubbling water wells to the surface, and the fires—ah! where are the fires? On another part of this desert are curious springs that look demure and innocuous enough most of the time, but occasionally they emit columns of spray and steam. It is related of these springs that once a party of emigrants passed by, and one of the men knelt down to take a drink of the clear, nice-looking water. At the instant he leaned over, the spring spurted a quantity of steam and spray all over him, scaring him nearly out of his wits. The man sprang up, and ran as if for his life, frantically beckoning the wagons to move on, at the same time shouting, at the top of his voice, "Drive on! drive on! hell's no great distance from here!"

From the Forty-mile Desert my road leads up the valley of the Humboldt River. On the shores of Humboldt Lake are camped a dozen Piute lodges, and I make a half-hour halt to pay them a visit. I shall never know whether I am a welcome visitor or not; they show no signs of pleasure or displeasure as I trundle the bicycle through the sage-brush toward them. Leaning it familiarly up against one of their *teepes*, I wander among them and pry into their domestic affairs like a health-officer in a New York tenement. I know I have no right to do this without saying, "By your leave," but item-hunters the world over do likewise, so I feel little squeamishness about it. Moreover, when I come back I find the Indians are playing "tit-for-tat" against me. Not only are they curiously examining the bicycle as a whole, but they have opened the tool-bag and are examining the tools, handing them around among themselves. I don't think these Piutes are smart or bold enough to steal nowadays; their intercourse with the whites along the railroad has, in a measure, relieved them of those aboriginal traits of character that would incite them to steal a brass button off their pale-faced brother's coat, or screw a nut off his bicycle ; but they have learned to beg ; the noble Piute of to-day is an incorrigible mendicant.   Gathering up my tools from among them, the monkey-wrench seems to have found favor in the eyes of a wrinkled-faced brave, who, it seems, is a chief.   He hands the wrench over with a smile that is meant to be captivating, and points at it as I am putting it back into the bag, and grunts, "Ugh! Piute likum! Piute likum!"   As I hold it up, and ask him if this is what he means, he again points and repeats, " Piute likum ; " and this time two others standing by point at *him* and also smile and say, " Him big chief ; big Piute chief, him ; " thinking, no doubt, this latter would be a clincher, and that I would at once recognize in "big Piute chief, him " a vastly superior being and hand him over the wrench.   In this, however, they are mistaken, for the wrench I cannot spare ; neither can I see any lingering trace of royalty about him, no kingliness of mien, or extra cleanliness ; nor is there anything winning about his smile—nor any of their smiles for that matter.   The Piute smile seems to me to be simply a cold, passionless expansion of the vast horizontal slit that reaches almost from one ear to the other, and separates the upper and lower sections of their expressionless faces.   Even the smiles of the squaws are of the same unlovely pattern, though they seem to be perfectly oblivious of any ugliness whatever, and whenever a pale-faced visitor appears near their *teepe* they straightway present him with one of those repulsive, unwinning smiles.

Sunday, May 4th, finds me anchored for the day at the village of Lovelocks, on the Humboldt River, where I spend quite a remarkable day.   Never before did such a strangely assorted crowd gather to see the first bicycle ride they ever saw, as the crowd that gathers behind the station at Lovelocks to-day to see me.   There are perhaps one hundred and fifty people, of whom a hundred are Piute and Shoshone Indians, and the remainder a mingled company of whites and Chinese railroaders ; and among them all it is difficult to say who are the most taken with the novelty of the exhibition—the red, the yellow, or the white.   Later in the evening I accept the invitation of a Piute brave to come out to their camp, behind the village, and witness rival teams of Shoshone and Piute squaws play a match-game of " Fi-re-fla," the national game of both the Shoshone and Piute tribes.   The principle of the game is similar to polo. The squaws are armed with long sticks, with which they endeavor to carry a shorter one to the goal. It is a picturesque and novel sight to see the squaws, dressed in costumes in which the garb of savagery and civilization is strangely mingled and the many colors of the rainbow are promiscuously blended, flitting about the field with the agility of a team of professional polo-players; while the bucks and old squaws, with their pappooses, sit around and watch the game with unmistakable enthusiasm. The Shoshone team wins and looks pleased.

Here, at Lovelocks, I fall in with one of those strange and seemingly incongruous characters that are occasionally met with in the West. He is conversing with a small gathering of Piutes in their own tongue, and I introduce myself by asking him the probable age of one of the Indians, whose wrinkled and leathery countenance would indicate unusual longevity. He tells me the Indian is probably ninety years old ; but the Indians themselves never know their age, as they count everything by the changes of the moon and the seasons, having no knowledge whatever of the calendar year. While talking on this subject, imagine my surprise to hear my informant—who looks as if the Scriptures are the last thing in the world for him to speak of—volunteer the information that our venerable and venerated ancestors, the antediluvians, used to count time in the same way as the Indians, and that instead of Methuselah being nine hundred and sixty-nine years of age, it ought to be revised so as to read " nine hundred and sixty-nine moons," which would bring that ancient and long-lived person—the oldest man that ever lived—down to the venerable but by no means extraordinary age of eighty years and nine months. This is the first time I have heard this theory, and my astonishment at hearing it from the lips of a rough-looking *habitué* of the Nevada plains, seated in the midst of a group of illiterate Indians, can easily be imagined.

On, up the Humboldt valley I continue, now riding over a smooth, alkali flat, and again slavishly trundling through deep sand, a dozen snowy mountain peaks round about, the Humboldt sluggishly winding its way through the alkali plain ; on past Rye Patch, to the right of which are more hot springs, and farther on mines of pure sulphur—all these things, especially the latter, unpleasantly suggestive of a certain place where the climate is popularly supposed to be uncomfortably warm ; on, past Humboldt Station, near which place I wantonly shoot a poor harmless badger, who peers inquisitively out of his hole as I ride past. There is something peculiarly pathetic about the actions of a dying badger, and no sooner has the thoughtless shot sped on its mission of death than I am sorry for doing it.

Going out of Mill City next morning I lose the way, and find myself up near a small mining camp among the mountains south of the railroad. Thinking to regain the road quickly by going across country through the sage-brush, I get into a place where that enterprising shrub is so thick and high that I have to hold the bicycle up overhead to get through.

At three o'clock in the afternoon I come to a railroad section-house. At the Chinese bunk-house I find a lone Celestial who, for some reason, is staying at home. Having had nothing to eat or drink since six o'clock this morning, I present the Chinaman with a smile that is intended to win his heathen heart over to any gastronomic scheme I may propose; but smiles are thrown away on John Chinaman.

"John, can you fix me up something to eat?"

"No; Chinaman no savvy whi' man eatee; bossee ow on thlack. Chinaman eatee nothing bu' licee [rice]; no licee cookee."

This sounds pretty conclusive; nevertheless I don't intend to be thus put off so easily. There is nothing particularly beautiful about a silver half-dollar, but in the almond-shaped eyes of the Chinaman scenes of paradisiacal loveliness are nothing compared to the dull surface of a twenty-year-old fifty-cent piece; and the jingle of the silver coins contains more melody for Chin Chin's unromantic ear than a whole musical festival.

"John, I'll give you a couple of two-bit pieces if you'll get me a bite of something," I persist. John's small, black eyes twinkle at the suggestion of two-bit pieces, and his expressive countenance assumes a commerical air as, with a ludicrous change of front, he replies:

"Wha'! You gib me flore bittee, me gib you bitee eatee?"

"That's what I said, John; and please be as lively as possible about it."

"All li; you gib me flore bittee me fly you Melican plan-cae."

"Yes, pancakes will do. Go ahead!"

Visions of pancakes and molasses flit before my hunger-distorted vision as I sit outside until he gets them ready. In ten minutes John calls me in. On a tin plate, that looks as if it has just been rescued from a barrel of soap-grease, reposes a shapeless mass of substance resembling putty—it is the "Melican plan-cae ;" and the Celestial triumphantly sets an empty box in front of it for me to sit on and extends his greasy palm for the stipulated price. May the reader never be ravenously hungry and have to choose between a "Melican plan-cae" and nothing ! It is simply a chunk of tenacious dough, made of flour and water only, and soaked for a few minutes in warm grease. I call for molasses ; he doesn't know what it is. I inquire for syrup, thinking he may recognize my want by that name. He brings a jar of thin Chinese catsup, that tastes something like Limburger cheese smells. I immediately beg of him to take it where its presumably benign influence will fail to reach me. He produces some excellent cold tea, however, by the aid of which I manage to "bolt" a portion of the "plan-cae." One doesn't look for a very elegant spread for fifty cents in the Sage-brush State ; but this "Melican plan-cae" is the worst fifty-cent meal I ever heard of.

To-night I stay in Winnemucca, the county seat of Humboldt County, and quite a lively little town of 1,200 inhabitants. "What'll yer have ?" is the first word on entering the hotel, and "Won't yer take a bottle of whiskey along ?" is the last word on leaving it next morning. There are Piutes and Piutes camped at Winnemucca, and in the morning I meet a young brave on horseback a short distance out of town and let him try his hand with the bicycle. I wheel him along a few yards and let him dismount ; and then I show him how to mount and invite him to try it himself. He gallantly makes the attempt, but springs forward with too much energy, and over he topples, with the bicycle cavorting around on top of him. This satisfies his aboriginal curiosity, and he smiles and shakes his head when I offer to swap the bicycle for his mustang. The road is heavy with sand all along by Winnemucca, and but little riding is to be done. The river runs through green meadows of rich bottom-land hereabouts ; but the meadows soon disappear as I travel eastward. Twenty miles east of Winnemucca the river and railroad pass through the cañon in a low range of mountains, while my route lies over the summit. It is a steep trundle up the mountains, but from the summit a broad view of the surrounding country is obtained. The Humboldt River is not a beautiful stream, and for the greater part of its length it meanders through alternate stretches of dreary sage-brush plain and low sand-hills, at long

The Piute's Header.

intervals passing through a cañon in some barren mountain chain. But "distance lends enchantment to the view," and from the summit of the mountain pass even the Humboldt looks beautiful. The sun shines on its waters, giving it a sheen, and for many a mile its glistening surface can be seen winding its serpentine course through the broad, gray-looking sage and grease-wood plains, while at occasional intervals narrow patches of green, in striking contrast to the surrounding gray, show where the hardy mountain grasses venturously endeavor to invade the domains of the autocratic sagebrush. What is that queer-looking little reptile, half lizard, half frog, that scuttles about among the rocks? It is different from anything I have yet seen. Around the back of its neck and along its sides, and, in a less prominent degree, all over its yellowish-gray body, are small, horn-like protuberances that give the little fellow a very peculiar appearance. Ah! I know who he is. I have heard of him, and have seen his picture in books. I am happy to make his acquaintance. He is "Prickey," the famed horned toad of Nevada. On this mountain spur, between the Golconda mining-camp and Iron Point, is the only place I have seen him on the tour. He is a very interesting little creature, more lizard than frog, perfectly harmless; and his little bead-like eyes are bright and fascinating as the eyes of a rattlesnake.

Alkali flats abound, and some splendid riding is to be obtained east of Iron Point. Just before darkness closes down over the surrounding area of plain and mountain I reach Stone-House section-house.

"Yes, I guess we can get you a bite of something; but it will be cold," is the answer vouchsafed in reply to my query about supper.

Being more concerned these days about the quantity of provisions I can command than the quality, the prospect of a cold supper arouses no ungrateful emotions. I would rather have a four-pound loaf and a shoulder of mutton for supper now than a smaller quantity of extra choice viands; and I manage to satisfy the cravings of my inner man before leaving the table. But what about a place to sleep? For some inexplicable reason these people refuse to grant me even the shelter of their roof for the night. They are not keeping hotel, they say, which is quite true; they have a right to refuse, even if it *is* twenty miles to the next place; and they *do* refuse.

"There's the empty Chinese bunk-house over there. You can

3

crawl in there, if you arn't afeerd of ghosts," is the parting remark, as the door closes and leaves me standing, like an outcast, on the dark, barren plain.

A week ago this bunk-house was occupied by a gang of Chinese railroaders, who got to quarrelling among themselves, and the quarrel wound up in quite a tragic poisoning affair, that resulted in the death of two, and nearly killed a third. The Chinese are nothing, if not superstitious, and since this affair no Chinaman would sleep in the bunk-house or work on this section; consequently the building remains empty. The "spooks" of murdered Chinese are everything but agreeable company; nevertheless they are preferable to inhospitable whites, and I walk over to the house and stretch my weary frame in—for aught I know—the same bunk in which, but a few days ago, reposed the ghastly corpses of the poisoned Celestials. Despite the unsavory memories clinging around the place, and my pillowless and blanketless couch, I am soon in the land of dreams. It is scarcely presumable that one would be blessed with rosy-hued visions of pleasure under such conditions, however, and near midnight I awake in a cold shiver. The snowy mountains rear their white heads up in the silent night, grim and ghostly all around, and make the midnight air chilly, even in midsummer. I lie there, trying in vain to doze off again, for it grows perceptibly cooler. At two o'clock I can stand it no longer, and so get up and strike out for Battle Mountain, twenty miles ahead.

The moon has risen; it is two-thirds full, and a more beautiful sight than the one that now greets my exit from the bunk-house it is scarcely possible to conceive. Only those who have been in this inter-mountain country can have any idea of a glorious moonlight night in the clear atmosphere of this dry, elevated region. It is almost as light as day, and one can see to ride quite well wherever the road is ridable. The pale moon seems to fill the whole broad valley with a flood of soft, silvery light; the peaks of many snowy mountains loom up white and spectral; the stilly air is broken by the excited yelping of a pack of coyotes noisily baying the pale-yellow author of all this loveliness, and the wild, unearthly scream of an unknown bird or animal coming from some mysterious, undefinable quarter completes an ideal Western picture, a poem, a dream, that fully compensates for the discomforts of the preceding hour. The inspiration of this beautiful scene awakes the slumbering poesy within, and I am inspired to compose a poem—"Moonlight in the Rockies"—that I expect some day to see the world go into raptures over !

A few miles from the Chinese shanty I pass a party of Indians

Ugh ! What is it ?

camped by the side of my road. They are squatting around the smouldering embers of a sage-brush fire, sleeping and dozing. I am riding slowly and carefully along the road that happens to be rida-ble just here, and am fairly past them before being seen. As I gradually vanish in the moonlit air I wonder what they think it was—that strange-looking object that so silently and mysteriously glided past. It is safe to warrant they think me anything but flesh and blood, as they rouse each other and peer at my shadowy form disappearing in the dim distance.

From Battle Mountain my route leads across a low alkali bottom, through which dozens of small streams are flowing to the Humboldt. Many of them are narrow enough to be jumped, but not with a bicycle on one's shoulder, for under such conditions there is always a disagreeable uncertainty that one may disastrously alight before he gets ready. But I am getting tired of partially undressing to ford streams that are little more than ditches, every little way, and so I hit upon the novel plan of using the machine for a vaulting-pole. Reaching it out into the centre of the stream, I place one hand on the head and the other on the saddle, and vault over, retaining my hold as I alight on the opposite shore. Pulling the bicycle out after me, the thing is done. There is no telling to what uses this two-wheeled "creature" could be put in case of necessity. Certainly the inventor never expected it to be used for a vaulting-pole in leaping across streams. Twenty-five miles east of Battle Mountain the valley of the Humboldt widens into a plain of some size, through which the river meanders with many a horseshoe curve, and maps out the pot-hooks and hangers of our childhood days in mazy profusion. Amid these innumerable curves and counter-curves, clumps of willows and tall blue-joint reeds grow thickly, and afford shelter to thousands of pelicans, that here make their homes far from the disturbing presence of man. All unconscious of impending difficulties, I follow the wagon trail leading through this valley until I find myself standing on the edge of the river, ruefully looking around for some avenue by which I can proceed on my way. I am in the bend of a horseshoe curve, and the only way to get out is to retrace my footsteps for several miles, which disagreeable performance I naturally feel somewhat opposed to doing. Casting about me I discover a couple of old fence-posts that have floated down from the Be-o-wa-we settlement above and lodged against the bank. I determine to try and utilize them in getting the machine across the river, which is not over thirty yards wide at this point. Swimming across with my clothes first, I tie the bicycle to the fence-posts, which barely keep it from sinking, and manage to navigate it successfully across. The village of Be-o-wa-we is full of cowboys, who are preparing for the annual

spring round-up. Whites, Indians, and Mexicans compose the
motley crowd. They look a wild lot, with their bear-skin *chaparejos*
and semi-civilized trappings, galloping to and fro in and about the
village. "I can't spare the time, or I would," is my slightly un-
truthful answer to an invitation to stop over for the day and have
some fun. Briefly told, this latter, with the cowboy, consists in
getting hilariously drunk, and then turning his "pop" loose at
anything that happens to strike his whiskey-bedevilled fancy as presenting a fitting target. Now a bicycle, above all things, would
intrude itself upon the notice of a cowboy on a "tear" as a peculiar
and conspicuous object, especially if it had a man on it ; so after
taking a "smile" with them for good-fellowship, and showing them
the *modus operandi* of riding the wheel, I consider it wise to push
on up the valley.

Three miles from Be-o-wa-we is seen the celebrated "Maiden's
Grave," on a low hill or bluff by the road-side ; and "thereby hangs
a tale." In early days, a party of emigrants were camped near by
at Gravelly Ford, waiting for the waters to subside, so that they
could cross the river, when a young woman of the party sickened
and died. A rudely carved head-board was set up to mark the spot
where she was buried. Years afterward, when the railroad was
being built through here, the men discovered this rude head-board
all alone on the bleak hill-top, and were moved by worthy sentiment
to build a rough stone wall around it to keep off the ghoulish coyotes ; and, later on, the superintendent of the division erected a
large white cross, which now stands in plain view of the railroad.
On one side of the cross is written the simple inscription, "Maid-
en's Grave ;" on the other, her name, "Lucinda Duncan." Leaving the bicycle by the road-side, I climb the steep bluff and examine
the spot with some curiosity. There are now twelve other graves
beside the original "Maiden's Grave," for the people of Be-o-wa-we
and the surrounding country have selected this romantic spot on
which to inter the remains of their departed friends. This afternoon I follow the river through Humboldt Cañon in preference to
taking a long circuitous route over the mountains. The first noticeable things about this cañon are the peculiar water-marks plainly
visible on the walls, high up above where the water could possibly
rise while its present channels of escape exist unobstructed. It is
thought that the country east of the spur of the Red Range, which
stretches clear across the valley at Be-o-wa-we, and through which

the Humboldt seems to have cut its way, was formerly a lake, and
that the water gradually wore a passage-way for itself through the
massive barrier, leaving only the high-water marks on the mountain sides to tell of the mighty change.   In this cañon the rocky
walls tower like gigantic battlements, grim and gloomy on either
side, and the seething, boiling waters of the Humboldt—that for
once awakens from its characteristic lethargy, and madly plunges
and splutters over a bed of jagged rocks which seem to have been
tossed into its channel by some Herculean hand—fill this mighty
"rift" in the mountains with a never-ending roar.   It has been
threatening rain for the last two hours, and now the first peal of
thunder I have heard on the whole journey awakens the echoing
voices of the cañon and rolls and rumbles along the great jagged
fissure like an angry monster muttering his mighty wrath.   Peal
after peal follow each other in quick succession, the vigorous, new-
born echoes of one peal seeming angrily to chase the receding
voices of its predecessor from cliff to cliff, and from recess to projection, along its rocky, erratic course up the cañon.   Vivid flashes
of forked lightning shoot athwart the heavy black cloud that seems
to rest on either wall, roofing the cañon with a ceiling of awful
grandeur.   Sheets of electric flame light up the dark, shadowy re-
cesses of the towering rocks as they play along the ridges and hover
on the mountain-tops ; while large drops of rain begin to patter
down, gradually increasing with the growing fury of their battling
allies above, until a heavy, drenching downpour of rain and hail
compels me to take shelter under an overhanging rock.

At 4 P.M. I reach Palisade, a railroad village situated in the most
romantic spot imaginable, under the shadows of the towering palisades that hover above with a sheltering care, as if their special
mission were to protect it from all harm.   Evidently these mountains have been rent in twain by an earthquake, and this great
gloomy chasm left open, for one can plainly see that the two walls
represent two halves of what was once a solid mountain.   Curious
caves are observed in the face of the cliffs, and one, more conspicuous than the rest, has been christened "Maggie's Bower," in honor
of a beautiful Scottish maiden who with her parents once lingered
in a neighboring creek-bottom for some time, recruiting their stock.
But all is not romance and beauty even in the glorious palisades of
the Humboldt ; for great, glaring, patent-medicine advertisements
are painted on the most conspicuously beautiful spots of the pali-

sades. Business enterprise is of course to be commended and encouraged ; but it is really annoying that one cannot let his æsthetic soul—that is constantly yearning for the sublime and beautiful—rest in gladsome reflection on some beautiful object without at the same time being reminded of " corns," and " biliousness," and all the multifarious evils that flesh is heir to.

It grows pitchy dark ere I leave the cañon on my way to Carlin. Farther on, the gorge widens, and thick underbrush intervenes between the road and the river. From out the brush I see peering two little round phosphorescent balls, like two miniature moons, turned in my direction. I wonder what kind of an animal it is, as I trundle along through the darkness, revolver in hand, ready to defend myself, should it make an attack. I think it is a mountain-lion, as they seem to be plentiful in this part of Nevada. Late as it is when I reach Carlin, the " boys " must see how a bicycle is ridden, and, as there is no other place suitable, I manage to circle around the pool-table in the hotel bar-room a few times, nearly scalping myself against the bronze chandelier in the operation. I hasten, however, to explain that these proceedings took place immediately after my arrival, lest some worldly wise, over-sagacious person should be led to suspect them to be the riotous undertakings of one who had " smiled with the boys once too often." Little riding is possible all through this section of Nevada, and, in order to complete the forty miles a day that I have rigorously imposed upon myself, I sometimes get up and pull out at daylight. It is scarce more than sunrise when, following the railroad through Five-mile Cañon—another rift through one of the many mountain chains that cross this part of Nevada in all directions under the general name of the Humboldt Mountains—I meet with a startling adventure. I am trundling through the cañon alongside the river, when, rounding the sharp curve of a projecting mountain, a tawny mountain lion is perceived trotting leisurely along ahead of me, not over a hundred yards in advance. He hasn't seen me yet ; he is perfectly oblivious of the fact that he is in " the presence." A person of ordinary discretion would simply have revealed his presence by a gentlemanly sneeze, or a slight noise of any kind, when the lion would have immediately bolted back into the underbrush. Unable to resist the temptation, I fired at him, and of course missed him, as a person naturally would at a hundred yards with a bull-dog revolver. The bullet must have singed him a little though, for, instead of wildly scooting for the brush, as I anticipated, he turns savagely round and comes bounding rapidly toward me, and at twenty paces crouches for a spring. Laying his cat-like head almost on the ground, his round eyes flashing fire, and his tail angrily waving to and fro, he looks savage and dangerous. Crouching behind the bicycle, I fire at him again. Nine times out of ten a person will overshoot the mark with a revolver under such circumstances, and, being anxious to avoid this, I do the reverse, and fire too low. The ball strikes the ground just in front of his head, and throws the sand and gravel in his face, and perhaps in his wicked round eyes ; for he shakes his head, springs up, and makes off into the brush. I shall shed blood of some sort yet before I leave Nevada ! There isn't a day that I don't shoot at something or other ; and all I ask of any animal is to come within two hundred yards and I will squander a cartridge on him, and I never fail to hit—the ground.

At Elko, where I take dinner, I make the acquaintance of an individual, rejoicing in the sobriquet of " Alkali Bill," who has the largest and most comprehensive views of any person I ever met. He has seen a paragraph, something about me riding round the world, and he considerately takes upon himself the task of summing up the few trifling obstacles that I shall encounter on the way round :

"There is only a small rise at Sherman," he rises to explain, " and another still smaller at the Alleghanies ; all the balance is downhill to the Atlantic. Of course you'll have to 'boat it' across the Frogpond ; then there's Europe—mostly level ; so is Asia, except the Himalayas—and you can soon cross them ; then you're all 'hunky,' for there's no mountains to speak of in China."

Evidently Alkali Bill is a person who points the finger of scorn at small ideas, and leaves the bothersome details of life to other and smaller-minded folks. In his vast and glorious imagery he sees a centaur-like 'cycler skimming like a frigate-bird across states and continents, scornfully ignoring sandy deserts and bridgeless streams, halting for nothing but oceans, and only slowing up a little when he runs up against a peak that bobs up its twenty thousand feet of snowy grandeur serenely in his path. What a Cæsar is lost to this benighted world, because in its blindness, it will not search out such men as Alkali and ask them to lead it onward to deeds of inconceivable greatness ! Alkali Bill can whittle more chips in an hour than some men could in a week.

Encounter with a Mountain Lion.

Much of the Humboldt Valley, through which my road now runs, is at present flooded from the vast quantities of water that are pouring into it from the Ruby Range of mountains now visible to the southeast, and which have the appearance of being the snowiest of any since leaving the Sierras. Only yesterday I threatened to shed blood before I left Nevada, and sure enough my prophecy is destined to speedy fulfilment. Just east of the Osino Cañon, and where the North Fork of the Humboldt comes down from the north and joins the main stream, is a stretch of swampy ground on which swarms of wild ducks and geese are paddling about. I blaze away at them, and a poor inoffensive gosling is no more!

While writing my notes this evening, in a room adjoining the "bar" at Halleck, near the United States fort of the same name, I overhear a boozy soldier modestly informing his comrades that forty-five miles an hour is no unusual speed to travel with a bicycle.

Gradually I am nearing the source of the Humboldt, and at the town of Wells I bid it farewell for good. Wells is named from a group of curious springs near the town. They are supposed to be extinct volcanoes, now filled with water ; and report says that no sounding-line has yet been found long enough to fathom the bottom. Some day when some poor, unsuspecting tenderfoot is peering inquisitively down one of these well-like springs, the volcano may suddenly come into play again and convert the water into steam that will shoot him clear up into the moon! These volcanoes may have been soaking in water for millions of years ; but they are not to be trusted on that account ; they can be depended upon to fill some citizen full of lively surprise one of these days. Everything here is surprising! You look across the desert and see flowing water and waving trees ; but when you get there, with your tongue hanging out and your fate wellnigh sealed, you are surprised to find nothing but sand and rocks. You climb a mountain expecting to find trees and birds' eggs, and you are surprised to find highwater marks and sea-shells. Finally, you look in the looking-glass and are surprised to find that the wind and exposure have transformed your nice blonde complexion to a semi-sable hue that would prevent your own mother from recognizing you.

The next day, when nearing the entrance to Montella Pass, over the Goose Creek Range, I happen to look across the mingled sagebrush and juniper-spruce brush to the right, and a sight greets my

eyes that causes me to instinctively look around for a tall tree,
though well knowing that there is nothing of the kind for miles ;
neither is there any ridable road near, or I might try my hand at
breaking the record for a few miles. Standing bolt upright on their
hind legs, by the side of a clump of juniper-spruce bushes and in-
tently watching my movements, are a pair of full-grown cinnamon
bears. When a bear sees a man before the man happens to descry
him, and fails to betake himself off immediately, it signifies that he
is either spoiling for a fight or doesn't care a continental password
whether war is declared or not. Moreover, animals recognize the
peculiar advantages of two to one in a fight equally with their human
inferi—superiors ; and those two over there are apparently in no particular hurry to move on. They don't seem awed at my presence. On
the contrary, they look suspiciously like being undecided and hesitative about whether to let me proceed peacefully on my way or not.
Their behavior is outrageous ; they stare and stare and stare, and
look quite ready for a fight. I don't intend one to come off, though,
if I can avoid it. I prefer to have it settled by arbitration. I haven't
lost these bears ; they aren't mine, and I don't want anything that
doesn't belong to me. I am not covetous ; so, lest I should be
tempted to shoot at them if I come within the regulation two hundred yards, I "edge off" a few hundred yards in the other direction,
and soon have the intense satisfaction of seeing them stroll off toward
the mountains. I wonder if I don't owe my escape on this occasion
to my bicycle ? Do the bright spokes glistening in the sunlight as
they revolve make an impression on their bearish intellects that
influences their decision in favor of a retreat. It is perhaps needless to add that, all through this mountain-pass, I keep a loose eye
busily employed looking out for bears.

But nothing more of a bearish nature occurs, and the early
gloaming finds me at Tacoma, a village near the Utah boundary
line. There is an awful calamity of some sort hovering over this
village. One can feel it in the air. The *habitués* of the hotel bar-
room sit around, listless and glum. When they speak at all it is to
predict all sorts of difficulties for me in my progress through Utah
and Wyoming Territories. "The black guats of the Salt Lake mud
flat'll eat you clean up," snarls one. "Bear River's flooding the hull
kintry up Weber Cañon way," growls another. "The slickest thing
you kin do, stranger, is to board the keers and git out of this,"
says a third, in a tone of voice and with an emphasis that plainly in-

dicates his great disgust at " this." By " this " he means the village of Tacoma ; and he is disgusted with it. They are all disgusted with it, and with the whole world this evening, because Tacoma is " out of whiskey." Yes, the village is destitute of whiskey ; it should have arrived yesterday, and hasn't shown up yet ; and the effect on the society of the bar-room is so depressing that I soon retire to my couch, to dream of Utah's strange intermingling of forbidding deserts and beautiful orchards through which my route now leads me.

CHAPTER III.

THROUGH MORMON-LAND AND OVER THE ROCKIES.

A DREARY-LOOKING country is the "Great American Desert," in Utah, the northern boundary line of which I traverse next morning. To the left of the road is a low chain of barren hills ; to the right, the uninviting plain, over which one's eye wanders in vain for some green object that might raise hopes of a less desolate region beyond ; and over all hangs an oppressive silence—the silence of a dead country—a country destitute of both animal and vegetable life. Over the great desert hangs a smoky haze, out of which Pilot Peak, thirty-eight miles away, rears its conical head 2,500 feet above the level plain at its base.

Some riding is obtained at intervals along this unattractive stretch of country, but there are no continuously ridable stretches, and the principal incentive to mount at all is a feeling of disgust at so much compulsory walking. A noticeable feature through the desert is the almost unquenchable thirst that the dry saline air inflicts upon one. Reaching a railway section-house, I find no one at home ; but there is a small underground cistern of imported water, in which "wrigglers" innumerable wriggle, but which is otherwise good and cool. There is nothing to drink out of, and the water is three feet from the surface ; while leaning down to try and drink, the wooden framework at the top gives way and precipitates me head first into the water. Luckily, the tank is large enough to enable me to turn round and reappear at the surface, head first, and with considerable difficulty I scramble out again, with, of course, not a dry thread on me.

At three in the afternoon I roll into Terrace, a small Mormon town. Here a rather tough-looking citizen, noticing that my garments are damp, suggests that 'cycling must be hard work to make a person perspire like that in this dry climate. At the Matlin' section-house I find accommodation for the night with a whole-souled section-house foreman, who is keeping bachelor's hall temporarily, as his wife is away on a visit at Ogden. From this house, which is situated on the table-land of the Red Dome Mountains, can be obtained a more comprehensive view of the Great American Desert than when we last beheld it. It has all the appearance of being the dry bed of an ancient salt lake or inland sea. A broad, level plain of white alkali, which is easily mistaken in the dim distance for smooth, still water, stretches away like a dead, motionless sea as far as human vision can penetrate, until lost in the haze; while, here and there, isolated rocks lift their rugged heads above the dreary level, like islets out of the sea. It is said there are many evidences that go to prove this desert to have once been covered by the waters of the great inland sea that still, in places, laves its eastern borders with its briny flood. I am informed there are many miles of smooth, hard, salt-flats, over which a 'cycler could skim like a bird; but I scarcely think enough of bird-like skimming to go searching for it on the American Desert. A few miles east of Matlin the road leads over a spur of the Red Dome Range, from whence I obtain my first view of the Great Salt Lake, and soon I am enjoying a long-anticipated bath in its briny waters. It is disagreeably cold, but otherwise an enjoyable bath. One can scarce sink beneath the surface, so strongly is the water impregnated with salt.

For dinner, I reach Kelton, a town that formerly prospered as the point from which vast quantities of freight were shipped to Idaho. Scores of huge freight-wagons are now bunched up in the corrals, having outlived their usefulness since the innovation from mules and "overland ships" to locomotives on the Utah Northern Railway. Empty stores and a general air of vanished prosperity are the main features of Kelton to-day; and the inhabitants seem to reflect in their persons the aspect of the town; most of them being freighters, who, finding their occupation gone, hang listlessly around, as though conscious of being fit for nothing else. From Kelton I follow the lake shore, and at six in the afternoon arrive at the salt-works, near Monument Station, and apply for accommodation, which is readily given. Here is erected a wind-mill, which pumps the water from the lake into shallow reservoirs, where it evaporates and leaves a layer of coarse salt on the bottom. These people drink water that is disagreeably brackish and unsatisfactory to one unaccustomed to it, but which they say has become more acceptable to them, from habitual use, than purely fresh water. This spot is the healthiest and most favorable for the prolific production of certain forms of insect-life I ever was in, and I spend the liveliest night here I ever spent anywhere. These people professed to give me a bed to myself, but no sooner have I laid my head on the pillow than I recognize the ghastly joke they are playing on me. The bed is already densely populated with guests, who naturally object to being ousted or overcrowded. They seem quite a kittenish and playful lot, rather inclined to accomplish their ends by playing wild pranks than by resorting to more austere measures. Watching till I have closed my eyes in an attempt to doze off, they slip up and playfully tickle me under the chin, or scramble around in my ear, and anon they wildly chase each other up and down my back, and play leap-frog and hide-and-go-seek all over my sensitive form, so that I arise in the morning anything but refreshed from my experience.

Still following the shores of the lake, for several miles, my road now leads over the northern spur of the Promontory Mountains. On these hills I find a few miles of hard gravel that affords the best riding I have experienced in Utah, and I speed along as rapidly as possible, for dark, threatening clouds are gathering overhead. But ere I reach the summit of the ridge a violent thunder-storm breaks over the hills, and I seem to be verily hobnobbing with the thunder and lightning, that appears to be round about me, rather than overhead. A troop of wild bronchos, startled and stampeded by the vivid lightning and sharp peals of thunder, come wildly charging down the mountain trail, threatening to run quite over me in their mad career. Pulling my six-shooter, I fire a couple of shots in the air to attract their attention, when they rapidly swerve to the left, and go tearing frantically over the rolling hills on their wild flight to the plains below.

Most of the rain falls on the plain and in the lake, and when I arrive at the summit I pause to take a view at the lake and surrounding country. A more auspicious occasion could scarcely have been presented. The storm has subsided, and far beneath my feet a magnificent rainbow spans the plain, and dips one end of its variegated beauty in the sky-blue waters of the lake. From this point the view to the west and south is truly grand—rugged, irregular mountain-chains traverse the country at every conceivable angle, and around among them winds the lake, filling with its blue waters the intervening spaces, and reflecting, impartially alike, their grand majestic beauty and their faults. What dreams of empire and white-winged commerce on this inland sea must fill the mind

A Stampede of Wild Mustangs,

and fire the imagery of the newly arrived Mormon convert who, standing on the commanding summit of these mountains, feasts his eyes on the glorious panorama of blue water and rugged mountains that is spread like a wondrous picture before him! Surely, if he be devotionally inclined, it fails not to recall to his mind another inland sea in far-off Asia Minor, on whose pebbly shores and by whose rippling waves the cradle of an older religion than Mormonism was rocked—but not rocked to sleep.

Ten miles farther on, from the vantage-ground of a pass over another spur of the same range, is obtained a widely extended view of the country to the east. For nearly thirty miles from the base of the mountains, low, level mud-flats extend eastward, bordered on the south by the marshy, sinuous shores of the lake, and on the north by the Blue Creek Mountains. Thirty miles to the east—looking from this distance strangely like flocks of sheep grazing at the base of the mountains—can be seen the white-painted houses of the Mormon settlements, that thickly dot the narrow but fertile strip of agricultural land between Bear River and the mighty Wahsatch Mountains, that, rearing their snowy crest skyward, shut out all view of what lies beyond. From this height the level mud-flats appear as if one could mount his wheel and bowl across at a ten-mile pace ; but I shall be agreeably surprised if I am able to aggregate ten miles of riding out of the thirty. Immediately after getting down into the bottom I make the acquaintance of the tiny black gnats that one of our whiskey-bereaved friends at Tacoma had warned me against. One's head is constantly enveloped in a black cloud of these little wretches. They are of infinitesimal proportions, and get into a person's ears, eyes, and nostrils, and if one so far forgets himself as to open his mouth, they swarm in as though they think it the "pearly gates ajar," and this their last chance of effecting an entrance. Mingled with them, and apparently on the best of terms, are swarms of mosquitoes, which appear perfect Jumbos in comparison with their disreputable associates.

As if partially to recompense me for the torments of the afternoon, Dame Fortune considerately provides me with two separate and distinct suppers this evening. I had intended, when I left Promontory Station, to reach Corinne for the night ; consequently I bring a lunch with me, knowing it will take me till late to reach there. These days, I am troubled with an appetite that makes me blush to speak of it, and about five o'clock I sit down—on the bleached skeleton of a defunct mosquito !—and proceed to eat my lunch of bread and meat—and gnats ; for I am quite certain of eating hundreds of these omnipresent creatures at every bite I take. Two hours afterward I am passing Quarry section-house, when the foreman beckons me over and generously invites me to remain over night. He brings out canned oysters and bottles of Milwaukee beer, and insists on my helping him discuss these acceptable viands ; to which invitation it is needless to say I yield without extraordinary pressure, the fact of having eaten two hours before being no obstacle whatever. So much for 'cycling as an aid to digestion. Arriving at Corinne, on Bear River, at ten o'clock next morning, I am accosted by a bearded, patriarchal Mormon, who requests me to constitute myself a parade of one, and ride the bicycle around the town for the edification of the people's minds.

"In course they knows what a ' perlocefede ' is, from seein' 'em in picturs ; but they never seed a real machine, and it'd be a ' hefty' treat fer 'em," is the eloquent appeal made by this person in behalf of the Corinnethians, over whose destinies and happiness he appears to preside with fatherly solicitude. As the streets of Corinne this morning consist entirely of black mud of uncertain depth, I am reluctantly compelled to say the elder nay, at the same time promising him that if he would have them in better condition next time I happened around, I would willingly second his brilliant idea of making the people happy by permitting them a glimpse of my " perlocefede " in action.

After crossing Bear River I find myself on a somewhat superior road leading through the Mormon settlements to Ogden. No greater contrast can well be imagined than that presented by this strip of country lying between the lake and the Wahsatch Mountains, and the desert country to the westward. One can almost fancy himself suddenly transported by some good genii to a quiet farming community in an Eastern State. Instead of untamed bronchos and wild-eyed cattle, roaming at their own free will over unlimited territory, are seen staid work-horses ploughing in the field, and the sleek milch-cow peacefully cropping tame grass in enclosed meadows. Birds are singing merrily in the willow hedges and the shade-trees ; green fields of alfalfa and ripening grain line the road and spread themselves over the surrounding country in

alternate squares, like those of a vast checker-board. Farms, on
the average, are small, and, consequently, houses are thick ; and
not a farm-house among them all but is embowered in an orchard
of fruit and shade-trees that mingle their green leaves and white
blossoms harmoniously. At noon I roll into a forest of fruit-trees,
among which, I am informed, Willard City is situated ; but one
can see nothing of any city. Nothing but thickets of peach, plum,
and apple trees, all in full bloom, surround the spot where I alight
and begin to look around for some indications of the city. "Where
is Willard City ?" I inquire of a boy who comes out from one of
the orchards carrying a can of kerosene in his hand, suggestive of
having just come from a grocery, and so he has. "This is Willard City, right here," replies the boy ; and then, in response to my
inquiry for the hotel, he points to a small gate leading into an
orchard, and tells me the hotel is in there.

The hotel—like every other house and store here—is embowered amid an orchard of blooming fruit-trees, and looks like anything but a public eating-house. No sign up, nothing to distinguish it from a private dwelling ; and I am ushered into a nicely
furnished parlor, on the neatly papered walls of which hang enlarged portraits of Brigham Young and other Mormon celebrities,
while a large-sized Mormon bible, expensively bound in morocco,
reposes on the centre-table. A charming Miss of —teen summers
presides over a private table, on which is spread for my material
benefit the finest meal I have eaten since leaving California. Such
snow-white bread ! Such delicious butter ! And the exquisite flavor
of "spiced peach-butter " lingers in my fancy even now ; and as if
this were not enough for "two bits " (a fifty per cent. come-down
from usual rates in the mountains), a splendid bouquet of flowers is
set on the table to round off the repast with their grateful perfume.
As I enjoy the wholesome, substantial food, I fall to musing on the
mighty chasm that intervenes between the elegant meal now before me and the "Melican plan-cae " of two weeks ago.

"You have a remarkably pleasant country here, Miss," I venture
to remark to the young lady who has presided over my table, and
whom I judge to be the daughter of the house, as she comes to the
door to see the bicycle.

"Yes ; we have made it pleasant by planting so many orchards,"
she answers, demurely.

"I should think the Mormons ought to be contented, for they

A Fair Young Mormon.

possess the only good piece of farming country between California and 'the States,'" I blunderingly continued.

"I never heard anyone say they are not contented, but their enemies," replies this fair and valiant champion of Mormonism in a voice that shows she quite misunderstands my meaning.

"What I intended to say was, that the Mormon people are to be highly congratulated on their good sense in settling here," I hasten to explain ; for were I to leave at this house, where my treatment has been so gratifying, a shadow of prejudice against the Mormons, I should feel like kicking myself all over the Territory. The women of the Mormon religion are instructed by the wiseacres of the church to win over strangers by kind treatment and by the charm of their conversation and graces; and *this* young lady has learned the lesson well; she has graduated with high honors. Coming from the barren deserts of Nevada and Western Utah—from the land where the irreverent and irrepressible " Old Timer " fills the air with a sulphurous odor from his profanity and where nature is seen in its sternest aspect, and then suddenly finding one's self literally surrounded by flowers and conversing with Beauty about Religion, is enough to charm the heart of a marble statue.

Ogden is reached for supper, where I quite expect to find a 'cycler or two (Ogden being a city of eight thousand inhabitants) ; but the nearest approach to a bicycler in Ogden is a gentleman who used to belong to a Chicago club, but who has failed to bring his " wagon " West with him. Twelve miles of alternate riding and walking eastwardly from Ogden bring me to the entrance of Weber Cañon, through which the Weber River, the Union Pacific Railroad, and an uncertain wagon-trail make their way through the Wahsatch Mountains on to the elevated table-lands of Wyoming Territory. Objects of interest follow each other in quick succession along this part of the journey, and I have ample time to examine them, for Weber River is flooding the cañon, and in many places has washed away the narrow space along which wagons are wont to make their way, so that I have to trundle slowly along the railway track. Now the road turns to the left, and in a few minutes the rugged and picturesque walls of the cañon are towering in imposing heights toward the clouds. The Weber River comes rushing—a resistless torrent—from under the dusky shadows of the mountains through which it runs for over fifty miles, and onward to the plain below, where it assumes a more moderate pace, as if conscious that it has at last escaped from the hurrying turmoil of its boisterous march down the mountain.

Advancing into the yawning jaws of the range, a continuously resounding roar is heard in advance, which gradually becomes louder as I proceed eastward; in a short time the source of the noise is discovered, and a weird scene greets my enraptured vision. At a place where the fall is tremendous, the waters are opposed in their mad march by a rough-and-tumble collection of huge, jagged rocks, that have at some time detached themselves from the walls above, and come crashing down into the bed of the stream. The rushing waters, coming with haste from above, appear to pounce with insane fury on the rocks that dare thus to obstruct their path; and then for the next few moments all is a hissing, seething, roaring caldron of strife, the mad waters seeming to pounce with ever-increasing fury from one imperturbable antagonist to another, now leaping clear over the head of one, only to dash itself into a cloud of spray against another, or pour like a cataract against its base in a persistent, endless struggle to undermine it; while over all tower the dark, shadowy rocks, grim witnesses of the battle. This spot is known by the appropriate name of "The Devil's Gate."

Wherever the walls of the cañon recede from the river's brink, and leave a space of cultivable land, there the industrious Mormons have built log or adobe cabins, and converted the circumscribed domain into farms, gardens, and orchards. In one of these isolated settlements I seek shelter from a passing shower at the house of a "three-ply Mormon" (a Mormon with three wives), and am introduced to his three separate and distinct better-halves; or, rather, one should say, "better-quarters," for how can anything have three halves? A noticeable feature at all these farms is the universal plurality of women around the house, and sometimes in the field. A familiar scene in any farming community is a woman out in the field, visiting her husband, or, perchance, assisting him in his labors. The same thing is observable at the Mormon settlements along the Weber River—only, instead of one woman, there are generally two or three, and perhaps yet another standing in the door of the house.

Passing through two tunnels that burrow through rocky spurs stretching across the cañon, as though to obstruct farther progress, across the river, to the right, is the "Devil's Slide"—two perpendicular walls of rock, looking strangely like man's handiwork, stretching in parallel lines almost from·base to summit of a slop-

ing, grass-covered mountain. The walls are but a dozen feet apart.
It is a curious phenomenon, but only one among many that are
scattered at intervals all through here. A short distance farther,
and I pass the famous "Thousand-mile Tree"—a rugged pine, that
stands between the railroad and the river, and which has won renown by springing up just one thousand miles from Omaha. This
tree is having a tough struggle for its life these days ; one side of its
honored trunk is smitten as with the leprosy. The fate of the Thousand-mile Tree is plainly sealed. It is unfortunate in being the
most conspicuous target on the line for the fe-ro-ci-ous youth who
comes West with a revolver in his pocket and shoots at things from
the car-window. Judging from the amount of cold lead contained
in that side of its venerable trunk next the railway few of these
thoughtless marksmen go past without honoring it with a shot.
Emerging from "the Narrows" of Weber Cañon, the route follows
across a less contracted space to Echo City, a place of two hundred
and twenty-five inhabitants, mostly Mormons, where I remain overnight. The hotel where I put up at Echo is all that can be desired,
so far as "provender" is concerned ; but the handsome and picturesque proprietor seems afflicted with sundry eccentric habits, his
leading eccentricity being a haughty contempt for fractional currency. Not having had the opportunity to test him, it is difficult
to say whether this peculiarity works both ways, or only when the
change is due his transient guests. However, we willingly give
him the benefit of the doubt.

Heavily freighted rain-clouds are hovering over the mountains
next morning and adding to the gloominess of the gorge, which,
just east of Echo City, contracts again and proceeds eastward under
the name of Echo Gorge. Turning around a bold rocky projection
to the left, the far-famed "Pulpit Rock" towers above, on which
Brigham Young is reported to have stood and preached to the Mormon host while halting over Sunday at this point, during their pilgrimage to their new home in the Salt Lake Valley below. Had
the redoubtable prophet turned "dizzy" while haranguing his followers from the elevated pinnacle of his novel pulpit, he would at
least have died a more romantic death than he is accredited with
—from eating too much green corn.

Fourteen miles farther brings me to "Castle Rocks," a name
given to the high sandstone bluffs that compose the left-hand side
of the cañon at this point, and which have been worn by the ele-

ments into all manner of fantastic shapes, many of them calling to mind the towers and turrets of some old-world castle so vividly, that one needs but the pomp and circumstance of old knight-errant days to complete the illusion. But, as one gazes with admiration on these towering buttresses of nature, it is easy to realize that the most massive and imposing feudal castle, or ramparts built with human hands, would look like children's toys beside them.

The weather is cool and bracing, and when, in the middle of the afternoon, I reach Evanston, Wyo. Terr., too late to get dinner at the hotel, I proceed to devour the contents of a bakery, filling the proprietor with boundless astonishment by consuming about two-thirds of his stock. When I get through eating, he bluntly refuses to charge anything, considering himself well repaid by having witnessed the most extraordinary gastronomic·feat on record—the swallowing of two-thirds of a bakery! Following the trail down Yellow Creek, I arrive at Hilliard after dark. The Hilliardites are "somewhat seldom," but they are made of the right material. The boarding-house landlady sets about preparing me supper, late though it be ; and the "boys" extend me a hearty invitation to turn in with them for the night. Here at Hilliard is a long V-shaped flume, thirty miles long, in which telegraph poles, ties, and cordwood are floated down to the railroad from the pineries of the Uintah Mountains, now plainly visible to the south. The "boys" above referred to are men engaged in handling ties thus floated down ; and sitting around the red-hot stove, they make the evening jolly with songs and yarns of tie-drives, and of wild rides down the long "V" flume. A happy, light-hearted set of fellows are these "tie-men," and not an evening but their rude shanty resounds with merriment galore. Fun is in the air to-night, and "Beaver" (so dubbed on account of an unfortunate tendency to fall into every hole of water he goes anywhere near) is the unlucky wight upon whom the rude witticisms concentrate ; for he has fallen into the water again to-day, and is busily engaged in drying his clothes by the stove. They accuse him of keeping up an uncomfortably hot fire, detrimental to everybody's comfort but his own, and threaten him with dire penalties if he doesn't let the room cool off; also broadly hinting their disapproval of his over-fondness for "Adam's ale," and threaten to make him "set 'em up" every time he tumbles in hereafter. In revenge for these remarks, "Beaver" piles more wood into the stove, and, with many a westernism—not permitted in print—threatens to keep up a fire that will drive them all out of the shanty if they persist in their persecutions.

Crossing next day the low, broad pass over the Uintah Mountains, some stretches of ridable surface are passed over, and at this point I see the first band of antelope on the tour ; but as they fail to come within the regulation two hundred yards they are graciously permitted to live.

At Piedmont Station I decide to go around by way of Fort Bridger and strike the direct trail again at Carter Station, twenty-four miles farther east.

A tough bit of Country.

The next day at noon finds me "tucked in my little bed" at Carter, decidedly the worse for wear, having experienced the toughest twenty-four hours of the entire journey. I have to ford no less than nine streams of ice-cold water ; get benighted on a rain-soaked adobe plain, where I have to sleep out all night in an abandoned freight-wagon ; and, after carrying the bicycle across seven miles of deep, sticky clay, I finally arrive at Carter, looking like the last sad remnant of a dire calamity—having had nothing to eat for twenty-four hours. From Carter my route leads through the Bad-Lands, amid buttes of mingled clay and rock, which the elements have worn into all conceivable shapes. and conspicuous among them

can be seen, to the south, "Church Buttes," so called from having
been chiselled by the dexterous hand of nature into a group of domes
and pinnacles, that, from a distance, strikingly resembles some
magnificent cathedral. High-water marks are observable on these ·
buttes, showing that Noah's flood, or some other aqueous calamity
once happened around here ; and one can easily imagine droves of
miserable, half-clad Indians, perched on top, looking with doleful,
melancholy expression on the surrounding wilderness of waters.
Arriving at Granger, for dinner, I find at the hotel a crest-fallen
state of affairs somewhat similar to the glumness of Tacoma. Tacoma had plenty of customers, but no whiskey ; Granger on the
contrary has plenty of whiskey, but no customers. The effect on
that marvellous, intangible something, the saloon proprietor's intellect, is the same at both places. Here is plainly a new field of re-
search for some ambitious student of psychology. Whiskey without
customers ! Customers without whiskey ! Truly all is vanity and
vexation of spirit.

Next day I pass the world-renowned castellated rocks of Green
River, and stop for the night at Rock Springs, where the Union
Pacific Railway Company has extensive coal mines. On calling for
my bill at the hotel here, next morning, the proprietor—a corpulent Teuton, whose thoughts, words, and actions, run entirely to
beer—replies, "Twenty-five cents a quart." Thinking my hearing
apparatus is at fault, I inquire again. "Twenty-five cents a quart
and vurnish yer own gan." The bill is abnormally large, but, as I
hand over the amount, a "loaded schooner " is shoved under my
nose, as though a glass of beer were a tranquillizing antidote for all
the ills of life. Splendid level alkali flats abound east of Rock
Springs, and I bowl across them at a lively pace until they terminate, and my route follows up Bitter Creek, where the surface is
just the reverse ; being seamed and furrowed as if it had just
emerged from a devastating flood. It is said that the teamster
who successfully navigated the route up Bitter Creek, considered
himself entitled to be called "a tough cuss from Bitter Creek, on
wheels, with a perfect education." A justifiable regard for individual rights would seem to favor my own assumption of this distinguished title after traversing the route with a bicycle.

Ten o'clock next morning finds me leaning on my wheel, surveying the scenery from the " Continental Divide "—the backbone
of the continent. Facing the north, all waters at my right hand

flow to the east, and all on my left flow to the west—the one eventually finding their way to the Atlantic, the other to the Pacific. This spot is a broad low pass through the Rockies, more plain than mountain, but from which a most commanding view of numerous mountain chains are obtained. To the north and northwest are the Seminole, Wind River, and Sweet-water ranges—bold, rugged mountain-chains, filling the landscape of the distant north with a mass of great, jagged, rocky piles, grand beyond conception ; their many snowy peaks peopling the blue ethery space above with ghostly, spectral forms well calculated to inspire with feelings of awe and admiration a lone cycler, who, standing in silence and solitude profound on the great Continental Divide, looks and meditates on what he sees. Other hoary monarchs are visible to the east, which, however, we shall get acquainted with later on. Down grade is the rule now, and were there a good road, what an enjoyable coast it would be, down from the Continental Divide! but half of it has to be walked. About eighteen miles from the divide I am greatly amused, and not a little astonished, at the strange actions of a coyote that comes trotting in a leisurely, confidential way toward me ; and when he reaches a spot commanding a good view of my. road he stops and watches my movements with an air of the greatest inquisitiveness and assurance. He stands and gazes as I trundle along, not over fifty yards away, and he looks so much like a well-fed collie, that I actually feel like patting my knee for him to come and make friends. Shoot at him ? Certainly not. One never abuses a confidence like that. He can come and rub his sleek coat up against the bicycle if he likes, and —blood-thirsty rascal though he no doubt is—I will never fire at him. He has as much right to gaze in astonishment at a bicycle as anybody else who never saw one before.

Staying over night and the next day at Rawlins, I make the sixteen miles to Fort Fred Steele next morning before breakfast, there being a very good road between the two places. This fort stands on the west bank of North Platte River, and a few miles west of the river I ride through the first prairie-dog town encountered in crossing the continent from the west, though I shall see plenty of these interesting little fellows during the next three hundred miles. These animals sit near their holes and excitedly bark at whatever goes past. Never before have they had an opportunity to bark at a bicycle, and they seem to be making the most of their opportunity. I see at this village none of the small speckled owls, which, with the rattlesnake, make themselves so much at home in the prairie-dogs' comfortable quarters, but I see them farther east. These three strangely assorted companions may have warm affections toward each other ; but one is inclined to think the great bond of sympathy that binds them together is the tender regard entertained by the owl and the rattlesnake for the nice, tender young prairie-pups that appear at intervals to increase the joys and cares of the elder animals.

I am now getting on to the famous Laramie Plains, and Elk Mountain looms up not over ten miles to the south—a solid, towery mass of black rocks and dark pine forests, that stands out bold and distinct from surrounding mountain chains as though some animate thing conscious of its own strength and superiority. A snow-storm is raging on its upper slopes, obscuring that portion of the mountain ; but the dark forest-clad slopes near the base are in plain view, and also the rugged peak which elevates its white-crowned head above the storm, and reposes peacefully in the bright sunlight in striking contrast to the warring elements lower down. I have heard old hunters assert that this famous "landmark of the Rockies" is hollow, and that they have heard wolves howling inside the mountain ; but some of these old western hunters see and hear strange things !

As I penetrate the Laramie Plains the persistent sage-brush, that has constantly hovered around my path for the last thousand miles, grows beautifully less, and the short, nutritious buffalo-grass is creeping everywhere. In Carbon, where I arrive after dark, I mention among other things in reply to the usual volley of questions, the fact of having to foot it so great a proportion of the way through the mountain country ; and shortly afterward, from among a group of men, I hear a voice, thick and husky with "valley tan," remark : "Faith, Oi cud roide a bicycle meself across the counthry av yeez ud lit me walluk it afut !" and straightway a luminous bunch of shamrocks dangled for a brief moment in the air, and then vanished. After passing Medicine Bow Valley and Como Lake I find some good ridable road, the surface being hard gravel and the plains high and dry. Reaching the brow of one of those rocky ridges that hereabouts divide the plains into so many shallow basins, I find myself suddenly within a few paces of a small herd of antelope peacefully grazing on the other side of the narrow

ridge, all unconscious of the presence of one of creation's alleged
proud lords. My ever handy revolver rings out clear and sharp on
the mountain air, and the startled antelope go bounding across the
plain in a succession of quick, jerky jumps peculiar to that nimble
animal ; but ere they have travelled a hundred yards one of them
lags behind and finally staggers and lays down on the grass. As I
approach him he makes a gallant struggle to rise and make off
after his companions, but the effort is too much for him, and coming up to him, I quickly put him out of pain by a shot behind the
ear. This makes a proud addition to my hitherto rather small list
of game, which now comprises jack-rabbits, a badger, a fierce gosling, an antelope, and a thin, attenuated coyote, that I bowled over
in Utah.

From this ridge an extensive view of the broad, billowy plains
and surrounding mountains is obtained. Elk Mountain still seems
close at hand, its towering form marking the western limits of the
Medicine Bow Range whose dark pine-clad slopes form the western
border of the plains. Back of them to the west is the Snowy
Range, towering in ghostly grandeur as far above the timber-clad
summits of the Medicine Bow Range as these latter are above the
grassy plains at their base. To the south more snowy mountains
stand out against the sky like white tracery on a blue ground, with
Long's Peak and Fremont's Peak towering head and shoulders
above them all. The Rattlesnake Range, with Laramie Peak rearing its ten thousand feet of rugged grandeur to the clouds, are
visible to the north. On the east is the Black Hills Range, the
last chain of the Rockies, and now the only barrier intervening between me and the broad prairies that roll away eastward to the
Missouri River and " the States."

A genuine Laramie Plains rain-storm is hovering overhead as I
pull out of Rock Creek, after dinner, and in a little while the performance begins. There is nothing of the gentle pattering shower
about a rain and wind storm on these elevated plains ; it comes on
with a blow and a bluster that threatens to take one off his feet.
The rain is dashed about in the air by the wild, blustering wind,
and comes from all directions at the same time. While you are
frantically hanging on to your hat, the wind playfully unbuttons
your rubber coat and lifts it up over your head and flaps the wet,
muddy corners about in your face and eyes ; and, ere you can dis-
entangle your features from the cold uncomfortable embrace of

the wet mackintosh, the rain—which "falls" upward as well as down, and sidewise, and every other way—has wet you through up as high as the armpits ; and then the gentle zephyrs complete your discomfiture by purloining your hat and making off across the sodden plain with it, at a pace that defies pursuit. The storm winds up in a pelting shower of hailstones—round chunks of ice that cause me to wince whenever one makes a square hit, and they strike the steel spokes of the bicycle and make them produce harmonious sounds. Trundling through Cooper Lake Basin, after dark, I get occasional glimpses of mysterious shadowy objects flitting hither and thither through the dusky pall around me. The basin is full of antelope, and my presence here in the darkness fills them with consternation ; their keen scent and instinctive knowledge of a strange presence warn them of my proximity ; and as they cannot see me in the darkness they are flitting about in wild alarm.

Stopping for the night at Lookout, I make an early start, in order to reach Laramie City for dinner. These Laramie Plains "can smile and look pretty" when they choose, and, as I bowl along over a fairly good road this sunny Sunday morning, they certainly choose. The Laramie River on my left, the Medicine Bow and Snowy ranges—black and white respectively—towering aloft to the right, and the intervening plains dotted with herds of antelope, complete a picture that can be seen nowhere save on the Laramie Plains. Reaching a swell of the plains, that almost rises to the dignity of a hill, I can see the nickel-plated wheels of the Laramie wheelmen glistening in the sunlight on the opposite side of the river several miles from where I stand. They have come out a few miles to meet me, but have taken the wrong side of the river, thinking I had crossed below Rock Creek. The members of the Laramie Bicycle Club are the first wheelmen I have seen since leaving California ; and, as I am personally acquainted at Laramie, it is needless to dwell on my reception at their hands. The rambles of the Laramie Club are well known to the cycling world from the many interesting letters from the graphic pen of their captain, Mr. Owen, who, with two other members, once took a tour on their wheels to the Yellowstone National Park. They have some very good natural roads around Laramie, but in their rambles over the mountains these "rough riders of the Rockies" necessarily take risks that are unknown to their fraternal brethren farther east.

Tuesday morning I pull out to scale the last range that separates me from " the plains "—popularly known as such—and, upon arriving at the summit, I pause to take a farewell view of the great and wonderful inter-mountain country, across whose mountains, plains, and deserts I have been travelling in so novel a manner for the last month. The view from where I stand is magnificent—ay, sublime beyond human power to describe—and well calculated to make an indelible impression on the mind of one gazing upon it, perhaps for the last time. The Laramie Plains extend northward and westward, like a billowy green sea. Emerging from a black cañon behind Jelm Mountain, the Laramie River winds its serpentine course in a northeast direction until lost to view behind the abutting mountains of the range, on which I now stand, receiving tribute in its course from the Little Laramie and numbers of smaller streams that emerge from the mountainous bulwarks forming the western border of the marvellous picture now before me. The unusual rains have filled the numberless depressions of the plains with ponds and lakelets that in their green setting glisten and glimmer in the bright morning sunshine like gems. A train is coming from the west, winding around among them as if searching out the most beautiful, and finally halts at Laramie City, which nestles in their midst—the fairest gem of them all—the "Gem of the Rockies." Sheep Mountain, the embodiment of all that is massive and indestructible, juts boldly and defiantly forward as though its mission were to stand guard over all that lies to the west. The Medicine Bow Range is now seen to greater advantage, and a bald mountain-top here and there protrudes above the dark forests, timidly, as if ashamed of its nakedness. Our old friend, Elk Mountain, is still in view, a stately and magnificent pile, serving as a land-mark for a hundred miles around. Beyond all this, to the west and south—a good hundred miles away—are the snowy ranges ; their hoary peaks of glistening purity penetrating the vast blue dome above, like monarchs in royal vestments robed. Still others are seen, white and shadowy, stretching away down into Colorado, peak beyond peak, ridge beyond ridge, until lost in the impenetrable distance.

As I lean on my bicycle on this mountain-top, drinking in the glorious scene, and inhaling the ozone-laden air, looking through the loop-holes of recent experiences in crossing the great wonderland to the west ; its strange intermingling of forest-clad hills and grassy valleys ; its barren, rocky mountains and dreary, desolate plains ; its vast, snowy solitudes and its sunny, sylvan nooks ; the no less strange intermingling of people ; the wandering red-skin with his pathetic history ; the feverishly hopeful prospector, toiling and searching for precious metals locked in the eternal hills ; and the wild and free cow-boy who, mounted on his wiry bronco, roams these plains and mountains, free as the Arab of the desert—I heave a sigh as I realize that no tongue or pen of mine can hope to do the subject justice.

My road is now over Cheyenne Pass, and from this point is mostly down-grade to Cheyenne. Soon I come to a naturally smooth granite surface which extends for twelve miles, where I have to keep the brake set most of the distance, and the constant friction heats the brake-spoon and scorches the rubber tire black. To-night I reach Cheyenne, where I find a bicycle club of twenty members, and where the fame of my journey from San Francisco draws such a crowd on the corner where I alight, that a blue-coated guardian of the city's sidewalks requests me to saunter on over to the hotel. Do I? Yes, I saunter over. The Cheyenne "cops" are bold, bad men to trifle with. They *have* to be "bold, bad men to trifle with," or the wild, wicked cow-boys would come in and "paint the city red" altogether too frequently.

It is the morning of June 4th as I bid farewell to the "Magic City," and, turning my back to the mountains, ride away over very fair roads toward the rising sun. I am not long out before meeting with that characteristic feature of a scene on the Western plains, a "prairie schooner ;" and meeting prairie schooners will now be a daily incident of my eastward journey. Many of these "pilgrims" come from the backwoods of Missouri and Arkansas, or the rural districts of some other Western State, where the persevering, but at present circumscribed, cycler has not yet had time to penetrate, and the bicycle is therefore to them a wonder to be gazed at and commented on, generally—it must be admitted—in language more fluent as to words than in knowledge of the subject discussed. Not far from where the trail leads out of Crow Creek bottom on to the higher table-land, I find the grassy plain smoother than the wagon-trail, and bowl along for a short distance as easily as one could wish. But not for long is this permitted ; the ground becomes covered with a carpeting of small, loose cacti that stick to the rubber tire with the clinging tenacity of a cuckle-burr to a

5

mule's tail. Of course they scrape off again as they come round
to the bridge of the fork, but it isn't the tire picking them up that
fills me with lynx-eyed vigilance and alarm ; it is the dreaded possibility of taking a header among these awful vegetables that un-
nerves one, starts the cold chills chasing each other up and down
my spinal column, and causes staring big beads of perspiration to
ooze out of my forehead. No more appalling physical calamity on
a small scale could befall a person than to take a header on to a
cactus-covered greensward ; millions of miniature needles would
fill his tender hide with prickly sensations, and his vision with
floating stars. It would perchance cast clouds of gloom over his
whole life. Henceforth he would be a solemn-visaged, bilious-eyed
needle-cushion among men, and would never smile again. I once
knew a young man named Whipple, who sat down on a bunch
of these cacti at a picnic in Virginia Dale, Wyo., and *he* never
smiled again. Two meek-eyed maidens of the Rockies invited him
to come and take a seat between them on a thin, innocuous-looking
layer of hay. Smilingly poor, unsuspecting Whipple accepted the
invitation ; jokingly he suggested that it would be a rose between
two thorns. But immediately he sat down he became convinced
that it was the liveliest thorn—or rather millions of thorns—between two roses. Of course the two meek-eyed maidens didn't
know it was there, how should they ? But, all the same, he never
smiled again—not on them.

At the section-house, where I call for dinner, I make the mistake of leaving the bicycle behind the house, and the woman takes
me for an uncommercial traveller—yes, a tramp. She snaps out,
"We can't feed everybody that comes along," and shuts the door
in my face. Yesterday I was the centre of admiring crowds in the
richest city of its size in America ; to-day I am mistaken for a hungry-eyed tramp, and spurned from the door by a woman with a
faded calico dress and a wrathy what-are-you-doing-here ? look in
her eye. Such is life in the Far West.

Gradually the Rockies have receded from my range of vision,
and I am alone on the boundless prairie. There is a feeling of
utter isolation at finding one's self alone on the plains that is not
experienced in the mountain country. There is something tangible and companionable about a mountain ; but here, where there
is no object in view anywhere—nothing but the boundless, level
plains, stretching away on every hand as far as the eye can reach,

and all around, whichever way one looks, nothing but the green
carpet below and the cerulean arch above—one feels that he is the
sole occupant of a vast region of otherwise unoccupied space.   This
evening, while fording Pole Creek with the bicycle, my clothes,

Fishing out my Clothes.

and shoes—all at the same time—the latter fall in the river ; and
in my wild scramble after the shoes I drop some of the clothes ;
then I drop the machine in my effort to save the clothes, and wind
up by falling down in the water with everything.   Everything is

fished out again all right, but a sad change has come over the
clothes and shoes. This morning I was mistaken for a homeless,
friendless wanderer; this evening as I stand on the bank of Pole
Creek with nothing over me but a thin mantle of native modesty,
and ruefully wring the water out of my clothes, I feel considerably
like one! Pine Bluffs provides me with shelter for the night, and a
few miles' travel next morning takes me across the boundary-line into
Nebraska. My route leads down Pole Creek, with ridable roads
probably half the distance, and low, rocky bluffs lining both sides of
the narrow valley, and leading up to high, rolling prairie beyond.
Over these rocky bluffs the Indians were wont to stampede herds
of buffalo, which falling over the precipitous bluffs, would be killed
by hundreds, thus procuring an abundance of beef for the long
winter. There are no buffalo here now—they have departed with
the Indians—and I shall never have a chance to add a bison to
my game-list on this tour. But they have left plenty of tangible
evidence behind, in the shape of numerous deeply worn trails leading from the bluffs to the creek.

The prairie hereabouts is spangled with a wealth of divers-col-
ored flowers that fill the morning air with gratifying perfume.
The air is soft and balmy, in striking contrast to the chilly atmosphere of early morning in the mountain country, where the accumulated snows of a thousand winters exert their chilling influence
in opposition to the benign rays of old Sol. This evening I pass
through "Prairie-dog City," the largest congregation of prairie-
dog dwellings met with on the tour. The "city" covers hundreds
of acres of ground, and the dogs come out in such multitudes to
present their noisy and excitable protests against my intrusion, that
I consider myself quite justified in shooting at them. I hit one
old fellow fair and square, but he disappears like a flash down his
hole, which now becomes his grave. The lightning-like movements
of the prairie-dog, and his instinctive inclination toward his home,
combine to perform the last sad rites of burial for his body at
death. As, toward dark, I near Potter Station, where I expect accommodation for the night, a storm comes howling from the west,
and it soon resolves into a race between me and the storm. With
a good ridable road I could win the race; but, being handicapped
with an unridable trail, nearly obscured beneath tall, rank grass,
the storm overtakes me, and comes in at Potter Station a winner
by about three hundred lengths.

In the morning I start out in good season, and, nearing Sidney, the road becomes better, and I sweep into that enterprising town at a becoming pace.   I conclude to remain at Sidney for dinner, and pass the remainder of the forenoon visiting the neighboring fort.

CHAPTER IV.

FROM THE GREAT PLAINS TO THE ATLANTIC.

THROUGH the courtesy of the commanding officer at Fort Sidney I am enabled to resume my journey eastward under the grateful shade of a military summer helmet in lieu of the semi-sombrero slouch that has lasted me through from San Francisco. Certainly it is not without feelings of compunction that one discards an old friend, that has gallantly stood by me through thick and thin throughout the eventful journey across the inter-mountain country; but the white helmet gives such a delightfully imposing air to my otherwise forlorn and woebegone figure that I ride out of Sidney feeling quite vain. The first thing done is to fill a poor yellow-spotted snake—whose head is boring in the sand—with lively surprise, by riding over his mottled carcass; and only the fact of the tire being rubber, and not steel, enables him to escape unscathed. This same evening, while halting for the night at Lodge Pole Station, the opportunity of observing the awe-inspiring aspect of a great thunder-storm on the plains presents itself. With absolutely nothing to obstruct the vision the Alpha and Omega of the whole spectacle are plainly observable. The gradual mustering of the forces is near the Rockies to the westward, then the skirmish-line of fleecy cloudlets comes rolling and tumbling in advance, bringing a current of air that causes the ponderous wind-mill at the railway tank to "about face" sharply, and sets its giant arms to whirling vigorously around. Behind comes the compact, inky veil that spreads itself over the whole blue canopy above, seemingly banishing all hope of the future; and athwart its Cimmerian surface shoot zigzag streaks of lightning, accompanied by heavy, muttering thunder that rolls and reverberates over the boundless plains seemingly conscious of the spaciousness of its play-ground. Broad sheets of electric flame play along the ground, filling the air with a strange, unnatural light; heavy, pattering raindrops begin to fall, and, ten minutes after, a pelting, pitiless down-pour is drenching the sod-cabin of the lonely rancher, and, for the time being, converting the level plain into a shallow lake.

A fleet of prairie schooners is anchored in the South Platte bottom, waiting for it to dry up, as I trundle down that stream— every mile made interesting by reminiscences of Indian fights and massacres—next day, toward Ogallala ; and one of the "Pilgrims" looks wise as I approach, and propounds the query, "Does it hev ter git very muddy afore yer kin ride yer verlocify, mister ? " "Ya-as, purty dog-goned muddy," I drawl out in reply ; for, although comprehending his meaning, I don't care to venture into

The First Homestead.

an explanatory lecture of uncertain length. Seven weeks' travel through bicycleless territory would undoubtedly convert an angel into a hardened prevaricator, so far as answering questions is concerned.

This afternoon is passed the first homestead, as distinguished from a ranch—consisting of a small tent pitched near a few acres of newly upturned prairie—in the picket-line of the great agricultural empire that is gradually creeping westward over the plains, crowding the autocratic cattle-kings and their herds farther west, even as the Indians and their still greater herds—buffaloes—have been crowded out by the latter. At Ogallala—which but a few years ago was *par excellence* the cow-boys' rallying point—"homesteads," "timber claims," and "pre-emption" now form the all-absorbing topic.

"The Platte's 'petered' since the hoosiers have begun to settle it up," deprecatingly reflects a bronzed cow-boy at the hotel supper-table ; and, from his standpoint, he is correct.

Passing the next night in the dug-out of a homesteader, in the forks of the North and South Platte, I pass in the morning Buffalo Bill's home ranch (the place where a ranch proprietor himself resides is denominated the "home ranch" as distinctive from a ranch presided over by employés only), the house and improvements of which are said to be the finest in Western Nebraska. Taking dinner at North Platte City, I cross over a substantial wagon-bridge, spanning the turgid yellow stream just below where the north and south branches fork, and proceed eastward as "the Platte" simply, reaching Brady Island for the night. Here I encounter extraordinary difficulties in getting supper. Four families, representing the Union Pacific force at this place, all living in separate houses, constitute the population of Brady Island. "All our folks are just recovering from the scarlet fever," is the reply to my first application ; "Muvver's down to ve darden on ve island, and we ain't dot no bread baked," says a barefooted youth at house No. 2 ; "Me ould ooman's across ter the naybur's, 'n' there ain't a boite av grub cooked in the shanty," answers the proprietor of No. 3, seated on the threshold, puffing vigorously at the traditional short clay ; "We all to Nord Blatte been to veesit, und shust back ter home got mit notings gooked," winds up the gloomy programme at No. 4. I am hesitating about whether to crawl in somewhere, supperless, for the night, or push on farther through the darkness, when, "I don't care, pa! it's a shame for a stranger to come here where there are four families and have to go without supper," greet my ears in a musical, tremulous voice. It is the convalescent daughter of house No. 1, valiantly championing my cause ; and so well does she succeed that her "pa" comes out, and notwithstanding my protests, insists on setting out the best they have cooked.

Homesteads now become more frequent, groves of young cottonwoods, representing timber claims, are occasionally encountered, and section-house accommodation becomes a thing of the past. Near Willow Island I come within a trifle of stepping on a belligerent rattlesnake, and in a moment his deadly fangs are hooked to one of the thick canvas gaiters I am wearing. Were my exquisitely outlined calves encased in cycling stockings only, I should have had a "heap sick foot" to amuse myself with for the next three weeks, though there is little danger of being "snuffed out" entirely by a rattlesnake favor these days ; an all-potent remedy is to drink plenty of whiskey as quickly as possible after being bitten, and whiskey is one of the easiest things to obtain in the West. Giving his snakeship to understand that I don't appreciate his "good intentions" by vigorously shaking him off, I turn my "barker" loose on him, and quickly convert him into a "goody-good snake ;" for if "the only good Indian is a dead one," surely the same terse remark applies with much greater force to the vicious and deadly rattler. As I progress eastward, sod-houses and dug-outs become less frequent, and at long intervals frame school-houses appear to remind me that I am passing through a civilized country. Stretches of sand alternate with ridable roads all down the Platte. Often I have to ticklishly wobble along a narrow space between two yawning ruts, over ground that is anything but smooth. I consider it a lucky day that passes without adding one or more to my long and eventful list of headers, and to-day I am fairly "un-horsed" by a squall of wind that—taking me unawares—blows me and the bicycle fairly over.

East of Plum Creek a greater proportion of ridable road is encountered, but they still continue to be nothing more than well-worn wagon-trails across the prairie, and when teams are met en route westward one has to give and the other take, in order to pass. It is doubtless owing to misunderstanding a cycler's capacities, rather than ill-nature, that makes these Western teamsters oblivious to the precept, "It is better to give than to receive ;" and if ignorance is bliss, an outfit I meet to-day ought to comprise the happiest mortals in existence. Near Elm Creek I meet a train of "schooners," whose drivers fail to recognize my right to one of the two wheel-tracks ; and in my endeavor to ride past them on the uneven greensward, I am rewarded by an inglorious header. A dozen freckled Arkansawish faces are watching my movements with undisguised astonishment ; and when my crest-fallen self is spread out on the prairie, these faces—one and all—resolve into expansive grins, and a squeaking female voice from out the nearest wagon, pipes : "La me ! that's a right smart chance of a travelling machine, but, if that's the way they stop 'em, I wonder they don't break every blessed bone in their body ! " But all sorts of people are mingled promiscuously here, for, soon after this incident, two young men come running across the prairie from a semi-dug-out, who prove to be college graduates from "the Hub," who are rooting prairie here in Nebraska, preferring the free, independent life of a Western farmer to the restraints of a position at an Eastern desk. They are more conversant with cycling affairs than myself, and, having heard of my tour, have been on the lookout, expecting I would pass this way.

At Kearney Junction the roads are excellent, and everything is satisfactory ; but an hour's ride east of that city I am shocked at the gross misconduct of a vigorous and vociferous young mule who is confined alone in a pasture, presumably to be weaned. He evidently mistakes the picturesque combination of man and machine for his mother, as, on seeing us approach, he assumes a thirsty, anxious expression, raises his unmusical, undignified voice, and endeavors to jump the fence. He follows along the whole length of the pasture, and when he gets to the end, and realizes that I am drawing away from him, perhaps forever, he bawls out in an agony of grief and anxiety, and, recklessly bursting through the fence, comes tearing down the road, filling the air with the unmelodious notes of his soul-harrowing music. The road is excellent for a piece, and I lead him a lively chase, but he finally overtakes me, and, when I slow-up, he jogs along behind quite contentedly.

East of Kearney the sod-houses disappear entirely, and the improvements are of a more substantial character. At Wood River I " make my bow " to the first growth of natural timber since leaving the mountains, which indicates my gradual advance off the vast timberless plains. Passing through Grand Island, Central City, and other towns, I find myself anchored Saturday evening, June 14th, at Duncan—a settlement of Polackers—an honest-hearted set of folks, who seem to thoroughly understand a cycler's digestive capacity, though understanding nothing whatever about the uses of the machine. Resuming my journey next morning, I find the roads fair. After crossing the Loup River, and passing through Columbus, I reach—about 11 A.M.—a country school-house, with a gathering of farmers hanging around outside, awaiting the arrival of the parson to open the meeting. Alighting, I am engaged in answering forty questions or thereabouts to the minute when that pious individual canters up, and, dismounting from his nag, comes forward and joins in the conversation. He invites me to stop over and hear the sermon ; and when I beg to be excused because desirous of pushing ahead while the weather is favorable His Reverence solemnly warns me against desecrating the Sabbath by going farther than the prescribed " Sabbath-day's journey."

At Fremont I bid farewell to the Platte—which turns south and joins the Missouri River at Plattsmouth—and follow the old military road through the Elkhorn Valley to Omaha. "Military road " sounds like music in a cycler's ear—suggestive of a well-kept and well-graded highway ; but this particular military road between Fremont and Omaha fails to awaken any blithesome sensations to-day, for it is almost one continuous mud-hole. It is called a military road simply from being the route formerly traversed by troops and supply trains bound for the Western forts. Resting a day in Omaha, I obtain a permit to trundle my wheel across the Union Pacific Bridge that spans the Missouri River— the "Big Muddy," toward which I have been travelling so long— between Omaha and Council Bluffs ; I bid farewell to Nebraska, and cross over to Iowa.

Heretofore I have omitted mentioning the tremendously hot weather I have encountered lately, because of my inability to produce legally tangible evidence ; but to-day, while eating dinner at a farm-house, I leave the bicycle standing against the fence, and old Sol ruthlessly unsticks the tire, so that, when I mount, it comes off, and gives me a gymnastic lesson all unnecessary. My first day's experience in the great " Hawkeye State " speaks volumes for the hospitality of the people, there being quite a rivalry between two neighboring farmers about which should take me in to dinner. A compromise is finally made, by which I am to eat dinner at one place, and be "turned loose" in a cherry orchard afterward at the other, to which happy arrangement I, of course, enter no objections. In striking contrast to these friendly advances is my own unpardonable conduct the same evening in conversation with an honest old farmer.

"I see you are taking notes. I suppose you keep track of the crops as you travel along ? " says the H. O. F.

" Certainly, I take more notice of the crops than anything ; I'm a natural born agriculturist myself."

" Well," continues the farmer, " right here where we stand is Carson Township."

"Ah! indeed! Is it possible that I have at last arrived at Carson Township?"

"You have heard of the township before, then, eh?"

"Heard of it! why, man alive, Carson Township is all the talk out in the Rockies; in fact, it is known all over the world as the finest Township for corn in Iowa!"

This sort of conduct is, I admit, unwarrantable in the extreme; but cycling is responsible for it all. If continuous cycling is productive of a superfluity of exhilaration, and said exhilaration bubbles over occasionally, plainly the bicycle is to blame. So forcibly does this latter fact intrude upon me as I shake hands with the farmer, and congratulate him on his rare good fortune in belonging to Carson Township that I mount, and with a view of taking a little of the shine out of it, ride down the long, steep hill leading to the bridge across the Nishnebotene River at a tremendous pace. The machine "kicks" against this treatment, however, and, when about half way down, it strikes a hole and sends me spinning and gyrating through space; and when I finally strike *terra firma*, it thumps me unmercifully in the ribs ere it lets me up.

"Variable" is the word descriptive of the Iowa roads; for seventy-five miles due east of Omaha the prairie rolls like a heavy Atlantic swell, and during a day's journey I pass through a dozen alternate stretches of muddy and dusky road; for like a huge watering-pot do the rain-clouds pass to and fro over this great garden of the West, that is practically one continuous fertile farm from the Missouri to the Mississippi.

Passing through Des Moines on the 23d, muddy roads and hot, thunder-showery weather characterize my journey through Central Iowa, aggravated by the inevitable question, "Why don't you ride?" one Solomon-visaged individual asking me if the railway company wouldn't permit me to ride along one of the rails. No base, unworthy suspicions of a cycler's inability to ride on a two-inch rail finds lodgement in the mind of this wiseacre; but his compassionate heart is moved with tender solicitude as to whether the soulless "company" will, or will not, permit it. Hurrying timorously through Grinnell—the city that was badly demolished and scattered all over the surrounding country by a cyclone in 1882—I pause at Victor, where I find the inhabitants highly elated over the prospect of building a new jail with the fines nightly in-

Germany Transplanted.

W. A. Rogers.

flicted on graders employed on a new railroad near by, who come to town and " hilare " every evening.

" What kind of a place do you call this ? " I inquire, on arriving at a queer-looking town twenty-five miles west of Iowa City.

" This is South Amana, one of the towns of the Amana Society," is the civil reply.

The Amana Society is found upon inquiry to be a communism of Germans, numbering 15,000 souls, and owning 50,000 acres of choice land in a body, with woollen factories, four small towns, and the best of credit everywhere. Everything is common property, and upon withdrawal or expulsion, a member takes with him only the value of what he brought in. The domestic relations are as usual; and while no person of ambition would be content with the conditions of life here, the slow, ease-loving, methodical people composing the society seem well satisfied with their lot, and are, perhaps, happier, on the whole, than the average outsider. I remain here for dinner, and take a look around. The people, the buildings, the language, the food, everything, is precisely as if it had been picked up bodily in some rural district in Germany, and set down unaltered here in Iowa. " Wie gehts," I venture, as I wheel past a couple of plump, rosy-cheeked maidens, in the quaint, old-fashioned garb of the German peasantry. " Wie gehts," is the demure reply from them, both at once; but not the shadow of a dimple responds to my unhappy attempt to win from them a smile. Pretty but not coquettish are these communistic maidens of Amana.

At Tiffin the stilly air of night is made joyous with the mellifluous voices of whip-poor-wills—the first I have heard on the tour—and their tuneful concert is impressed on my memory in happy contrast to certain other concerts, both vocal and instrumental, endured *en route*. Passing through Iowa City, crossing Cedar River at Moscow, nine days after crossing the Missouri, I hear the distant whistle of a Mississippi steamboat. Its hoarse voice is sweetest music to me, heralding the fact that two-thirds of my long tour across the continent is completed. Crossing the " Father of Waters " over the splendid government bridge between Davenport and Rock Island, I pass over into Illinois. For several miles my route leads up the Mississippi River bottom, over sandy roads ; but nearing Rock River, the sand disappears, and, for some distance, an excellent road winds through the oak-groves lining

this beautiful stream. The green woods are free from underbrush, and a cool undercurrent of air plays amid the leafy shades,
which, if not ambrosial, are none the less grateful, as it registers
over 100° in the sun ; without, the silvery sheen of the river glimmers through the interspaces ; the dulcet notes of church-bells
come floating on the breeze from over the river, seeming to proclaim, with their melodious tongues, peace and good-will to all.
Rock River, with its 300 yards in width of unbridged waters, now
obstructs my path, and the ferryboat is tied up on the other shore.
" Whoop-ee," I yell at the ferryman's hut opposite, but without
receiving any response. " Wh-o-o-p-e-ee," I repeat in a gentle,
civilized voice—learned, by the by, two years ago on the Crow res-
ervation in Montana, and which sets the surrounding atmosphere
in a whirl and drowns out the music of the church-bells—but it
has no effect whatever on the case-hardened ferryman in the hut ;
he pays no heed whatever until my persuasive voice is augmented by
the voices of two new arrivals in a buggy, when he sallies serenely
forth and slowly ferries us across. Riding along rather indifferent
roads, between farms worth $100 an acre, through the handsome
town of Geneseo, stopping over night at Atkinson, I resume my journey next morning through a country abounding in all that goes to
make people prosperous, if not happy.   Pretty names are given to
places hereabouts, for on my left I pass " Pink Prairie, bordered
with Green River." Crossing over into Bureau County, I find
splendid gravelled roads, and spend a most agreeable hour with
the jolly Bicycle Club, of Princeton, the handsome county seat of
Bureau County.   Pushing on to Lamoille for the night, the enterprising village barber there hustles me into his cosey shop,
and shaves, shampoos, shingles, bay-rums, and otherwise manipu-
lates me, to the great enhancement of my personal appearance, all,
so he says, for the honor of having lathered the chin of the " great
and only——" In fact, the Illinoisians seem to be most excellent
folks.

   After three days' journey through the great Prairie State my
head is fairly turned with kindness and flattery ; but the third
night, as if to rebuke my vanity, I am bluntly refused shelter at
three different farm-houses.   I am benighted, and conclude to make
the best of it by "turning in" under a hay-cock ; but the Fox
River mosquitoes oust me in short order, and compel me to "mosey"
along through the gloomy night to Yorkville.   At Yorkville a stout

German, on being informed that I am going to ride to Chicago, replies, "What! Ghigago mit dot? Why, mine dear vellow, Ghigago's more as vorty miles; you gan't ride mit dot to Ghigago;" and the old fellow's eyes fairly bulge with astonishment at the bare idea of riding forty miles "mit dot." I considerately refrain from telling him of my already 2,500-mile jaunt "mit dot," lest an apoplectic fit should waft his Teutonic soul to realms of sauer-kraut bliss and Limburger happiness forever. On the morning of July 4th I roll into Chicago, where, having persuaded myself that I deserve a few days' rest, I remain till the Democratic Convention winds up on the 13th.

Fifteen miles of good riding and three of tough trundling, through deep sand, brings me into Indiana, which for the first thirty-five miles around the southern shore of Lake Michigan is simply and solely sand. Finding it next to impossible to traverse the wagon-roads, I trundle around the water's edge, where the sand is firmer because wet. After twenty miles of this I have to shoulder the bicycle and scale the huge sand-dunes that border the lake here, and after wandering for an hour through a bewildering wilderness of swamps, sand-hills, and hickory thickets, I finally reach Miller Station for the night. This place is enough to give one the yellow-edged blues: nothing but swamps, sand, sad-eyed turtles, and ruthless, relentless mosquitoes. At Chesterton the roads improve, but still enough sand remains to break the force of headers, which, notwithstanding my long experience on the road, I still manage to execute with undesirable frequency. To-day I take one, and while unravelling myself and congratulating my lucky stars at being in a lonely spot where none can witness my discomfiture, a gruff, sarcastic "haw-haw" falls like a funeral knell on my ear, and a lanky "Hoosier" rides up on a diminutive pumpkin-colored mule that looks a veritable pygmy between his hoop-pole legs. It is but justice to explain that this latter incident did not occur in "Posey County."

At La Porte the roads improve for some distance, but once again I am benighted, and sleep under a wheat-shock. Traversing several miles of corduroy road, through huckleberry swamps, next morning, I reach Crum's Point for breakfast. A remnant of some Indian tribe still lingers around here and gathers huckleberries for the market, two squaws being in the village purchasing supplies for their camp in the swamps. "What's the name of these Indians here?" I ask.

"One of em's Blinkie, and t'other's Seven-up," is the reply, in a
voice that implies such profound knowledge of the subject that I

Jumbo comes out to meet me.

forbear to investigate further. Splendid gravel roads lead from
Crum's Point to South Bend, and on through Mishawaka, alternating with sandy stretches to Goshen, which town is said—by the

6

Goshenites—to be the prettiest in Indiana ; but there seems to be considerable pride of locality in the great Hoosier State, and I venture there are scores of "prettiest towns in Indiana." Nevertheless, Goshen is certainly a very handsome place, with unusually broad, well-shaded streets ; the centre of a magnificent farming country, it is romantically situated on the banks of the beautiful Elkhart River. At Wawaka I find a corpulent 300-pound cycler, who, being afraid to trust his jumbolean proportions on an ordinary machine, has had an extra stout bone-shaker made to order, and goes out on short runs with a couple of neighbor wheelmen, who, being about fifty per cent. less bulky, ride regulation wheels. "Jumbo" goes all right when mounted, but, being unable to mount without aid, he seldom ventures abroad by himself for fear of having to foot it back. Ninety-five degrees in the shade characterizes the weather these days, and I generally make a few miles in the gloaming—not, of course, because it is cooler, but because the "gloaming" is so delightfully romantic.

At ten o'clock in the morning, July 17th, I bowl across the boundary line into Ohio. Following the Merchants' and Bankers' Telegraph road to Napoleon, I pass through a district where the rain has overlooked them for two months ; the rear wheel of the bicycle is half buried in hot dust ; the blackberries are dead on the bushes, and the long-suffering corn looks as though afflicted with the yellow jaundice. I sup this same evening with a family of Germans, who have been settled here forty years, and scarcely know a word of English yet. A fat, phlegmatic-looking baby is peacefully reposing in a cradle, which is simply half a monster pumpkin scooped out and dried ; it is the most intensely rustic cradle in the world. Surely, this youngster's head ought to be level on agricultural affairs, when he grows up, if anybody's ought !

From Napoleon my route leads up the Maumee River and canal, first trying the tow-path of the latter, and then relinquishing it for the very fair wagon-road. The Maumee River, winding through its splendid rich valley, seems to possess a peculiar beauty all its own, and my mind, unbidden, mentally compares it with our old friend, the Humboldt. The latter stream traverses dreary plains, where almost nothing but sage-brush grows ; the Maumee waters a smiling valley, where orchards, fields, and meadows alternate with sugar-maple groves, and in its fair bosom reflects beautiful landscape views, that are changed and rebeautified by the master-hand of the sun every hour of the day, and doubly embellished at night by the moon. It is whispered that during "the late unpleasantness" the Ohio regiments could out-yell the Louisiana tigers, or any other Confederate troops, two to one. Who has not heard the "Ohio yell?" Most people are magnanimously inclined to regard this rumor as simply a "gag" on the Buckeye boys ; but it isn't. The Ohioans are to the manner born ; the "Buckeye yell" is a tangible fact. All along the Maumee it resounds in my ears ; nearly every man or boy, who from the fields, far or near, sees me bowling along the road, straightway delivers himself of a yell, pure and simple. At Perrysburg I strike the famous "Maumee pike"—forty miles of stone road, almost a dead level. The western half is kept in rather poor repair these days ; but from Fremont eastward it is splendid wheeling. The atmosphere of Bellevue is blue with politics, and myself and another innocent, unsuspecting individual, hailing from New York, are enticed into a political meeting by a wily politician, and dexterously made to pose before the assembled company as two gentlemen who have come—one from the Atlantic, the other from the Pacific—to witness the overwhelming success of the only honest, horny-handed, double-breasted patriots—the . . . party. The roads are found rather sandy east of the pike, and the roadful of wagons going to the circus, which exhibits to-day at Norwalk, causes considerable annoyance.

Erie County, through which I am now passing, is one of the finest fruit countries in the world, and many of the farmers keep open orchard. Staying at Ridgeville overnight, I roll into Cleveland, and into the out-stretched arms of a policeman, at 10 o'clock, next morning. "He was violating the city ordinance by riding on the sidewalk," the arresting policeman informs the captain. "Ah ! he was, hey !" thunders the captain, in a hoarse, bass voice that causes my knees to knock together with fear and trembling ; and the captain's eye seems to look clear through my trembling form. "P-l-e-a-s-e, s-i-r, I d-i-d-n't t-r-y t-o d-o i-t," I falter, in a weak, gasping voice that brings tears to the eyes of the assembled officers and melts the captain's heart, so that he is already wavering between justice and mercy when a local wheelman comes gallantly to the rescue, and explains my natural ignorance of Cleveland's city laws, and I breathe the joyous air of freedom once again.

Three members of the Cleveland Bicycle Club and a visiting

wheelman accompany me ten miles out, riding down far-famed Euclid Avenue, and calling at Lake View Cemetery to pay a visit to
Garfield's tomb.   I bid them farewell at Euclid village.   Following
the ridge road leading along the shore of Lake Erie to Buffalo, I
ride through a most beautiful farming country, passing through
Willoughby and Mentor—Garfield's old home.   Splendidly kept
roads pass between avenues of stately maples, that cast a grateful
shade athwart the highway, both sides of which are lined with
magnificent farms, whose fields and meadows fairly groan beneath
their wealth of produce, whose fructiferous orchards are marvels
of productiveness, and whose barns and stables would be veritable
palaces to the sod-housed homesteaders on Nebraska's frontier
prairies.   Prominent among them stands the old Garfield homestead—a fine farm of one hundred and sixty-five acres, at present
managed by Mrs. Garfield's brother.   Smiling villages nestling
amid stately groves, rearing white church-spires from out their
green, bowery surroundings, dot the low, broad, fertile shore-land
to the left ; the gleaming waters of Lake Erie here and there glisten
like burnished steel through the distant interspaces, and away beyond stretches northward, like a vast mirror, to kiss the blue Canadian skies.

Near Conneaut I whirl the dust of the Buckeye State from my
tire and cross over into Pennsylvania, where, from the little hamlet
of Springfield, the roads become good, then better, and finally best
at Girard—the home of the veteran showman, Dan Rice, the beau-
tifying works of whose generous hand are everywhere visible in his
native town.   Splendid is the road and delightful the country coming east from Girard ; even the red brick school-houses are embowered amid leafy groves ; and so it continues with ever-varying, ever-
pleasing beauty to Erie, after which the highway becomes hardly
so good.

Twenty-four hours after entering Pennsylvania I make my exit
across the boundary into the Empire State.   The roads continue
good, and after dinner I reach Westfield, six miles from the famous
Lake Chautauqua, which beautiful hill and forest embowered sheet
of water is popularly believed by many of its numerous local admirers
to be the highest navigable lake in the world.   If so, however, Lake
Tahoe in the Sierra Nevada Mountains comes next, as it is about six
thousand feet above the level of the sea, and has three steamers plying on its waters !   At Fredonia I am shown through the celebrated

watch-movement factory here, by the captain of the Fredonia Club, who accompanies me to Silver Creek, where we call on another enthusiastic wheelman—a physician who uses the wheel in preference to a horse, in making professional calls throughout the surrounding country. Taking supper with the genial "Doc.," they both accompany me to the summit of a steep hill leading up out of the creek bottom. No wheelman has ever yet rode up this hill, save the muscular and gritty captain of the Fredonia Club, though several have attempted the feat. From the top my road ahead is plainly visible for miles, leading through the broad and smiling Cattaraugus Valley that is spread out like a vast garden below, through which Cattaraugus Creek slowly winds its tortuous way. Stopping over night at Angola I proceed to Buffalo next morning, catching the first glimpse of that important "seaport of the lakes," where, fifteen miles across the bay, the wagon-road is almost licked by the swashing waves; and entering the city over a "misfit" plankroad, off which I am almost upset by the most audaciously indifferent woman in the world. A market woman homeward bound with her empty truck-wagon, recognizes my road-rights to the extent of barely room to squeeze past between her wagon and the ditch ; and holds her long, stiff buggy-whip so that it "swipes" me viciously across the face, knocks my helmet off into the mud ditch, and well-nigh upsets me into the same. The woman—a crimson-crested blonde —jogs serenely along without even deigning to turn her head.

Leaving the bicycle at "Isham's "—who volunteers some slight repairs—I take a flying visit by rail to see Niagara Falls, returning the same evening to enjoy the proffered hospitality of a genial member of the Buffalo Bicycle Club. Seated on the piazza of his residence, on Delaware Avenue, this evening, the symphonious voice of the club-whistle is cast adrift whenever the glowing orb of a cycle-lamp heaves in sight through the darkness, and several members of the club are thus rounded up and their hearts captured by the witchery of a smile—a "smile" in Buffalo, I hasten to explain, is no kin whatever to a Rocky Mountain "smile"—far be it from it ! This club-whistle of the Buffalo Bicycle Club happens to sing the same melodious song as the police-whistle at Washington, D.C.; and the Buffalo cyclers who graced the national league-meet at the Capital with their presence took a folio of club music along. A small but frolicsome party of them on top of the Washington monument, "heaved a sigh " from their whistles, at a comrade passing along the street

below, when a corpulent policeman, naturally mistaking it for a
signal from a brother "cop," hastened to climb the five hundred
feet or thereabouts of ascent up the monument. When he arrived,
puffing and perspiring, to the summit, and discovered his mistake,
the wheelmen say he made such awful use of the Queen's English
that the atmosphere had a blue, sulphurous tinge about it for some
time after.

Leaving Buffalo next morning I pass through Batavia, where
the wheelmen have a most æsthetic little club-room. Besides being jovial and whole-souled fellows, they are awfully æsthetic ; and
the sweetest little Japanese *curios* and *bric-à-brac* decorate the walls
and tables.

Stopping over night at LeRoy, in company with the president
and captain of the LeRoy Club, I visit the State fish-hatchery at
Mumford next morning, and ride on through the Genesee Valley,
finding fair roads through the valley, though somewhat hilly and
stony toward Canandaigua. Inquiring the best road to Geneva I
am advised of the superiority of the one leading past the poor-
house. Finding them somewhat intricate, and being too super-
sensitive to stop people and ask them the road to the poor-house,
I deservedly get lost, and am wandering erratically eastward
through the darkness, when I fortunately meet a wheelman in
a buggy, who directs me to his mother's farm-house near by,
with instructions to that most excellent lady to accommodate me
for the night. Nine o'clock next morning I reach fair Geneva, so
beautifully situated on Seneca's silvery lake, passing the State agricultural farm *en route ;* continuing on up the Seneca River, passing
through Waterloo and Seneca Falls to Cayuga, and from thence to
Auburn and Skaneateles, where I heave a sigh at the thoughts of
leaving the last—I cannot say the loveliest, for all are equally lovely
—of that beautiful chain of lakes that transforms this part of New
York State into a vast and delightful summer resort.

"Down a romantic Swiss glen, where scores of sylvan nooks
and rippling rills invite one to cast about for fairies and sprites," is
the word descriptive of my route from Marcellus next morning.
Once again, on nearing the Camillus outlet from the narrow vale, I
hear the sound of Sunday bells, and after the church-bell-less
Western wilds, it seems to me that their notes have visited me
amid beautiful scenes, strangely often of late. Arriving at Camillus, I ask the name of the sparkling little stream that dances along

Amenities of the Erie Tow-path.

this fairy glen like a child at play, absorbing the sun-rays and coquettishly reflecting them in the faces of the venerable oaks that bend over it like loving guardians protecting it from evil. My ears are prepared to hear a musical Indian name—"Laughing-Waters" at least ; but, like a week's washing ruthlessly intruding upon love's young dream, falls on my waiting ears the unpoetic misnomer, "Nine-Mile Creek."

Over good roads to Syracuse, and from thence my route leads down the Erie Canal, alternately riding down the canal tow-path, the wagon-roads, and between the tracks of the New York Central Railway. On the former, the greatest drawback to peaceful cycling is the towing-mule and his unwarrantable animosity toward the bicycle, and the awful, unmentionable profanity engendered thereby in the utterances of the boatmen. Sometimes the burden of this sulphurous profanity is aimed at me, sometimes at the inoffensive bicycle, or both of us collectively, but oftener is it directed at the unspeakable mule, who is really the only party to blame. A mule scares, not because he is really afraid, but because he feels skittishly inclined to turn back, or to make trouble between his enemies—the boatmen, his task-master, and the cycler, an intruder on his exclusive domain, the Erie tow-path. A span of mules will pretend to scare, whirl around, and jerk loose from the driver, and go ":scooting" back down the tow-path in a manner indicating that nothing less than a stone wall would stop them ; but, exactly in the nick of time to prevent the tow-line jerking them sidewise into the canal, they stop. Trust a mule for never losing his head when he runs away, as does his hot-headed relative, the horse ; he never once allows surrounding circumstances to occupy his thoughts to an extent detrimental to his own self-preservative interests. The Erie Canal mule's first mission in life is to engender profanity and strife between boatmen and cyclists, and the second is to work and chew hay, which brings him out about even with the world all round.

At Rome I enter the famous and beautiful Mohawk Valley, a place long looked forward to with much pleasurable anticipation, from having heard so often of its natural beauties and its interesting historical associations. "It's the garden spot of the world ; and travellers who have been all over Europe and everywhere, say there's nothing in the world to equal the quiet landscape beauty of the Mohawk Valley," enthusiastically remarks an old gentleman

in spectacles, whom I chance to encounter on the heights east of
Herkimer. Of the first assertion I have nothing to say, having
passed through a dozen " garden spots of the world " on this tour
across America ; but there is no gainsaying the fact that the Mohawk
Valley, as viewed from this vantage spot, is wonderfully beautiful.
I think it must have been on this spot that the poet received inspiration to compose the beautiful song that is sung alike in the
quiet homes of the valley itself and in the trapper's and hunter's
tent on the far off Yellowstone—

> " Fair is the vale where the Mohawk gently glides,
> On its clear, shining way to the sea."

The valley is one of the natural gateways of commerce, for, at Little Falls—where it contracts to a mere pass between the hills—one
can almost throw a stone across six railway tracks, the Erie Canal
and the Mohawk River. Spending an hour looking over the magnificent Capitol building at Albany, I cross the Hudson, and
proceed to ride eastward between the two tracks of the Boston &
Albany Railroad, finding the riding very fair. From the elevated
road-bed I cast a longing, lingering look down the Hudson Valley,
that stretches away southward like a heaven-born dream, and
sigh at the impossibility of going two ways at once. " There's
$50 fine for riding a bicycle along the B. & A. Railroad," I am
informed at Albany, but risk it to Schodack, where I make inquiries
of a section foreman. " No ; there's no foine ; but av yeez are run
over an' git killed, it'll be useless for yeez to inther suit agin the
company for damages," is the reassuring reply ; and the unpleasant
visions of bankrupting fines dissolve in a smile at this characteristic
Milesian explanation.

Crossing the Massachusetts boundary at the village of State
Line, I find the roads excellent ; and, thinking that the highways
of the " Old Bay State " will be good enough anywhere, I grow
careless about the minute directions given me by Albany wheelmen, and, ere long, am laboriously toiling over the heavy roads
and steep grades of the Berkshire Hills, endeavoring to get what
consolation I can, in return for unridable roads, out of the charming
scenery, and the many interesting features of the Berkshire-Hill
country. It is at Otis, in the midst of these hills, that I first become acquainted with the peculiar New England dialect in its native home.

The widely heralded intellectual superiority of the Massachusetts fair ones asserts itself even in the wildest parts of these wild hills ; for at small farms—that, in most States, would be characterized by bare-footed, brown-faced housewives—I encounter spectacled ladies whose fair faces reflect the encyclopædia of knowledge within, and whose wise looks naturally fill me with awe. At Westfield I learn that Karl Kron, the author and publisher of the American road-book, "Ten Thousand Miles on a Bicycle"—not to be outdone by my exploit of floating the bicycle across the Humboldt—undertook the perilous feat of swimming the Potomac with his bicycle suspended at his waist, and had to be fished up from the bottom with a boat-hook. Since then, however, I have seen the gentleman himself, who assures me that the whole story is a canard. Over good roads to Springfield—and on through to Palmer ; from thence riding the whole distance to Worcester between the tracks of the railway, in preference to the variable country roads.

On to Boston next morning, now only forty miles away, I pass venerable weather-worn mile-stones, set up in old colonial days, when the Great West, now trailed across with the rubber hoof-marks of "the popular steed of to-day," was a pathless wilderness, and on the maps a blank. Striking the famous "sand-papered roads" at Framingham—which, by the by, ought to be pumice-stoned a little to make them as good for cycling as stretches of gravelled road near Springfield, Sandwich, and Plano, Ill. ; La Porte, and South Bend, Ind. ; Mentor, and Willoughby, O. ; Girard, Penn. ; several places on the ridge road between Erie and Buffalo, and the alkali flats of the Rocky Mountain territories. Soon the blue intellectual haze hovering over "the Hub" heaves in sight, and, at two o'clock in the afternoon of August 4th, I roll into Boston, and whisper to the wild waves of the sounding Atlantic what the sad sea-waves of the Pacific were saying when I left there, just one hundred and three and a half days ago, having wheeled about 3,700 miles to deliver the message.

Passing the winter of 1884-85 in New York, I became acquainted with the *Outing Magazine*, contributed to it sketches of my tour across America, and in the Spring of 1885 continued around the world as its special correspondent ; embarking April 9th from New York, for Liverpool, aboard the City of Chicago.

CHAPTER V.

At one P.M., on that day, the ponderous but shapely hull of the City of Chicago, with its living and lively freight, moves from the dock as though it, too, were endowed with mind as well as matter ; the crowds that a minute ago disappeared down the gang-plank are now congregated on the outer end of the pier, a compact mass of waving handkerchiefs, and anxious-faced people shouting out signs of recognition to friends aboard the departing steamer.

From beginning to end of the voyage across the Atlantic the weather is delightful ; and the passengers—well, half the cabin-passengers are members of Henry Irving's Lyceum Company en route home after their second successful tour in America ; and old voyagers abroad who have crossed the Atlantic scores of times pronounce it altogether the most enjoyable trip they ever experienced. The third day out we encountered a lonesome-looking iceberg—an object that the captain seemed to think would be better appreciated, and possibly more affectionately remembered, if viewed at the respectful distance of about four miles. It proves a cold, unsympathetic berg, yet extremely entertaining in its own way, since it accommodates us by neutralizing pretty much all the surplus caloric in the atmosphere around for hours after it has disappeared below the horizon of our vision.

I am particularly fortunate in finding among my fellow-passengers Mr. Harry B. French, the traveller and author, from whom I obtain much valuable information, particularly of. China. Mr. French has travelled some distance through the Flowery Kingdom himself, and thoughtfully forewarns me to anticipate a particularly lively and interesting time in invading that country with a vehicle so strange and incomprehensible to the Celestial mind as a bicycle. This experienced gentleman informs me, among other interesting things, that if five hundred chattering Celestials batter down the door and swarm unannounced at midnight into the apartment where

I am endeavoring to get the first wink of sleep obtained for a whole week, instead of following the natural inclinations of an Anglo-Saxon to energetically defend his rights with a stuffed club, I shall display Solomon-like wisdom by quietly submitting to the invasion, and deferentially bowing to Chinese inquisitiveness. If, on an occasion of this nature, one stationed himself behind the door, and, as a sort of preliminary warning to the others, greeted the first interloper with the business end of a boot-jack, he would be morally certain of a lively one-sided misunderstanding that might end disastrously to himself; whereas, by meekly submitting to a critical and exhaustive examination by the assembled company, he might even become the recipient of an apology for having had to batter down the door in order to satisfy their curiosity. One needs more discretion than valor in dealing with the Chinese.

At noon on the 19th we reach Liverpool, where I find a letter awaiting me from A. J. Wilson (Faed), inviting me to call on him at Powerscroft House, London, and offering to tandem me through the intricate mazes of the West End ; likewise asking whether it would be agreeable to have him, with others, accompany me from London down to the South coast—a programme to which, it is needless to say, I entertain no objections. As the custom-house officer wrenches a board off the broad, flat box containing my American bicycle, several fellow-passengers, prompted by their curiosity to obtain a peep at the machine which they have learned is to carry me around the world, gather about ; and one sympathetic lady, as she catches a glimpse of the bright nickeled forks, exclaims, " Oh, what a shame that they should be allowed to wrench the planks off ! They might injure it ;" but a small tip thoroughly convinces the individual prying off the board that, by removing one section and taking a conscientious squint in the direction of the closed end, his duty to the British government would be performed as faithfully as though everything were laid bare ; and the kind-hearted lady's apprehensions of possible injury are thus happily allayed. In two hours after landing, the bicycle is safely stowed away in the underground store-rooms of the Liverpool & Northwestern Railway Company, and in two hours more I am wheeling rapidly toward London, through neatly cultivated fields, and meadows and parks of that intense greenness met with nowhere save in the British Isles, and which causes a couple of native Americans, riding in the same compartment, and who are visiting England for the first time, to express their admiration of it all in the unmeasured language of the genuine Yankee when truly astonished and delighted.

Arriving in London I lose no time in seeking out Mr. Bolton, a well-known wheelman, who has toured on the continent probably as extensively as any other English cycler, and to whom I bear a letter of introduction. Together, on Monday afternoon, we ruthlessly invade the sanctums of the leading cycling papers in London. Mr. Bolton is also able to give me several useful hints concerning wheeling through France and Germany. Then comes the application for a passport, and the inevitable unpleasantness of being suspected by every policeman and detective about the government buildings of being a wild-eyed dynamiter recently arrived from America with the fell purpose of blowing up the place.

On Tuesday I make a formal descent on the Chinese Embassy, to seek information regarding the possibility of making a serpentine trail through the Flowery Kingdom *via* Upper Burmah to Hong-Kong or Shanghai. Here I learn from Dr. McCarty, the interpreter at the Embassy, as from Mr. French, that, putting it as mildly as possible, I must expect a wild time generally in getting through the interior of China with a bicycle. The Doctor feels certain that I may reasonably anticipate the pleasure of making my way through a howling wilderness of hooting Celestials from one end of the country to the other. The great danger, he thinks, will be not so much the well-known aversion of the Chinese to having an " outer barbarian " penetrate the sacred interior of their country, as the enormous crowds that would almost constantly surround me out of curiosity at both rider and wheel, and the moral certainty of a foreigner unwittingly doing something to offend the Chinamen's peculiar and deep-rooted notions of propriety. This, it is easily seen, would be a peculiarly ticklish thing to do when surrounded by surging masses of dangling pig-tails and cerulean blouses, the wearers of which are from the start predisposed to make things as unpleasant as possible. My own experience alone, however, will prove the kind of reception I am likely to meet with among them ; and if they will only considerately refrain from impaling me on a bamboo, after a barbarous and highly ingenious custom of theirs, I little reck what other unpleasantries they have in store. After one remains in the world long enough to find it out, he usually becomes less fastidious about the future of things in general, than when in the hopeful days of boyhood every prospect ahead was fringed with the golden expectations of a budding and inexperienced imagery ; nevertheless, a thoughtful, meditative person, who realizes the necessity of drawing the line somewhere, would naturally draw it at impalation. Not being conscious of any presentiment savoring of impalation, however, the only request I make of the Chinese, at present, is to place no insurmountable obstacle against my pursuing the even—or uneven, as the case may be—tenor of my way through their country. China, though, is several revolutions of my fifty-inch wheel away to the eastward, at this present time of writing, and speculations in regard to it are rather premature.

Soon after reaching London I have the pleasure of meeting "Faed," a gentleman who carries his cycling enthusiasm almost where some people are said to carry their hearts—on his sleeve ; so that a very short acquaintance only is necessary to convince one of being in the company of a person whose interest in whirling wheels is of no ordinary nature. When I present myself at Powers-croft House, Faed is busily wandering around among the curves and angles of no less than three tricycles, apparently endeavoring to encompass the complicated mechanism of all three in one grand comprehensive effort of the mind, and the addition of as many tricycle crates standing around makes the premises so suggestive of a flourishing tricycle agency that an old gentleman, happening to pass by at the moment, is really quite excusable in stopping and inquiring the prices, with a view to purchasing one for himself. Our tandem ride through the West End has to be indefinitely postponed, on account of my time being limited, and our inability to procure readily a suitable machine ; and Mr. Wilson's bump of discretion would not permit him to think of allowing me to attempt the feat of manœuvring a tricycle myself among the bewildering traffic of the metropolis, and risk bringing my " wheel around the world " to an inglorious conclusion before being fairly begun. While walking down Parliament Street my attention is called to a venerable-looking gentleman wheeling briskly along among the throngs of vehicles of every description, and I am informed that the bold tricycler is none other than Major Knox Holmes, a vigorous youth of some seventy-eight summers, who has recently accomplished the feat of riding one hundred and fourteen miles in ten hours ; for a person nearly eighty years of age this is really quite a promising performance, and there is small doubt but that when the gallant

Major gets a little older—say when he becomes a centenarian—he
will develop into a veritable prodigy on the cinder-path !

Having obtained my passport, and got it *viséd* for the Sultan's
dominions at the Turkish consulate, and placed in Faed's possession a bundle of maps, which he generously volunteers to forward
to me, as I require them in the various countries it is proposed to
traverse, I return on April 30th to Liverpool, from which point the
formal start on the wheel across England is to be made. Four
o'clock in the afternoon of May 2d is the time announced, and
Edge Hill Church is the appointed place, where Mr. Lawrence
Fletcher, of the Anfield Bicycle Club, and a number of other Liverpool wheelmen, have volunteered to meet and accompany me some
distance out of the city. Several of the Liverpool daily papers have
made mention of the affair. Accordingly, upon arriving at the appointed place and time, I find a crowd of several hundred people
gathered to satisfy their curiosity as to what sort of a looking individual it is who has crossed America awheel, and furthermore proposes to accomplish the greater feat of the circumlocution of the
globe. A small sea of hats is enthusiastically waved aloft ; a ripple
of applause escapes from five hundred English throats as I mount
my glistening bicycle ; and, with the assistance of a few policemen,
the twenty-five Liverpool cyclers who have assembled to accompany
me out, extricate themselves from the crowd, mount and fall into
line two abreast ; and merrily we wheel away down Edge Lane and
out of Liverpool.

English weather at this season is notoriously capricious, and the
present year it is unusually so, and ere the start is fairly made we
are pedaling along through quite a pelting shower, which, however,
fails to make much impression on the roads beyond causing the
flinging of more or less mud. The majority of my escort are members of the Anfield Club, who have the enviable reputation of being
among the hardest road-riders in England, several members having
accomplished over two hundred miles within the twenty-four hours ;
and I am informed that Mr. Fletcher is soon to undertake the task
of beating the tricycle record over that already well-contested route,
from John o' Groat's to Land's End. Sixteen miles out I become the
happy recipient of hearty well-wishes innumerable, with the accompanying hand-shaking, and my escort turn back toward home and
Liverpool—all save four, who wheel on to Warrington and remain
overnight, with the avowed intention of accompanying me twenty-

five miles farther to-morrow morning. Our Sunday morning experience begins with a shower of rain, which, however, augurs well
for the remainder of the day ; and, save for a gentle head wind, no
reproachful remarks are heard about that much-criticised individual, the clerk of the weather ; especially as our road leads through
a country prolific of everything charming to one's sense of the beautiful. Moreover, we are this morning bowling along the self-same
highway that in days of yore was among the favorite promenades
of a distinguished and enterprising individual known to every British juvenile as Dick Turpin—a person who won imperishable renown, and the undying affection of the small Briton of to-day, by
making it unsafe along here for stage-coaches and travellers indis-
creet enough to carry valuables about with them.

"Think I'll get such roads as this all through England ? " I ask
of my escort as we wheel joyously southward along smooth, macadamized highways that would make the "sand-papered roads"
around Boston seem almost unfit for cycling in comparison, and
that lead through picturesque villages and noble parks ; occasionally catching a glimpse of a splendid old manor among venerable
trees, that makes one unconsciously begin humming :—

> "The ancient homes of England,
>   How beautiful they stand
> Amidst the tall ancestral trees
>   O'er all the pleasant land ! "

"Oh, you'll get much better roads than this in the southern
counties," is the reply ; though, fresh from American roads, one
can scarce see what shape the improvements can possibly take.
Out of Lancashire into Cheshire we wheel, and my escort, after
wishing me all manner of good fortune in hearty Lancashire style,
wheel about and hie themselves back toward the rumble and roar
of the world's greatest sea-port, leaving me to pedal pleasantly
southward along the green lanes and amid the quiet rural scenery
of Staffordshire to Stone, where I remain Sunday night. The country is favored with another drenching down-pour of rain during the
night, and moisture relentlessly descends at short, unreliable intervals on Monday morning, as I proceed toward Birmingham.
Notwithstanding the superabundant moisture the morning ride is
a most enjoyable occasion, requiring but a dash of sunshine to
make everything perfect. The mystic voice of the cuckoo is heard

from many an emerald copse around; songsters that inhabit only the green hedges and woods of "Merrie England" are carolling their morning vespers in all directions; skylarks are soaring, soaring skyward, warbling their unceasing pæans of praise as they gradually ascend into cloudland's shadowy realms; and occasionally I bowl along beneath an archway of spreading beeches that are colonized by crowds of noisy rooks incessantly "cawing" their approval or disapproval of things in general. Surely England, with its wellnigh perfect roads, the wonderful greenness of its vegetation, and its roadsters that meet and regard their steel-ribbed rivals with supreme indifference, is the natural paradise of 'cyclers. There is no annoying dismounting for frightened horses on these happy highways, for the English horse, though spirited and brimful of fire, has long since accepted the inevitable, and either has made friends with the wheelman and his swift-winged steed, or, what is equally agreeable, maintains a haughty reserve.

Pushing along leisurely, between showers, into Warwickshire, I reach Birmingham about three o'clock, and, after spending an hour or so looking over some tricycle works, and calling for a leather writing-case they are making especially for my tour, I wheel on to Coventry, having the company of Mr. Priest, Jr., of the tricycle works, as far as Stonehouse. Between Birmingham and Coventry the recent rainfall has evidently been less, and I mentally note this fifteen-mile stretch of road as the finest traversed since leaving Liverpool, both for width and smoothness of surface, it being a veritable boulevard. Arriving at Coventry I call on "Brother Sturmey," a gentleman well and favorably known to readers of 'cycling literature everywhere; and, as I feel considerably like deserving reasonably gentle treatment after perseveringly pressing forward sixty miles in spite of the rain, I request him to steer me into the Cyclists' Touring Club Hotel—an office which he smilingly performs, and thoughtfully admonishes the proprietor to handle me as tenderly as possible. I am piloted around to take a hurried glance at Coventry, visiting, among other objects of interest, the Starley Memorial. This memorial is interesting to 'cyclers from having been erected by public subscription in recognition of the great interest Mr. Starley took in the 'cycle industry, he having been, in fact, the father of the interest in Coventry, and, consequently, the direct author of the city's present prosperity.

The mind of the British small boy along my route has been

7

taxed to its utmost to account for my white military helmet, and
various and interesting are the passing remarks heard in consequence. The most general impression seems to be that I am direct
from the Soudan, some youthful Conservatives blandly intimating

The Starley Memorial, Coventry.

that I am the advance-guard of a general scuttle of the army out
of Egypt, and that presently whole regiments of white-helmeted
wheelmen will come whirling along the roads on nickel-plated
steeds, some even going so far as to do me the honor of calling

me General Wolseley ; while others—rising young Liberals, probably—recklessly call me General Gordon, intimating by this that the hero of Khartoum was not killed, after all, and is proving it by sweeping through England on a bicycle, wearing a white helmet to prove his identity !

A pleasant ride along a splendid road, shaded for miles with rows of spreading elms, brings me to the charming old village of Dunchurch, where everything seems moss-grown and venerable with age. A squatty, castle-like church-tower, that has stood the brunt of

Resting in an English Village.

many centuries, frowns down upon a cluster of picturesque, thatched cottages of primitive architecture, and ivy-clad from top to bottom ; while, to make the picture complete, there remain even the old wooden stocks, through the holes of which the feet of boozy unfortunates were wont to be unceremoniously thrust in the good old times of rude simplicity ; in fact, the only really unprimitive building about the place appears to be a newly erected Methodist chapel.   It couldn't be—no, of course it couldn't be possible, that there is any connecting link between the American peculiarity of

elevating the feet on the window-sill or the drum of the heating-
stove and this old-time custom of elevating the feet of those of our
ancestors possessed of boozy, hilarious proclivities !

At Weedon Barracks I make a short halt to watch the soldiers •
go through the bayonet exercises, and suffer myself to be persuaded into quaffing a mug of delicious, creamy stout at the can-
teen with a genial old sergeant, a bronzed veteran who has seen
active service in several of the tough expeditions that England
seems ever prone to undertake in various uncivilized quarters of
the world ; after which I wheel away over old Roman military
roads, through Northamptonshire and Buckinghamshire, reaching
Fenny Stratford just in time to find shelter against the machina-
tions of the weather-clerk, who, having withheld rain nearly all the
afternoon, begins dispensing it again in the gloaming. It rains
uninterruptedly all night ; but, although my route for some miles is
now down cross-country lanes, the rain has only made them rather
disagreeable, without rendering them in any respect unridable ;
and although I am among the slopes of the Chiltern Hills, scarcely
a dismount is necessary during the forenoon.   Spending the night
at Berkhamstead, Hertfordshire, I pull out toward London on
Thursday morning, and near Watford am highly gratified at meeting Faed and the captain of the North London Tricycle Club, who
have come out on their tricycles from London to meet and escort
me into the metropolis.   At Faed's suggestion I decide to remain
over in London until Saturday, to be present at the annual tricycle
meet on Barnes Common, and together we wheel down the Edge-
ware Road, Park Road, among the fashionable turnouts of Pic-
cadilly, past Knightsbridge and Brompton to the "Inventories"
Exhibition, where we spend a most enjoyable afternoon inspecting
the thousand and one material evidences of inventive genius from
the several countries represented.

Five hundred and twelve 'cyclers, including forty-one tandem
tricycles and fifty ladies, ride in procession at the Barnes Common
meet, making quite an imposing array as they wheel two abreast
between rows of enthusiastic spectators.   Here, among a host of
other wheeling celebrities, I am introduced to Major Knox Holmes,
before mentioned as being a gentleman of extraordinary powers of
endurance, considering his advanced age.   After tea a number of •
tricyclers accompany me down as far as Croydon, which place we
enter to the pattering music of a drenching rain-storm, experienc-

ing the accompanying pleasure of a wet skin, etc. The threatening aspect of the weather on the following morning causes part of our company to hesitate about venturing any farther from London ; but Faed and three companions wheel with me toward Brighton through a gentle morning shower, which soon clears away, however, and, before long, the combination of the splendid Sussex roads, fine breezy weather, and lovely scenery, amply repays us for the discomforts of yester-eve. Fourteen miles from Brighton we are met by eight members of the Kempton Rangers Bicycle Club, who have sallied forth thus far northward to escort us into town ; having done which, they deliver us over to Mr. C———, of the Brighton Tricycle Club, and brother-in-law to the mayor of the city. It is two in the afternoon. This gentleman straightway ingratiates himself into our united affections, and wins our eternal gratitude, by giving us a regular wheelman's dinner, after which he places us under still further obligations by showing us as many of the lions of Brighton as are accessible on Sunday, chief among which is the famous Brighton Aquarium, where, by his influence, he kindly has the diving-birds and seals fed before their usual hour, for our especial delectation—a proceeding which naturally causes the barometer of our respective self-esteems to rise several notches higher than usual, and doubtless gives equal satisfaction to the seals and diving-birds. We linger at the aquarium until near sun-down, and it is fifteen miles by what is considered the smoothest road to Newhaven. Mr. C——— declares his intention of donning his riding-suit and, by taking a shorter, though supposably rougher, road, reach Newhaven as soon as we. As we halt at Lewes for tea, and ride leisurely, likewise submitting to being photographed *en route*, he actually arrives there ahead of us.

It is Sunday evening, May 10th, and my ride through " Merrie England " is at an end. Among other agreeable things to be ever remembered in connection with it is the fact that it is the first three hundred miles of road I ever remember riding over without scoring a header—a circumstance that impresses itself none the less favorably perhaps when viewed in connection with the solidity of the average English road. It is not a very serious misadventure to take a flying header into a bed of loose sand on an American country road ; but the prospect of rooting up a flint-stone with one's nose, or knocking a curb-stone loose with one's bump of cautiousness, is an entirely different affair ; consequently, the universal smoothness of the surface of the English highways is appreciated at its full value by at least one wheelman whose experience of roads is nothing if not varied. Comfortable quarters are assigned me on board the Channel steamer, and a few minutes after bidding friends and England farewell, at Newhaven, at 11.30 P.M., I am gently rocked into unconsciousness by the motion of the vessel, and remain happily and restfully oblivious to my surroundings until awakened next morning at Dieppe, where I find myself, in a few minutes, on a foreign shore. All the way from San Francisco to Newhaven there is a consciousness of being practically in one country and among one people—people who, though acknowledging separate governments, are bound so firmly together by the ties of common instincts and interests, and the mystic brotherhood of a common language and a common civilization, that nothing of a serious nature can ever come between them. But now I am verily among strangers, and the first thing talked of is to make me pay duty on the bicycle.

The captain of the vessel, into whose hands Mr. C—— assigned me at Newhaven, protests on my behalf, and I likewise enter a gentle demurrer; but the custom-house officer declares that a duty will have to be forthcoming, saying that the amount will be returned again when I pass over the German frontier. The captain finally advises the payment of the duty and the acceptance of a receipt for the amount, and takes his leave. Not feeling quite satisfied as yet about paying the duty, I take a short stroll about Dieppe, leaving my wheel at the custom-house; and when I shortly return, prepared to pay the assessment, whatever it may be; the officer who, but thirty minutes since, declared emphatically in favor of a duty, now answers, with all the politeness imaginable: "Monsieur is at liberty to take the velocipede and go whithersoever he will." It is a fairly prompt initiation into the impulsiveness of the French character. They don't accept bicycles as baggage, though, on the Channel steamers, and six shillings freight, over and above passage-money, has to be yielded up.

Although upon a foreign shore, I am not yet, it seems, to be left entirely alone to the tender mercies of my own lamentable inability to speak French. Fortunately there lives at Dieppe a gentleman named Mr. Parkinson, who, besides being an Englishman to the backbone, is quite an enthusiastic wheelman, and, among other things, considers it his solemn duty to take charge of visiting

The Dieppe Milkman.

'cyclers from England and America and see them safely launched along the magnificent roadways of Normandy, headed fairly toward their destination. Faed has thoughtfully notified Mr. Parkinson of my approach, and he is watching for my coming as tenderly as though I were a returning prodigal and he charged with my welcoming home. Close under the frowning battlements of Dieppe Castle—a once wellnigh impregnable fortress that was some time in possession of the English—romantically nestles Mr. Parkinson's studio, and that genial gentleman promptly proposes accompanying me some distance into the country. On our way through Dieppe I notice blue-bloused peasants guiding small flocks of goats through the streets, calling them along with a peculiar, tuneful instrument that sounds somewhat similar to a bagpipe. I learn that they are Normandy peasants, who keep their flocks around town all summer, goat's milk being considered beneficial for infants and invalids. They lead the goats from house to house, and milk whatever quantity their customers want at their own door—a custom that we can readily understand will never become widely popular among Anglo-Saxon milkmen, since it leaves no possible chance for pump-handle combinations and corresponding profits. The morning is glorious with sunshine and the carols of feathered songsters as together we speed away down the beautiful Arques Valley, over roads that are simply perfect for wheeling ; and, upon arriving at the picturesque ruins of the Château d'Arques, we halt and take a casual peep at the crumbling walls of this once famous fortress, which the trailing ivy of Normandy now partially covers with a dark-green mantle of charity, as though its purpose and its mission were to hide its fallen grandeur from the rude gaze of the passing stranger.

All along the roads we meet happy-looking peasants driving into Dieppe market with produce. They are driving Normandy horses —and that means fine, large, spirited animals—which, being unfamiliar with bicycles, almost invariably take exception to ours, prancing about after the usual manner of high-strung steeds. Unlike his English relative, the Norman horse looks not supinely upon the whirling wheel, but arrays himself almost unanimously against us, and usually in the most uncompromising manner, similar to the phantom-eyed roadster of the United States agriculturist. The similarity between the turnouts of these two countries I am forced to admit, however, terminates abruptly with the horse itself, and does not by any means extend to the driver ; for, while the Nor-

mandy horse capers about and threatens to upset the vehicle into
the ditch, the Frenchman's face is wreathed in apologetic smiles;
and, while he frantically endeavors to keep the refractory horse
under control, he delivers himself of a whole dictionary of apologies
to the wheelman for the animal's foolish conduct, touches his cap
with an air of profound deference upon noticing that we have considerately slowed up, and invariably utters his *Bon jour, monsieur*,
as we wheel past, in a voice that plainly indicates his acknowledgment of the wheelman's—or anybody else's—right to half the roadway.   A few days ago I called the English roads perfect, and England the paradise of 'cyclers ; and so it is ; but the Normandy roads
are even superior, and the scenery of the Arques Valley is truly
lovely.   There is not a loose stone, a rut, or depression anywhere
on these roads, and it is little exaggeration to call them veritable
billiard-tables for smoothness of surface.   As one bowls smoothly
along over them he is constantly wondering how they can possibly
keep them in such condition.   Were these fine roads in America
one would never be out of sight of whirling wheels.

A luncheon of Normandy cheese and cider at Clères, and then
onward to Rouen is the word.   At every cross-roads is erected an
iron guide-post, containing directions to several of the nearest
towns, telling the distances in kilometres and yards ; and small
stone pillars are set up alongside the road, marking every hundred
yards.   Arriving at Rouen at four o'clock, Mr. Parkinson shows me
the famous old Rouen Cathedral, the Palace of Justice, and such
examples of old mediæval Rouen as I care to visit, and, after inviting me to remain and take dinner with him by the murmuring
waters of the historic Seine, he bids me *bon voyage*, turns my head
southward, and leaves me at last a stranger among strangers, to
"*comprendre Français*" unassisted.   Some wiseacre has placed it
on record that too much of a good thing is worse than none at all ;
however that may be, from having concluded that the friendly iron
guide-posts would be found on every corner where necessary,
pointing out the way with infallible truthfulness, and being doubtless influenced by the superior levelness of the road leading down
the valley of the Seine in comparison with the one leading over the
bluffs, I wander toward eventide into Elbeuf, instead of Pont de
l'Arques, as I had intended ; but it matters little, and I am content to make the best of my surroundings.   Wheeling along the
crooked, paved streets of Elbeuf, I enter a small hotel, and, after

the customary exchange of civilities, I arch my eyebrows at an in-
telligent-looking madame, and inquire, " *Comprendre Anglais ?*"—
" *Non,*" replies the lady, looking puzzled, while I proceed to venti-
late my pantomimic powers to try and make my wants understood.
After fifteen minutes of despairing effort, mademoiselle, the daughter, is despatched to the other side of the town, and presently returns with a bewhiskered Frenchman, who, in very much broken
English, accompanying his words with wondrous gesticulations,
gives me to understand that he is the only person in all Elbeuf
capable of speaking the English language, and begs me to unbur-
den myself to him without reserve.   He proves himself useful and
obliging, kindly interesting himself in obtaining me comfortable
accommodation at reasonable rates.   This Elbeuf hotel, though, is
anything but an elegant establishment, and *le propriétaire,* though
seemingly intelligent enough, brings me out a bottle of the inevitable *vin ordinaire* (common red wine) at breakfast-time, instead of
the coffee for which my opportune interpreter said he had given
the order yester-eve.   If a Frenchman only sits down to a bite of
bread and cheese he usually consumes a pint bottle of *vin ordinaire*
with it.   The loaves of bread here are rolls three and four feet long,
and frequently one of these is laid across—or rather along, for it is
oftentimes longer than the table is wide—the table for you to
hack away at during your meal, according to your bread-eating
capacity or inclination.

Monsieur, the accomplished, comes down to see his *Anglais*
friend and *protégé* next morning, a few minutes after his *Anglais*
friend and *protégé* has started off toward a distant street called Rue
Poussen, which *le garçon* had unwittingly directed him to when he
inquired the way to the *bureau de poste ;* the natural result, I suppose, of the difference between Elbeuf pronunciation and mine.
Discovering my mistake upon arriving at the Rue Poussen, I am
more fortunate in my attack upon the interpreting abilities of a
passing citizen, who sends an Elbeuf *gamin* to guide me to the
post-office.

Post-office clerks are proverbially intelligent people in any country, consequently it doesn't take me long to transact my business
at the *bureau de poste ;* but now—shades of Cæsar !—I have
thoughtlessly neglected to take down either the name of the hotel
or the street in which it is located, and for the next half-hour go
wandering about as helplessly as the "babes in the wood."   Once,

twice I fancy recognizing the location ; but the ordinary Elbeuf house is not easily recognized from its neighbors, and I am standing looking around me in the bewildered attitude of one uncertain of his bearings, when, lo ! the landlady, who has doubtless been wondering whatever has become of me, appears at the door of a building which I should certainly never have recognized as my hotel, besom in hand, and her pleasant, "*Oui, monsieur,*" sounds cheery and welcome enough, under the circumstances, as one may readily suppose.

Fine roads continue, and between Gaillon and Vernon one can see the splendid highway, smooth, straight, and broad, stretching ahead for miles between rows of stately poplars, forming magnificent avenues that add not a little to the natural loveliness of the country. Noble chateaus appear here and there, oftentimes situated upon the bluffs of the Seine, and forming the background to a long avenue of chestnuts, maples, or poplars, running at right angles to the main road and principal avenue. The well-known thriftiness of the French peasantry is noticeable on every hand, and particularly away off to the left yonder, where their small, well-cultivated farms make the sloping bluffs resemble huge log-cabin quilts in the distance. Another glaring and unmistakable evidence of the Normandy peasants' thriftiness is the remarkable number of patches they manage to distribute over the surface of their pantaloons, every peasant hereabouts averaging twenty patches, more or less, of all shapes and sizes. When the British or United States Governments impose any additional taxation on the people, the people grumblingly declare they won't put up with it, and then go ahead and pay it; but when the Chamber of Deputies at Paris turns on the financial thumb-screw a little tighter, the French peasant simply puts yet another patch on the seat of his pantaloons, and smilingly hands over the difference between the patch and the new pair he intended to purchase !

Huge cavalry barracks mark the entrance to Vernon, and, as I watch with interest the manœuvring of the troops going through their morning drill, I cannot help thinking that with such splendid roads as France possesses she might take many a less practical measure for home defence than to mount a few regiments of light infantry on bicycles ; infantry travelling toward the front at the rate of seventy-five or a hundred miles a day would be something of an improvement, one would naturally think. Every few miles my

road leads through the long, straggling street of a village, every
building in which is of solid stone, and looks at least a thousand
years old ; while at many cross-roads among the fields, and in all
manner of unexpected nooks and corners of the villages, crucifixes
are erected to accommodate the devotionally inclined. Most of
the streets of these interior villages are paved with square stones
which the wear and tear of centuries have generally rendered too
rough for the bicycle ; but occasionally one is ridable, and the astonishment of the inhabitants as I wheel leisurely through, whistling the solemn strains of "Roll, Jordan, roll," is really quite
amusing. Every village of any size boasts a church that, for fine-
ness of architecture and apparent costliness of construction, looks
out of all proportion to the straggling street of shapeless structures
that it overtops. Everything here seems built as though intended
to last forever, it being no unusual sight to see a ridiculously small
piece of ground surrounded by a stone wall built as though to resist a bombardment ; an enclosure that must have cost more to
erect than fifty crops off the enclosed space could repay.

The important town of Mantes is reached early in the evening,
and a good inn found for the night.

The market-women are arraying their varied wares all along
the main street of Mantes as I wheel down toward the banks of
the Seine this morning. I stop to procure a draught of new milk,
and, while drinking it, point to sundry long rows of light, flaky-
looking cakes strung on strings, and motion that I am desirous of
sampling a few at current rates ; but the good dame smiles and
shakes her head vigorously, as well enough she might, for I learn
afterward that the cakes are nothing less than dried yeast-cakes, a
breakfast off which would probably have produced spontaneous
combustion. Getting on to the wrong road out of Mantes, I find
myself at the river's edge down among the Seine watermen. I am
shown the right way, but from Mantes to Paris they are not Normandy roads ; from Mantes southward they gradually deteriorate
until they are little or no better than the "sand-papered roads of
Boston." Having determined to taboo *vin ordinaire* altogether I
astonish the *restaurateur* of a village where I take lunch by motion-
ing away the bottle of red wine and calling for "*de l'eau*," and the
glances cast in my direction by the other customers indicate plainly
enough that they consider the proceeding as something quite extraordinary.

Rolling through Saint Germain, Chalon Pavèy, and Nanterre, the magnificent Arc de Triomphe looms up in the distance ahead, and at about two o'clock, Wednesday, May 13th, I wheel into the gay capital through the Porte Maillott. Asphalt pavement now takes the place of macadam, and but a short distance inside the city limits I notice the 'cycle depot of Renard Fèrres. Knowing instinctively that the fraternal feelings engendered by the magic wheel reaches to wherever a wheelman lives, I hesitate not to dismount and present my card. Yes, Jean Glinka, apparently an *employé* there, comprehends *Anglais;* they have all heard of my tour, and wish me *bon voyage,* and Jean and his bicycle is forthwith produced and delegated to accompany me into the interior of the city and find me a suitable hotel. The streets of Paris, like the streets of other large cities, are paved with various compositions, and they have just been sprinkled. French-like, the luckless Jean is desirous of displaying his accomplishments on the wheel to a visitor so *distingué;* he circles around on the slippery pavement in a manner most unnecessary, and in so doing upsets himself while crossing a car-track, rips his pantaloons, and injures his wheel. At the Hotel du Louvre they won't accept bicycles, having no place to put them ; but a short distance from there we find a less pretentious establishment, where, after requiring me to fill up a formidable-looking blank, stating my name, residence, age, occupation, birthplace, the last place I lodged at, etc., they finally assign me quarters.

From Paul Devilliers, to whom I bring an introduction, I learn that by waiting here till Friday evening, and repairing to the rooms of the Société Vélocipédique Métropolitaine, the president of that club can give me the best bicycle route between Paris and Vienna ; accordingly I domicile myself at the hotel for a couple of days. Many of the lions of Paris are within easy distance of my hotel. The reader, however, probably knows more about the sights of Paris than one can possibly find out in two days ; therefore I refrain from any attempt at describing them ; but my hotel is worthy of remark.

Among other agreeable and sensible arrangements at the Hotel du Loiret, there is no such thing as opening one's room-door from the outside save with the key ; and unless one thoroughly understands this handy peculiarity, and has his wits about him continually, he is morally certain, sometime when he is leaving his room, absent-mindedly to shut the door and leave the key inside. This is, of course, among the first things that happen to me, and it costs me half a franc and three hours of wretchedness before I see the interior of my room again. The hotel keeps a rude skeleton-key on hand, presumably for possible emergencies of this nature ; but in manipulating this uncouth instrument *le portier* actually locks the door, and as the skeleton-key is expected to manage the catch only, and not the lock, this, of course, makes matters infinitely worse. The keys of every room in the house are next brought into requisition and tried in succession, but not a key among them all is a duplicate of mine. What is to be done? *Le portier* looks as dejected as though Paris was about to be bombarded, as he goes down and breaks the dreadful news to *le propriétaire*. Up comes *le propriétaire*—avoirdupois three hundred pounds—sighing like an exhaust-pipe at every step. For fifteen unhappy minutes the skeleton-key is wriggled and twisted about again in the key-hole, and the fat *propriétaire* rubs his bald head impatiently, but all to no purpose. Each returns to his respective avocation. Impatient to get at my writing materials, I look up at the iron bars across the fifth-story windows above, and motion that if they will procure a rope I will descend from thence and enter the window. They one and all point out into the street; and, thinking they have sent for something or somebody, I sit down and wait with Job-like patience for something to turn up. Nothing, however, turns up, and at the expiration of an hour I naturally begin to feel neglected and impatient, and again suggest the rope ; when, at a motion from *le propriétaire*, *le portier* pilots me around a neighboring corner to a locksmith's establishment, where, voluntarily acting the part of interpreter, he engages on my behalf, for half a franc, a man to come with a bunch of at least a hundred skeleton-keys of all possible shapes to attack the refractory key-hole. After trying nearly all the keys, and disburdening himself of whole volumes of impulsive French ejaculations, this man likewise gives it up in despair ; but, now everything else has been tried and failed, the countenance of *le portier* suddenly lights up, and he slips quietly around to an adjoining room, and enters mine inside of two minutes by simply lifting a small hook out of a staple with his knife-blade. There appears to be a slight coolness, as it were, between *le propriétaire* and me after this incident, probably owing to the intellectual standard of each becoming somewhat lowered in the other's estimation in consequence of it. *Le pro-*

The Champs Elysée at 10 P.M.

priétaire, doubtless, thinks a man capable of leaving the key inside of the door must be the worst type of an ignoramus ; and certainly my opinion of him for leaving such a diabolical arrangement unchanged in the latter half of the nineteenth century is not far removed from the same.

Visiting the headquarters of the Société Vélocipédique Métropolitaine on Friday evening, I obtain from the president the desired directions regarding the route, and am all prepared to continue eastward in the morning.   Wheeling down the famous Champs Elysées at eleven at night, when the concert gardens are in full blast and everything in a blaze of glory, with myriads of electric lights festooned and in long brilliant rows among the trees, is something to be remembered for a lifetime.   Before breakfast I leave the city by the Porte Daumesiul, and wheel through the environments toward Vincennes and Joinville, pedalling, to the sound of martial music, for miles beyond the Porte.

The roads for thirty miles east of Paris are not Normandy roads, but the country for most of the distance is fairly level, and for mile after mile, and league beyond league, the road is beneath avenues of plane and poplar, which, crossing the plain in every direction like emerald walls of nature's own building, here embellish and beautify an otherwise rather monotonous stretch of country.   The villages are little different from the villages of Normandy, but the churches have not the architectural beauty of the Normandy churches, being for the most part massive structures without any pretence to artistic embellishment in their construction.   Monkish-looking priests are a characteristic feature of these villages, and when, on passing down the narrow, crooked streets of Fontenay, I wheel beneath a massive stone archway, and looking around, observe cowled priests and everything about the place seemingly in keeping with it, one can readily imagine himself transported back to medieval times.   One of these little interior French villages is the most unpromising looking place imaginable for a hungry person to ride into ; often one may ride the whole length of the village expectantly looking around for some visible evidence of wherewith to cheer the inner man, and all that greets the hungry vision is a couple of four-foot sticks of bread in one dust-begrimed window, and a few mournful-looking crucifixes and Roman Catholic paraphernalia in another.   Neither are the peasants hereabouts to be compared with the Normandy peasantry in

.8

personal appearance. True, they have as many patches on their
pantaloons, but they don't seem to have acquired the art of at-
taching them in a manner to produce the same picturesque effect
as does the peasant of Normandy ; the original garment is almost
invariably a shapeless corduroy, of a bagginess and an o'er-ample-
ness most unbeautiful to behold.

The well-known axiom about fair paths leading astray holds
good with the high-ways and by-ways of France, as elsewhere, and
soon after leaving the ancient town of Provins, I am tempted by a
splendid road, following the windings of a murmuring brook, that
appears to be going in my direction, in consequence of which I
soon find myself among cross-country by-ways, and among peasant
proprietors who apparently know little of the world beyond their
native villages. Four o'clock finds me wheeling through a hilly
vineyard district toward Villenauxe, a town several kilometres off
my proper route, from whence a dozen kilometres over a very good
road brings me to Sezanne, where the Hotel de France affords excellent accommodation. After the *table d'hôte* the clanging bells of
the old church hard by announce services of some kind, and having a natural *penchant* when in strange places from wandering
whithersoever inclination leads, in anticipation of the ever possible
item of interest, I meander into the church and take a seat. There
appears to be nothing extraordinary about the service, the only
unfamiliar feature to me being a man wearing a uniform similar to
the *gendarmerie* of Paris : cockade, sash, sword, and everything
complete ; in addition to which he carries a large cane and a long
brazen-headed staff resembling the boarding-pike of the last century.

It has rained heavily during the night, but the roads around
here are composed mainly of gravel, and are rather improved than
otherwise by the rain ; and from Sezanne, through Champenoise
and on to Vitry le François, a distance of about sixty-five kilometres, is one of the most enjoyable stretches of road imaginable.
The contour of the country somewhat resembles the swelling
prairies of Western Iowa, and the roads are as perfect for most
of the distance as an asphalt boulevard. The hills are gradual acclivities, and, owing to the good roads, are mostly ridable, while
the declivities make the finest coasting imaginable ; the exhilaration of gliding down them in the morning air, fresh after the rain,
can be compared only to Canadian tobogganing. Ahead of you

A Glimpse of Medieval France.

stretches a gradual downward slope, perhaps two kilometres long. Knowing full well that from top to bottom there exists not a loose stone or a dangerous spot, you give the ever-ready steel-horse the rein ; faster and faster whirl the glistening wheels until objects by the road-side become indistinct phantoms as they glide instantaneously by, and to strike a hole or obstruction is to be transformed into a human sky-rocket, and, later on, into a new arrival in another world. A wild yell of warning at a blue-bloused peasant in the road ahead, shrill screams of dismay from several females at a cluster of cottages, greet the ear as you sweep past like a whirlwind, and the next moment reach the bottom at a rate of speed that would make the engineer of the Flying Dutchman green with envy. Sometimes, for the sake of variety, when gliding noiselessly along on the ordinary level, I wheel unobserved close up behind an unsuspecting peasant walking on ahead, without calling out, and when he becomes conscious of my presence and looks around and sees the strange vehicle in such close proximity it is well worth the price of a new hat to see the lively manner in which he hops out of the way, and the next moment becomes fairly rooted to the ground with astonishment ; for bicycles and bicycle riders are less familiar objects to the French peasant, outside of the neighborhood of a few large cities, than one would naturally suppose.

Vitry le François is a charming old town in the beautiful valley of the Marne ; in the middle ages it was a strongly fortified city ; the moats and earth-works are still perfect. The only entrance to the town, even now, is over the old draw-bridges, the massive gates, iron wheels, chains, etc., still being intact, so that the gates can yet be drawn up and entrance denied to foes, as of yore ; but the moats are now utilized for the boats of the Marne and Rhine Canal, and it is presumable that the old draw-bridges are nowadays always left open. To-day is Sunday—and Sunday in France is equivalent to a holiday—consequently Vitry le François, being quite an important town, and one of the business centres of the prosperous and populous Marne Valley, presents all the appearance of circus-day in an American agricultural community. Several booths are erected in the market square, the proprietors and *attachés* of two peregrinating theatres, several peep-shows, and a dozen various games of chance, are vying with each other in the noisiness of their demonstrations to attract the attention and small change of the crowd to their respective enterprises. Like every other highway in this part of France the Marne and Rhine Canal is fringed with an avenue of poplars, that from neighboring elevations can be seen winding along the beautiful valley for miles, presenting a most pleasing effect.

East of Vitry le François the roads deteriorate, and from thence to Bar-le-Duc they are inferior to any hitherto encountered in France ; nevertheless, from the American standpoint they are very good roads, and when, at five o'clock, I wheel into Bar-le-Duc and come to sum up the aggregate of the day's journey I find that, without any undue exertion, I have covered very nearly one hundred and sixty kilometres, or about one hundred English miles, since 8.30 A.M., notwithstanding a good hour's halt at Vitry le François for dinner. Bar-le-Duc appears to be quite an important business centre, pleasantly situated in the valley of the Ornain River, a tributary of the Marne ; and the stream, in its narrow, fertile valley, winds around among hills from whose sloping sides, every autumn, fairly ooze the celebrated red wines of the Meuse and Moselle regions.

The valley has been favored with a tremendous downpour of rain and hail during the night, and the partial formation of the road leading along the level valley eastward being a light-colored, slippery clay, I find it anything but agreeable wheeling this morning ; moreover, the Ornain Valley road is not so perfectly kept as it might be. As in every considerable town in France, so also in Bar-le-Duc, the military element comes conspicuously to the fore. Eleven kilometres of slipping and sliding through the greasy clay brings me to the little village of Tronville, where I halt to investigate the prospect of obtaining something to eat. As usual, the prospect, from the street, is most unpromising, the only outward evidence being a few glass jars of odds and ends of candy in one small window. Entering this establishment, the only thing the woman can produce besides candy and raisins is a box of brown, wafer-like biscuits, the unsubstantial appearance of which is, to say the least, most unsatisfactory to a person who has pedalled his breakfastless way through eleven kilometres of slippery clay. Uncertain of their composition, and remembering my unhappy mistake at Mantes in desiring to breakfast off yeast-cakes, I take the precaution of sampling one, and in the absence of anything more substantial conclude to purchase a few, and so motion to the woman to hand me the box in order that I can show her how many I want. But the o'er-careful Frenchwoman, mistaking my meaning, and fearful that I only want to sample yet another one, probably feeling uncertain of whether I might not wish to taste a whole handful this time, instead of handing it over moves it out of my reach altogether, meanwhile looking quite angry, and not a little mystified at her mysterious, pantomimic customer. A half-franc is produced, and, after taking the precaution of putting it away in advance, the cautious female weighs me out the current quantity of her ware ; and I notice that, after giving lumping weight, she throws in a few extra, presumably to counterbalance what, upon sober second thought, she perceives to have been an unjust suspicion.

While I am extracting what satisfaction my feathery purchase contains, it begins to rain and hail furiously, and so continues with little interruption all the forenoon, compelling me, much against my inclination, to search out in Tronville, if possible, some accommodation till to-morrow morning. The village is a shapeless cluster of stone houses and stables, the most prominent feature of the streets being huge heaps of manure and grape-vine prunings ; but I manage to obtain the necessary shelter, and such other accommodations as might be expected in an out-of-the-way village, unfrequented by visitors from one year's end to another. The following morning is still rainy, and the clayey roads of the Ornain Valley are anything but inviting wheeling ; but a longer stay in Tronville is not to be thought of, for, among other pleasantries of the place here, the chief table delicacy appears to be boiled *escargots*, a large, ungainly snail procured from the neighboring hills. Whilst fond of table delicacies, I emphatically draw the line at *escargots*.

Pulling out toward Toul I find the roads, as expected, barely ridable ; but the vineyard-environed little valley, lovely in its tears, wrings from one praise in spite of muddy roads and lowering weather. *En route* down the valley I meet a battery of artillery travelling from Toul to Bar-le Duc or some other point to the westward ; and if there is any honor in throwing a battery of French artillery into confusion, and wellnigh routing them, then the bicycle and I are fairly entitled to it.

As I ride carelessly toward them, the leading horses suddenly wheel around and begin plunging about the road. The officers' horses, and, in fact, the horses of the whole company, catch the infection, and there is a plunging and a general confusion all along the line, seeing which I, of course, dismount and retire—but not discomfited—from the field until they have passed. These French horses are certainly not more than half-trained. I passed a battery of English artillery on the road leading out of Coventry, and had I wheeled along under the horses' noses there would have been no confusion whatever.

On the divide between the Ornain and Moselle Valleys the roads are hillier, but somewhat less muddy. The weather continues showery and unsettled, and a short distance beyond Void I find myself once again wandering off along the wrong road. The peasantry hereabout seem to have retained a lively recollection of the Prussians, my helmet appearing to have the effect of jogging their memory, and frequently, when stopping to inquire about the roads, the first word in response will be the pointed query, "Prussian?" By following the directions given by three different peasants, I wander along the muddy by-roads among the vineyards for two wet, unhappy hours ere I finally strike the main road to Toul again. After floundering along the wellnigh unimproved by-ways for two hours one thoroughly appreciates how much he is indebted to the military necessities of the French Government for the splendid highways of France, especially among these hills and valleys, where natural roadways would be anything but good. Following down the Moselle Valley, I arrive at the important city of Nancy in the eventide, and am fortunate, I suppose, in discovering a hotel where a certain, or, more properly speaking, an uncertain, quantity and quality of English are spoken. Nancy is reputed to be one of the loveliest towns in France. But I merely remained in it over night, and long enough next morning to exchange for some German money, as I cross over the frontier to-day.

Lunéville is a town I pass through, some distance nearer the border, and the military display here made is perfectly overshadowing. Even the scarecrows in the fields are military figures, with wooden swords threateningly waving about in their hands with every motion of the wind, and the most frequent sound heard along the route is the sharp bang! bang! of muskets, where companies of soldiers are target-practising in the woods. There seems to be a bellicose element in the very atmosphere ; for every dog in every village I ride through verily takes after me, and I run clean over one bumptious cur, which, miscalculating the speed at which I am coming, fails to get himself out of the way in time. It is the narrowest escape from a header I have had since starting from Liverpool; although both man and dog were more scared than hurt. Sixty-five kilometres from Nancy, and I take lunch at the frontier town of Blamont. The road becomes more hilly, and a short distance out of Blamont, behold, it is as though a chalk-line were made across the roadway, on the west side of which it had been swept with scrupulous care, and on the east side not swept at all ; and when, upon passing the next roadman, I notice that he bears not upon his cap the brass stencil-plate bearing the inscription, "Cantonnier," I know that I have passed over the frontier into the territory of Kaiser Wilhelm.

My journey through fair France has been most interesting, and perhaps instructive, though I am afraid that the lessons I have taken in French politeness are altogether too superficial to be lasting. The " *Bon jour, monsieur,*" and " *Bon voyage,*" of France, may not mean any more than the "If I don't see you again, why, hello ! " of America, but it certainly sounds more musical and pleasant. It is at the *table d'hôte*, however, that I have felt myself to have invariably shone superior to the natives ; for, lo! the Frenchman eats soup from the end of his spoon. True, it is more convenient to eat soup from the prow of a spoon than from the larboard ; nevertheless, it is when eating soup that I instinctively feel my superiority. The French peasants, almost without exception, conclude that the bright-nickelled surface of the bicycle is silver, and presumably consider its rider nothing less than a millionnaire in consequence ; but it is when I show them the length of time the rear wheel or a pedal will spin round that they manifest their greatest surprise. The crowning glory of French landscape is the magnificent avenues of poplars that traverse the country in every direction, winding with the roads, the railways, and canals along the valleys, and marshalled like sentinels along the brows of the distant hills ; without them French scenery would lose half its charm.

CHAPTER VI.

GERMANY, AUSTRIA, AND HUNGARY.

NOTWITHSTANDING Alsace was French territory only fourteen years ago (1871) there is a noticeable difference in the inhabitants, to me the most acceptable being their great linguistic superiority over the people on the French side of the border. I linger in Saarburg only about thirty minutes, yet am addressed twice by natives in my own tongue ; and at Pfalzburg, a smaller town, where I remain over night, I find the same characteristic. Ere I penetrate thirty kilometres into German territory, however, I have to record what was never encountered in France ; an insolent teamster, who, having his horses strung across a narrow road-way in the suburbs of Saarburg, refuses to turn his leaders' heads to enable me to ride past, thus compelling me to dismount. Soldiers drilling, soldiers at target practice, and soldiers in companies marching about in every direction, greet my eyes upon approaching Pfalzburg ; and although there appears to be less beating of drums and blare of trumpets than in French garrison towns, one seldom turns a street corner without hearing the measured tramp of a military company receding or approaching. These German troops appear to march briskly and in a business-like manner in comparison with the French, who always seem to carry themselves with a tired and dejected deportment ; but the over-ample and rather slouchy-looking pantaloons of the French are probably answerable, in part, for this impression. One cannot watch these sturdy-looking German soldiers without a conviction that for the stern purposes of war they are inferior only to the soldiers of our own country.

At the little gasthaus at Pfalzburg the people appear to understand and anticipate an Englishman's gastronomic peculiarities, and for the first time since leaving England I am confronted at the supper-table with excellent steak and tea.

It is raining next morning as I wheel over the rolling hills toward Saverne, a city nestling pleasantly in a little valley beyond

those dark wooded heights ahead that form the eastern boundary
of the valley of the Rhine. The road is good but hilly, and for
several kilometres, before reaching Saverne, winds its way among
the pine forests tortuously and steeply down from the elevated divide. The valley, dotted here and there with pleasant villages, is
spread out like a marvellously beautiful picture, the ruins of several old castles on neighboring hill-tops adding a charm, as well as
a dash of romance.

The rain pours down in torrents as I wheel into Saverne. I
pause long enough to patronize a barber shop ; also to procure
an additional small wrench. Taking my nickelled monkey-wrench
into a likely-looking hardware store, I ask the proprietor if he
has anything similar. He examines it with lively interest, for, in
comparison with the clumsy tools comprising his stock-in-trade,
the wrench is as a watch-spring to an old horse-shoe. I purchase a
rude tool that might have been fashioned on the anvil of a village
blacksmith. From Saverne my road leads over another divide
and down into the glorious valley of the Rhine, for a short distance
through a narrow defile that reminds me somewhat of a cañon in
the Sierra Nevada foot-hills ; but a fine, broad road, spread with a
coating of surface-mud only by this morning's rain, prevents the
comparison from assuming definite shape for a cycler. Extensive
and beautifully terraced vineyards mark the eastern exit.

The road-beds of this country are hard enough for anything ;
but a certain proportion of clay in their composition makes a slippery coating in rainy weather. I enter the village of Marlenheim
and observe the first stork's nest, built on top of a chimney, that I
have yet seen in Europe, though I saw plenty of them afterward.
The parent stork is perched solemnly over her youthful brood,
which one would naturally think would get smoke-dried. A short
distance from Marlenheim I descry in the hazy distance the famous
spire of Strasburg cathedral looming conspicuously above everything else in all the broad valley ; and at 1.30 P.M. I wheel through
the massive arched gateway forming part of the city's fortifications,
and down the broad but roughly paved streets, the most mud-be-
spattered object in all Strasburg. The fortifications surrounding
the city are evidently intended strictly for business, and not merely
for outward display. The railway station is one of the finest in
Europe, and among other conspicuous improvements one notices
steam tram-cars. While trundling through the city I am impera-

lively ordered off the sidewalk by the policeman ; and when stopping to inquire of a respectable-looking Strasburger for the Appenweir road, up steps an individual with one eye and a cast off military cap three sizes too small. After querying, " *Appenweir? Englander?*" he wheels "about face" with military precision—doubtless thus impelled by the magic influence of his headgear—and beckons me to follow. Not knowing what better course to pursue I obey, and after threading the mazes of a dozen streets, composed of buildings ranging in architecture from the much gabled and not unpicturesque structures of mediæval times to the modern brown-stone front, he pilots me outside the fortifications again, points up the Appenweir road, and after the never neglected formality of touching his cap and extending his palm, returns city-ward.

Crossing the Rhine over a pontoon bridge, I ride along level and, happily, rather less muddy roads, through pleasant suburban villages, near one of which I meet a company of soldiers in undress uniform, strung out carelessly along the road, as though returning from a tramp into the country. As I approach them, pedalling laboriously against a stiff head wind, both myself and the bicycle fairly yellow with clay, both officers and soldiers begin to laugh in a good-natured, bantering sort of manner, and a round dozen of them sing out in chorus "*Ah! ah! der Englander!*" and as I reply, "Yah! yah!" in response, and smile as I wheel past them, the laughing and banter go all along the line. The sight of an "*Englander*" on one of his rambling expeditions of adventure furnishes much amusement to the average German, who, while he cannot help admiring the spirit of enterprise that impels him, fails to comprehend where the enjoyment can possibly come in. The average German would much rather loll around, sipping wine or beer, and smoking cigarettes, than impel a bicycle across a continent.

A few miles eastward of the Rhine another grim fortress frowns upon peaceful village and broad, green meads, and off yonder to the right is yet another ; sure enough, this Franco-German frontier is one vast military camp, with forts, and soldiers, and munitions of war everywhere! When I crossed the Rhine I left Lower Alsace, and am now penetrating the middle Rhine region, where villages are picturesque clusters of gabled cottages—a contrast to the shapeless and ancient-looking stone structures of the French villages. The difference also extends to the inhabitants; the peasant women of France, in either real or affected modesty, would usually pretend not to notice anything extraordinary as I wheeled past, but upon looking back they would almost invariably be seen standing and gazing after my receding figure with unmistakable interest; but the women of these Rhine villages burst out into merry peals of laughter.

Rolling over fair roads into the village of Oberkirch, I conclude to remain for the night, and the first thing undertaken is to disburden the bicycle of its covering of clay. The awkward-looking hostler comes around several times and eyes the proceedings with glances of genuine disapproval, doubtless thinking I am cleaning it myself instead of letting him swab it with a besom with the single purpose in view of dodging the inevitable tip. The proprietor can speak a few words of English. He puts his bald head out of the window above, and asks : "Pe you Herr Shtevens?"

"Yah, yah," I reply.

"Do you go mit der veld around?"

"Yah; I goes around mit the world."

"I shoust read about you mit der noospaper."

"Ah, indeed! what newspaper?"

"*Die Frankfurter Zeitung.* You go around mit der veld."

The landlord looks delighted to have for a guest the man who goes "mit der veld around," and spreads the news. During the evening several people of importance and position drop in to take a curious peep at me and my wheel.

A dampness about the knees, superinduced by wheeling in rubber leggings, causes me to seek the privilege of the kitchen fire upon arrival. After listening to the incessant chatter of the cook for a few moments, I suddenly dispense with all pantomime, and ask in purest English the privilege of drying my clothing in peace and tranquillity by the kitchen fire. The poor woman hurries out, and soon returns with her highly accomplished master, who, comprehending the situation, forthwith tenders me the loan of his Sunday pantaloons for the evening; which offer I gladly accept, notwithstanding the wide disproportion in their size and mine, the landlord being, horizontally, a very large person.

Oberkirch is a pretty village at the entrance to the narrow and charming valley of the River Rench, up which my route leads, into the fir-clad heights of the Black Forest. A few miles farther up

Borrowed Plumage.

the valley I wheel through a small village that nestles amid surroundings the loveliest I have yet seen. Dark, frowning firs intermingled with the lighter green of other vegetation crown the surrounding spurs of the Knibis Mountains ; vineyards, small fields of
waving rye, and green meadow cover the lower slopes with variegated beauty, at the foot of which huddles the cluster of pretty
cottages amid scattered orchards of blossoming fruit-trees. The
cheery lute of the herders on the mountains, the carol of birds,
and the merry music of dashing mountain-streams fill the fresh
morning air with melody. All through this country there are
apple-trees, pear-trees, cherry-trees—everywhere. In the fruit
season one can scarce open his mouth out-doors without having
the goddess Pomona pop in some delicious morsel. The poplar
avenues of France have disappeared, but the road is frequently
shaded for miles with fruit-trees. I never before saw a spot so
lovely—certainly not in combination with a wellnigh perfect road
for wheeling. On through Oppenau and Petersthal my way leads
—this latter a place of growing importance as a summer resort,
several commodious hotels with swimming-baths, mineral waters,
etc., being already prepared to receive the anticipated influx of
health and pleasure-seeking guests this coming summer—and then
up, up, up among the dark pines leading over the Black Forest
Mountains. Mile after mile of steep incline has now been trundled,
following the Rench River to its source. Ere long the road I have
lately traversed is visible far below, winding and twisting up the
mountain-slopes. Groups of swarthy peasant women are carrying
on their heads baskets of pine cones to the villages below. At a distance the sight of their bright red dresses among the sombre green
of the pines is suggestive of the fairies with which legend has peopled the Black Forest.

The summit is reached at last, and two boundary posts apprise
the traveller that on this wooded ridge he passes from Baden into
Würtemberg. The descent for miles is agreeably smooth and
gradual ; the mountain air blows cool and refreshing, with an odor
of the pines ; the scenery is Black Forest scenery, and what more
could be possibly desired than this happy combination of circumstances ?

Reaching Freudenstadt about noon, the mountain-climbing, the
bracing air, and the pine fragrance cause me to give the good people at the gasthaus an impressive lesson in the effect of cycling on

the human appetite. At every town and village I pass through in Würtemberg the whole juvenile population collects around me in an incredibly short time. The natural impulse of the German small boy appears to be to start running after me, shouting and laughing immoderately, and when passing through some of the larger villages, it is no exaggeration to say that I have had two hundred small Germans, noisy and demonstrative, clattering along behind in their heavy wooden shoes.

Würtemburg, by this route at least, is a decidedly hilly country, and the roads are far inferior to those of both England and France. There will be, perhaps, three kilometres of trundling up through wooded heights leading out of a small valley, then, after several kilometres over undulating, stony upland roads, a long and not always smooth descent into another small valley, this programme, several times repeated, constituting the journey of the day. The small villages of the peasantry are frequently on the uplands, but the larger towns are invariably in the valleys, sheltered by wooded heights, perched among the crags of the most inaccessible of which are frequently seen the ruins of an old castle. Scores of little boys of eight or ten are breaking stones by the road-side, at which I somewhat marvel, since there is a compulsory school law in Germany ; but perhaps to-day is a holiday ; or maybe, after school hours, it is customary for these unhappy youngsters to repair to the road-sides and blister their hands with cracking flints.

"Hungry as a buzz-saw" I roll into the sleepy old town of Rothenburg at six o'clock, and, repairing to the principal hotel, order supper. Several flunkeys of different degrees of usefulness come in and bow obsequiously from time to time, as I sit around, expecting supper to appear every minute. At seven o'clock the waiter comes in, bows profoundly, and lays the table-cloth ; at 7.15 he appears again, this time with a plate, knife, and fork, doing more bowing and scraping as he lays them on the table. Another half-hour rolls by, when, doubtless observing my growing impatience as he happens in at intervals to close a shutter or re-regulate the gas, he produces a small illustrated paper, and, bowing profoundly, lays it before me. I feel very much like making him swallow it, but resigning myself to what appears to be inevitable fate, I wait and wait, and at precisely 8.15 he produces a plate of soup ; at 8.30 the *kalbscotolet* is brought on, and at 8.45 a small plate of mixed biscuits. During the meal I call for another piece of bread, and behold there is a hurrying to and fro, and a resounding of feet scurrying along the stone corridors of the rambling old building, and ten minutes later I receive a small roll. At the opposite end of the long table upon which I am writing some half-dozen ancient and honorable Rothenburgers are having what they doubtless consider a " howling time." Confronting each is a huge tankard of foaming lager, and the one doubtless enjoying himself the most and making the greatest success of exciting the envy and admiration of those around him is a certain ponderous individual who sits from hour to hour in a half comatose condition, barely keeping a large porcelain pipe from going out, and at fifteen-minute intervals taking a telling pull at the lager. Were it not for an occasional blink of the eyelids and the periodical visitation of the tankard to his lips, it would be difficult to tell whether he were awake or sleeping, the act of smoking being barely perceptible to the naked eye.

In the morning I am quite naturally afraid to order anything to eat here for fear of having to wait until mid-day, or thereabouts, before getting it; so, after being the unappreciative recipient of several more bows, more deferential and profound if anything than the bows of yesterday eve, I wheel twelve kilometres to Tübingen for breakfast. It showers occasionally during the forenoon, and after about thirty-five kilometres of hilly country it begins to descend in torrents, compelling me to follow the example of several peasants in seeking the shelter of a thick pine copse. We are soon driven out of it, however, and donning my gossamer rubber suit, I push on to Alberbergen, where I indulge in rye bread and milk, and otherwise while away the hours until three o'clock, when, the rain ceasing, I pull out through the mud for Blaubeuren.

Down the beautiful valley of one of the Danube's tributaries I ride on Sunday morning, pedalling to the music of Blaubeuren's church-bells. After waiting until ten o'clock, partly to allow the roads to dry a little, I conclude to wait no longer, and so pull out toward the important and quite beautiful city of Ulm. The character of the country now changes, and with it likewise the characteristics of the people, who verily seem to have stamped upon their features the peculiarities of the region they inhabit. My road eastward of Blaubeuren follows down a narrow, winding valley, beside the rippling head-waters of the Danube, and eighteen kilometres of variable road brings me to the strongly fortified city of

Ulm, the place I should have reached yesterday, except for the inclemency of the weather, and where I cross from Würtemberg into Bavaria. On the uninviting uplands of Central Würtemberg one looks in vain among the peasant women for a prepossessing countenance or a graceful figure, but along the smiling valleys of Bavaria, the women, though usually with figures disproportionately broad, nevertheless carry themselves with a certain gracefulness; and, while far from the American or English idea of beautiful, are several degrees more so than their relatives of the part of Würtemberg I have traversed. I stop but a few minutes at Ulm, to test a mug of its lager and inquire the details of the road to Augsburg, yet during that short time I find myself an object of no little curiosity to the citizens, for the fame of my undertaking has pervaded Ulm.

The roads of Bavaria possess the one solitary merit of hardness, otherwise they would be simply abominable, the Bavarian idea of road-making evidently being to spread unlimited quantities of loose stones over the surface. For miles a wheelman is compelled to follow along narrow, wheel-worn tracks, incessantly dodging loose stones, or otherwise to pedal his way cautiously along the edges of the roadway. I am now wheeling through the greatest beer-drinking, sausage-consuming country in the world; hop-gardens are a prominent feature of the landscape, and long links of sausages are dangling in nearly every window. The quantities of these viands I see consumed to-day are something astonishing, though the celebration of the Whitsuntide holidays is probably augmentative of the amount.

The strains of instrumental music come floating over the level bottom of the Lech valley as, toward eventide, I approach the beautiful environs of Augsburg, and ride past several beer-gardens, where merry crowds of Augsburgers are congregated, quaffing foaming lager, eating sausages, and drinking inspiration from the music of military bands. "Where is the headquarters of the Augsburg Velocipede Club?" I inquire of a promising-looking youth as, after covering one hundred and twenty kilometres since ten o'clock, I wheel into the city. The club's headquarters are at a prominent café and beer-garden in the south-eastern suburbs, and repairing thither I find an accommodating individual who can speak English, and who willingly accepts the office of interpreter between me and the proprietor of the garden. Seated amid

9

hundreds of soldiers, Augsburg civilians, and peasants from the surrounding country, and with them extracting genuine enjoyment from a tankard of foaming Augsburg lager, I am informed that most of the members of the club are celebrating the Whitsuntide holidays by touring about the surrounding country, but that I am very welcome to Augsburg, and I am conducted to the Hotel Mohrenkopf (Moor's Head Hotel), and invited to consider myself the guest of the club as long as I care to remain in Augsburg—the Bavarians are nothing if not practical.

Mr. Josef Kling, the president of the club, accompanies me as far out as Friedburg on Monday morning; it is the last day of the holidays, and the Bavarians are apparently bent on making the most of it. The suburban beer-gardens are already filled with people, and for some distance out of the city the roads are thronged with holiday-making Augsburgers repairing to various pleasure resorts in the neighboring country, and the peasantry streaming cityward from the villages, their faces beaming in anticipation of unlimited quantities of beer. About every tenth person among the outgoing Augsburgers is carrying an accordion ; some playing merrily as they walk along, others preferring to carry theirs in blissful meditation on the good time in store immediately ahead, while a thoughtful majority have large umbrellas strapped to their backs. Music and song are heard on every hand, and as we wheel along together in silence, enforced by an ignorance of each other's language, whichever way one looks, people in holiday attire and holiday faces are moving hither and thither.

Some of the peasants are fearfully and wonderfully attired : the men wear high top-boots, polished from the sole to the uppermost hair's-breadth of leather ; black, broad-brimmed felt hats, frequently with a peacock's feather a yard long stuck through the band, the stem protruding forward, and the end of the feather behind ; and their coats and waistcoats are adorned with long rows of large, ancestral buttons. I am now in the Swabian district, and these buttons that form so conspicuous a part of the holiday attire are made of silver coins, and not infrequently have been handed down from generation to generation for several centuries, they being, in fact, family heirlooms. The costumes of the Swabish peasant women are picturesque in the extreme : their finest dresses and that wondrous head-gear of brass, silver, or gold—the *Schwä-*

bische Bauernfrauenhaube (Swabish farmer-woman hat)—being, like the buttons of the men, family heirlooms.   Some of these wonderful ancestral dresses, I am told, contain no less than one hundred and fifty yards of heavy material, gathered and closely pleated in innumerable perpendicular folds, frequently over a foot thick, making the form therein incased appear ridiculously broad and squatty. The waistbands of the dresses are up in the region of the shoulder-blades ; the upper portion of the sleeves are likewise padded out to fearful proportions.

The day is most lovely, the fields are deserted, and the roads and villages are alive with holiday-making peasants.   In every village a tall pole is erected, and decorated from top to bottom with small flags and evergreen wreaths.   The little stone churches and the adjoining cemeteries are filled with worshippers chanting in solemn chorus ; not so preoccupied with their devotional exercises and spiritual meditations, however, as to prevent their calling one another's attention to me as I wheel past, craning their necks to obtain a better view, and, in one instance, an o'er-inquisitive worshipper even beckons for me to stop—this person both chanting and beckoning vigorously at the same time.

Now my road leads through forests of dark firs ; and here I overtake a procession of some fifty peasants, the men and women alternately chanting in weird harmony as they trudge along the road. The men are bareheaded, carrying their hats in hand. Many of the women are barefooted, and the pedal extremities of others are incased in stockings of marvellous pattern ; not any are wearing shoes.   All the colors of the rainbow are represented in their respective costumes, and each carries a large umbrella strapped at his back ; they are trudging along at quite a brisk pace, and altogether there is something weird and fascinating about the whole scene : the chanting and the surroundings.   The variegated costumes of the women are the only bright objects amid the gloominess of the dark green pines.   As I finally pass ahead, the unmistakable expressions of interest on the faces of the men, and the even rows of ivories displayed by the women, betray a diverted attention.

Near noon I arrive at the antiquated town of Dachau, and upon repairing to the gasthaus, an individual in a last week's paper collar, and with general appearance in keeping, comes forward and addresses me in quite excellent English, and during the dinner

hour answers several questions concerning the country and the
natives so intelligently that, upon departing, I ungrudgingly offer
him the small tip customary on such occasions in Germany.   " No,

Whitsuntide in Bavaria.

I thank you, very muchly," he replies, smiling, and shaking his
head.   " I am not an employé of the hotel, as you doubtless think ;
I am a student of modern languages at the Munich University,
visiting Dachau for the day."   Several soldiers playing billiards in

the room grin broadly in recognition of the ludicrousness of the situation ; and I must confess that for the moment I feel like asking one of them to draw his sword and charitably prod me out of the room. The unhappy memory of having, in my ignorance, tendered a small tip to a student of the Munich University will cling around me forever. Nevertheless, I feel that after all there are extenuating circumstances—he ought to change his paper collar occasionally.

An hour after noon I am industriously dodging loose flints on the level road leading across the Isar River Valley toward Munich ; the Tyrolese Alps loom up, shadowy and indistinct, in the distance to the southward, their snowy peaks recalling memories of the Rockies through which I was wheeling exactly a year ago. While wending my way along the streets toward the central portion of the Bavarian capital the familiar sign, "American Cigar Store," looking like a ray of light penetrating through the gloom and mystery of the multitudinous unreadable signs that surround it, greets my vision, and I immediately wend my footsteps thitherward. I discover in the proprietor, Mr. Walsch, a native of Munich, who, after residing in America for several years, has returned to dream away declining years amid the smoke of good cigars and the quaffing of the delicious amber beer that the brewers of Munich alone know how to brew. Then who should happen in but Mr. Charles Buscher, a thorough-going American, from Chicago, who is studying art here at the Royal Academy of Fine Arts, and who straightway volunteers to show me Munich.

Nine o'clock next morning finds me under the pilotage of Mr. Buscher, wandering through the splendid art galleries. We next visit the Royal Academy of Fine Arts, a magnificent building, being erected at a cost of 7,000,000 marks.

We repair at eleven o'clock to the royal residence, making a note by the way of a trifling mark of King Ludwig's well-known eccentricity. Opposite the palace is an old church, with two of its four clocks facing the King's apartments. The hands of these clocks are, according to my informant, made of gold. Some time since the King announced that the sight of these golden hands hurt his eyesight, and ordered them painted black. It was done, and they are black to-day. Among the most interesting objects in the palace are the room and bed in which Napoleon I. slept in 1809, and which has since been occupied by no other person ; the " rich bed," a gorgeous affair of pink and scarlet satin-work, on which forty women wove, with gold thread, daily, for ten years, until 1,600,000 marks were expended.

At one of the entrances to the royal residence, and secured with iron bars, is a large bowlder weighing three hundred and sixty-three pounds ; in the wall above it are driven three spikes, the highest spike being twelve feet from the ground ; and Bavarian historians have recorded that Earl Christoph, a famous giant, tossed this bowlder up to the mark indicated by the highest spike, with his foot. ·

After this I am kindly warned by both Messrs. Buscher and Walsch not to think of leaving the city without visiting the *König-liche Hofbräuhaus* (Royal Court Brewery) the most famous place of its kind in all Europe. For centuries Munich has been famous for the excellent quality of its beer, and somewhere about four centuries ago the king founded this famous brewery for the charitable purpose of enabling his poorer subjects to quench their thirst with the best quality of beer, at prices within their means, and from generation to generation it has remained a favorite resort in Munich for lovers of good beer. In spite of its remaining, as of yore, a place of rude benches beneath equally rude, open sheds, with cobwebs festooning the rafters and a general air of dilapidation about it ; in spite of the innovation of dozens of modern beer-gardens with waving palms, electric lights, military music, and all modern improvements, the *Königliche Hofbräuhaus* is daily and nightly thronged with thirsty visitors, who for the trifling sum of twenty-two pfennigs (about five cents) obtain a quart tankard of the most celebrated brew in all Bavaria.

"Munich is the greatest art-centre of the world, the true hub of the artistic universe," Mr. Buscher enthusiastically assures me as we wander together through the sleepy old streets, and he points out a bright bit of old frescoing, which is already partly obliterated by the elements, and compares it with the work of recent years ; calls my attention to a piece of statuary, and anon pilots me down into a restaurant and beer-hall in some ancient, underground vaults and bids me examine the architecture and the frescoing. The very custom-house of Munich is a glorious old church, that would be carefully preserved as a relic of no small interest and importance in cities less abundantly blessed with antiquities, but which is here piled with the cases and boxes and bags of commerce.

One other conspicuous feature of Munich life must not be overlooked ere I leave it, viz., the hackmen.   Unlike their Transatlantic
brethren, they appear supremely indifferent about whether they pick
up any fares or not.   Whenever one comes to a hack-stand it is a
pretty sure thing to bet that nine drivers out of every ten are taking a quiet snooze, reclining on their elevated boxes, entirely oblivious of their surroundings, and a timid stranger would almost
hesitate about disturbing their slumbers.   But the Munich cabby
has long since got hardened to the disagreeable process of being
wakened up.   Nor does this lethargy pervade the ranks of hackdom
only : at least two-thirds of the teamsters one meets on the roads,
hereabouts, are stretched out on their respective loads, contentedly
sleeping while the horses or oxen crawl leisurely along toward their
goal.

Munich is visited heavily with rain during the night, and for
several kilometres, next morning, the road is a horrible waste of
loose flints and mud-filled ruts, along which it is all but impossible
to ride ; but after leaving the level bottom of the Isar River the
road improves sufficiently to enable me to take an occasional, admiring glance at the Bavarian and Tyrolese Alps, towering cloud-
ward on the southern horizon, their shadowy outlines scarcely distinguishable in the hazy distance from the fleecy clouds their peaks
aspire to invade.   While absentmindedly taking a more lingering
look than is consistent with safety when picking one's way along
the narrow edge of the roadway between the stone-strewn centre
and the ditch, I run into the latter, and am rewarded with my first
Cis-atlantic header, but fortunately both myself and the bicycle
come up uninjured.   Unlike the Swabish peasantry, the natives east
of Munich appear as prosy and unpicturesque in dress as a Kansas
homesteader.

Ere long there is noticeable a decided change in the character
of the villages, they being no longer clusters of gabled cottages,
but usually consist of some three or four huge, rambling buildings, at one of which I call for a drink and observe that brewing
and baking are going on as though they were expecting a whole
regiment to be quartered on them.   Among other things I mentally
note this morning is that the men actually seem to be bearing the
drudgery of the farm equally with the women ; but the favorable
impression becomes greatly imperilled upon meeting a woman harnessed to a small cart, heavily laboring along, while her husband—

kind man—is walking along-side, holding on to a rope, upon which he considerately pulls to assist her along and lighten her task. Nearing Hoag, and thence eastward, the road becomes greatly improved, and along the Inn River Valley, from Mühldorf to Alt Oetting, where I remain for the night, the late rain-storm has not reached, and the wheeling is superior to any I have yet had in Germany. Mühldorf is a curious and interesting old town. The sidewalks of Mühldorf are beneath long arcades from one end of the principal street to the other; not modern structures either, but massive archways that are doubtless centuries old, and that support the front rooms of the buildings that tower a couple of stories above them.

As toward dusk I ride into the market square of Alt Oetting, it is noticeable that nearly all the stalls and shops remaining open display nothing but rosaries, crucifixes, and other paraphernalia of the prevailing religion. Through Eastern Bavaria the people seem pre-eminently devotional; church-spires dot the landscape at every point of the compass. At my hotel in Alt Oetting, crucifixes, holy water, and burning tapers are situated on the different stairway landings. I am sitting in my room, penning these lines to the music of several hundred voices chanting in the old stone church near by, and can look out of the window and see a number of peasant women taking turns in dragging themselves on their knees round and round a small religious edifice in the centre of the market square, carrying on their shoulders huge, heavy wooden crosses, the ends of which are trailing on the ground.

All down the Inn River Valley, there is many a picturesque bit of intermingled pine-copse and grassy slopes; but admiring scenery is anything but a riskless undertaking along here, as I quickly discover. On the Inn River I find a primitive ferry-boat operated by a *fac-simile* of the Ancient Mariner, who takes me and my wheel across for the consideration of five pfennigs—a trifle over one cent—and when I refuse the tiny change out of a ten-pfennig piece the old fellow touches his cap as deferentially, and favors me with a look of gratitude as profound, as though I were bestowing a pension upon him for life. My arrival at a broad, well-travelled highway at once convinces me that I have again been unwittingly wandering among the comparatively untravelled by-ways as the result of following the kindly meant advice of people whose knowledge of bicycling requirements is of the slimmest nature. The Inn River has a warm, rich vale; haymaking is already in full progress, and the delightful perfume is wafted on the fresh morning air from meadows where scores of barefooted Maud Müllers are raking hay, ay, and mowing it too, swinging scythes side by side with the men. Some of the out-door crucifixes and shrines (small, substantial buildings containing pictures, images, and all sorts of religious emblems) along this valley are really quite elaborate affairs. All through Roman Catholic Germany these emblems of religion are very elaborate, or the reverse, according to the locality, the chosen spot in rich and fertile valleys generally being favored with better and more artistic affairs, and more of them, than the comparatively unproductive uplands. This is evidently because the inhabitants of the latter regions are either less wealthy, and consequently cannot afford it, or otherwise realize that they have really much less to be thankful for than their comparatively fortunate neighbors in the more productive valleys.

At the town of Simbach I cross the Inn River again on a substantial wooden bridge, and on the opposite side pass under an old stone archway bearing the Austrian coat-of-arms. Here I am conducted into the custom-house by an officer wearing the sombre uniform of Franz Josef, and required, for the first time in Europe, to produce my passport. After a critical and unnecessarily long examination of this document I am graciously permitted to depart. In an adjacent money-changer's office I exchange what German money I have remaining for the paper currency of Austria, and once more pursue my way toward the Orient, finding the roads rather better than the average German ones, the Austrians, hereabouts at least, having had the goodness to omit the loose flints so characteristic of Bavaria. Once out of the valley of the Inn River, however, I find the uplands intervening between it and the valley of the Danube aggravatingly hilly.

While eating my first luncheon in Austria, at the village of Altheim, the village pedagogue informs me in good English that I am the first Briton he has ever had the pleasure of conversing with. He learned the language entirely from books, without a tutor, he says, learning it for pleasure solely, never expecting to utilize the accomplishment in any practical way. One hill after another characterizes my route to-day; the weather, which has hitherto remained reasonably mild, is turning hot and sultry, and, arriving at Hoag about five o'clock, I feel that I have done sufficient hill-

climbing for one day. I have been wheeling through Austrian
territory since 10.30 this morning, and, with observant eyes the
whole distance, I have yet to see the first native, male or female,
possessing in the least degree either a graceful figure or a prepossessing face. There has been à great horse-fair at Hoag to-day ;
the business of the day is concluded, and the principal occupation
of the men, apart from drinking beer and smoking, appears to be
frightening the women out of their wits by leading prancing horses
as near them as possible.

My road, on leaving Hoag, is hilly, and the snowy heights of the
Nordliche Kalkalpen (North Chalk Mountains), a range of the Austrian Alps, loom up ahead at an uncertain distance.   To-day is what
Americans call a "scorcher," and climbing hills among pine-woods,
that shut out every passing breeze, is anything but exhilarating exercise with the thermometer hovering in the vicinity of one hundred degrees.   The peasants are abroad in their fields as usual,
but a goodly proportion are reclining beneath the trees.   Reclining is, I think, a favorite pastime with the Austrian.   The teamster, who happens to be wide awake and sees me approaching,
knows instinctively that his team is going to scare at the bicycle,
yet he makes no precautionary movements whatever, neither does
he arouse himself from his lolling position until the horses or oxen
begin to swerve around.   As a usual thing the teamster is filling
his pipe, which has a large, ungainly-looking, porcelain bowl, a
long, straight wooden stem, and a crooked mouth-piece.   Almost
every Austrian peasant from sixteen years old upward carries one
of these uncomely pipes.

The men here seem to be dull, uninteresting mortals, dressed
in tight-fitting, and yet, somehow, ill-fitting, pantaloons, usually
about three sizes too short, a small apron of blue ducking—an un-
becoming garment that can only be described as a cross between a
short jacket and a waistcoat—and a narrow-rimmed, prosy-looking
billycock hat.   The peasant women are the poetry of Austria, as
of any other European country, and in their short red dresses and
broad-brimmed, gypsy hats, they look picturesque and interesting
in spite of homely faces and ungraceful figures.   Riding into Lam·
bach this morning, I am about wheeling past a horse and drag that,
careless and Austrian-like, has been left untied and unwatched in
the middle of the street, when the horse suddenly scares, swerves
around just in front of me, and dashes, helter-skelter, down the

street. The horse circles around the market square and finally stops of his own accord without doing any damage. Runaways, like other misfortunes, it seems, never come singly, and ere I have left Lambach an hour I am the innocent cause of yet another one; this time it is a large, powerful work-dog, who becomes excited upon meeting me along the road, and upsets things in the most lively manner. Small carts pulled by dogs are common vehicles here, and this one is met coming up an incline, the man considerately giving the animal a lift. A life of drudgery breaks the spirit of these work-dogs and makes them cowardly and cringing. At my approach this one howls, and swerves suddenly around with a rush that upsets both man and cart, topsy-turvy, into the ditch, and the last glimpse of the rumpus obtained, as I sweep past and down the hill beyond, is the man pawing the air with his naked feet and the dog struggling to free himself from the entangling harness.

Up among the hills, at the village of Strenburg, night arrives at a very opportune moment to-day, for Strenburg proves a nice, sociable sort of village, where the doctor can speak good English and plays the rôle of interpreter for me at the gasthaus. The school-ma'am, a vivacious Italian lady, in addition to French and German, can also speak a few words of English, though she persistently refers to herself as the " school-master."• She boards at the same gasthaus, and all the evening long I am favored by the liveliest prattle and most charming gesticulations imaginable, while the room is half filled with her class of young lady aspirants to linguistic accomplishments, listening to our amusing, if not instructive, efforts to carry on a conversation. It is altogether a most enjoyable evening, and on parting I am requested to write when I get around the world and tell the Strenburgers all that I have seen and experienced. On top of the gasthaus is a rude observatory, and before starting I take a view of the country. The outlook is magnificent ; the Austrian Alps are towering skyward to the southeast, rearing snow-crowned heads out from among a billowy sea of pine-covered hills, and to the northward is the lovely valley of the Danube, the river glistening softly through the morning haze.

On yonder height, overlooking the Danube on the one hand and the town of Mölk on the other, is the largest and most imposing edifice I have yet seen in Austria ; it is a convent of the

Benedictine monks; and though Mölk is a solid, substantially built town, of perhaps a thousand inhabitants, I should think there is more material in the immense convent building than in the whole town besides, and one naturally wonders whatever use the monks can possibly have for a building of such enormous dimensions.

Entering a barber's shop here for a shave, I find the barber fol-

The Barber of Mölk.

lowing the example of so many of his countrymen by snoozing the mid-day hours happily and unconsciously away. One could easily pocket and walk off with his stock-in-trade, for small is the danger of his awakening. Waking him up, he shuffles mechanically over to his razor and lathering apparatus, this latter being a soup-plate with a semicircular piece chipped out to fit, after a fashion, the contour of the customers' throats. Pressing this jagged edge of queen's-ware against your windpipe, the artist alternately rubs the water and a cake of soap therein contained about your face with his hands, the water meanwhile passing freely between the ill-fitting soup-plate and your throat, and running down your breast ; but don't complain ; be reasonable : no reasonable-minded person could expect one soup-plate, however carefully chipped out, to fit the throats of the entire male population of Mölk, besides such travellers as happen along.

Spending the night at Neu Lengbach, I climb hills and wabble along, over rough, lumpy roads, toward Vienna, reaching the Austrian capital Sunday morning, and putting up at the *Englischer Hof* about noon. At Vienna I determine to make a halt of two days, and on Tuesday pay a visit to the headquarters of the Vienna Wanderers' Bicycle Club, away out on a suburban street called *Schwimmschulenstrasse ;* and the club promises that if I will delay my departure another day they will get up a small party of wheelmen to escort me seventy kilometres, to Presburg. The bicycle clubs of Vienna have, at the Wanderers' headquarters, constructed an excellent race-track, three and one-third laps to the English mile, at an expense of 2,000 gulden, and this evening several of Austria's fliers are training upon it for the approaching races. English and American wheelmen little understand the difficulties these Vienna cyclers have to contend with : all the city inside the Ringstrasse, and no less than fifty streets outside, are forbidden to the mounted cyclers, and they are required to ticket themselves with big, glaring letters, as also their lamps at night, so that, in case of violating any of these regulations, they can by their number be readily recognized by the police. Self-preservation compels the clubs to exercise every precaution against violating the police regulations, in order not to excite popular prejudice overwhelmingly against bicycles, and ere a new rider is permitted to venture outside their own grounds he is hauled up before a regularly organized committee, consisting of officers from each club in Vienna, and required to go through a regular examination in mounting, dismounting, and otherwise proving to their entire satisfaction his proficiency in managing and manœuvring his wheel ; besides which every cycler is provided with a pamphlet containing a list of the streets he may and may not frequent. In spite of all these harassing regulations, the Austrian capital has already two hundred riders.

The Viennese impress themselves upon me as being possessed

of more than ordinary individuality. Yonder comes a man, walking languidly along, and carrying his hat in his hand, because it is warm, and just behind him comes a fellow-citizen muffled up in an overcoat because—because of Viennese individuality. The people seem to walk the streets with a swaying, happy-go-anyhow sort of gait, colliding with one another and jostling together on the sidewalk in the happiest manner imaginable.

At five o'clock on Thursday morning I am dressing, when I am notified that two cyclers are awaiting me below. Church-bells are clanging joyously all over Vienna as we meander toward suburbs, and people are already streaming in the direction of the St. Stephen's Church, near the centre of the city, for to-day is *Frohnleichnam* (Corpus Christi), and the Emperor and many of the great ecclesiastical, civil, and military personages of the empire will pass in procession with all pomp and circumstance ; and the average Viennese is not the person to miss so important an occasion. Three other wheelmen are awaiting us in the suburbs, and together we ride through the waving barley-fields of the Danube bottom to Schwechat, for the light breakfast customary in Austria, and thence onward to Petronelle, thirty kilometres distant, where we halt a few minutes for a Corpus Christi procession, and drink a glass of white Hungarian wine. Near Petronelle are the remains of an old Roman wall, extending from the Danube to a lake called the *Neusiedler See.* My companions say it was built 2,000 years ago, when the sway of the Romans extended over such parts of Europe as were worth the trouble and expense of swaying. The roads are found rather rough and inferior, on account of loose stones and uneven surface, as we push forward toward Presburg, passing through a dozen villages whose streets are carpeted with fresh-cut grass, and converted into temporary avenues, with branches stuck in the ground, in honor of the day they are celebrating. At Hamburg we pass beneath an archway nine hundred years old, and wheel on through the grass-carpeted streets between rows of Hungarian soldiers drawn up in line, with green oak-sprigs in their hats ; the villagers are swarming from the church, whose bells are filling the air with their clangor, and on the summit of an over-shadowing cliff are the massive ruins of an ancient castle. Near about noon we roll into Presburg, warm and dusty, and after dinner take a stroll through the Jewish quarter of the town up to the height upon which Presburg castle is situated, and from which a most extensive

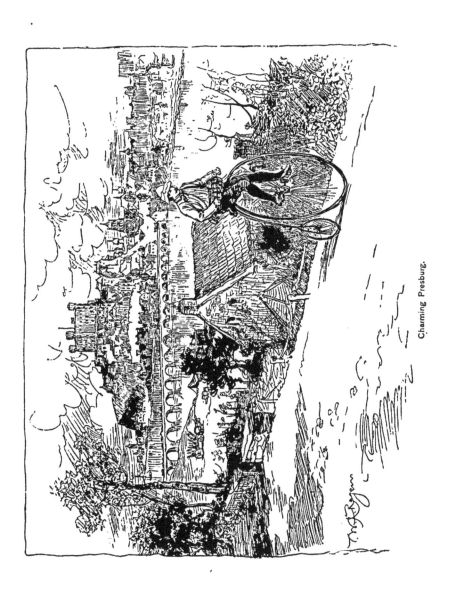

Charming Presburg.

and beautiful view of the Danube, its wooded bluffs and broad, rich bottom-lands, is obtainable. At dinner the waiter hands me a card, which reads : " Pardon me, but I believe you are an Englishman, in which case I beg the privilege of drinking a glass of wine with you." The sender is an English gentleman residing at Budapest, Hungary, who, after the requested glass of wine, tells me that he guessed who I was when he first saw me enter the garden with the five Austrian wheelmen.

My Austrian escort rides out with me to a certain cross-road, to make sure of heading me direct toward Budapest, and as we part they bid me good speed, with a hearty " *Eljen !* "—the Hungarian "*Hip, hip, hurrah.*" After leaving Presburg and crossing over into Hungary the road-bed is of a loose gravel that, during the dry weather this country is now experiencing, is churned up and loosened by every passing vehicle, until one might as well think of riding over a ploughed field. But there is a fair proportion of ridable side-paths, so that I make reasonably good time. Altenburg, my objective point for the night, is the centre of a sixty-thousand-acre estate belonging to the Archduke Albrecht, uncle of the present Emperor of Austro-Hungary, and one of the wealthiest land-owners in the empire. Ere I have been at the gasthaus an hour I am honored by a visit from Professor Thallmeyer, of the Altenburg Royal Agricultural School, who invites me over to his house to spend an hour in conversation, and in the discussion of a bottle of Hungary's best vintage, for the learned professor can talk very good English, and his wife is of English birth and parentage. Although Frau Thallmeyer left England at the tender age of two years, she calls herself an Englishwoman, speaks of England as " home," and welcomes to her house as a countryman any wandering Briton happening along. I am no longer in a land of small peasant proprietors, and there is a noticeably large proportion of the land devoted to grazing purposes, that in France or Germany would be found divided into small farms, and every foot cultivated. Villages are farther apart, and are invariably adjacent to large commons, on which roam flocks of noisy geese, herds of ponies, and cattle with horns that would make a Texan blush—the long-horned roadsters of Hungary. The costumes of the Hungarian peasants are both picturesque and novel, the women and girls wearing top-boots and short dresses on holiday occasions and Sundays, and at other times short dresses without any boots at all ; the men wear loose-flowing pantaloons of white, coarse linen that reach just below the knees, and which a casual observer would unhesitatingly pronounce a short skirt, the material being so ample. Hungary is still practically a land of serfs and nobles, and nearly every peasant encountered along the road touches his cap respectfully, in instinctive acknowledgment, as it were, of his inferiority. Long rows of women are seen hoeing in the fields with watchful overseers standing over them—a scene not unsuggestive of plantation life in the Southern States in the days of slavery. If these gangs of women are not more than about two hundred yards from the road their inquisitiveness overcomes every other consideration, and dropping everything, the whole crowd comes helter-skelter across the field to obtain a closer view of the strange vehicle; for it is only in the neighborhood of one or two of the principal cities of Hungary that one ever sees a bicycle.

Gangs of gypsies are now frequently met with ; they are dark-skinned, interesting people, and altogether different-looking from those occasionally encountered in England and America, where, although swarthy and dark-skinned, they bear no comparison in that respect to these, whose skin is wellnigh black, and whose gleaming white teeth and brilliant, coal-black' eyes stamp them plainly as alien to the race around them. Ragged, unwashed, happy gangs of vagabonds these stragglers appear, and regular droves of partially or wholly naked youngsters come racing after me, calling out "kreuzer! kreuzer! kreuzer!" and holding out hand or tattered hat in a supplicating manner as they run along-side. Unlike the peasantry, none of these gypsies touch their hats ; indeed, yon swarthy-faced vagabond, arrayed mainly in gewgaws, and eying me curiously with his piercing black eyes, may be priding himself on having royal blood in his veins ; and, unregenerate chicken-lifter though he doubtless be, would scarce condescend to touch his tattered tile even to the Emperor of Austria. The black eyes scintillate as they take notice of what they consider the great wealth of sterling silver about the machine I bestride. Eastward from Altenburg the main portion of the road continues for the most part unridably loose and heavy.

For some kilometres out of Raab the road presents a far better surface, and I ride quite a lively race with a small Danube passenger steamer that is starting down-stream. The steamboat toots and forges ahead, and in answer to the waving of hats and exclamations

10

of encouragement from the passengers, I likewise forge ahead, and
although the boat is going down-stream with the strong current
of the Danube, as long as the road continues fairly good I manage
to keep in advance ; but soon the loose surface reappears, and when
I arrive at Gonys, for lunch, I find the steamer already tied up, and
the passengers and officers greet my appearance with shouts of recognition.   My route along the Danube Valley leads through broad,
level wheat-fields that recall memories of the Sacramento Valley,
California.   Geese appear as the most plentiful objects around the
villages : there are geese and goslings everywhere ; and this evening, in a small village, I wheel quite over one, to the dismay of the
maiden driving them homeward, and the unconcealed delight of
several small Hungarians.

At the village of Nezmely I am to-night treated to a foretaste of
what is probably in store for me at a goodly number of places
ahead by being consigned to a bunch of hay and a couple of sacks
in the stable as the best sleeping accommodations the village gasthaus affords.   True, I am assigned the place of honor in the manger, which, though uncomfortably narrow and confining, is perhaps
better accommodation, after all, than the peregrinating tinker and
three other likely-looking characters are enjoying on the bare floor.
Some of these companions, upon retiring, pray aloud at unseemly
length, and one of them, at least, keeps it up in his sleep at frequent
intervals through the night ; horses and work-cattle are rattling
chains and munching hay, and an uneasy goat, with a bell around
his neck, fills the stable with an incessant tinkle till dawn.   Black
bread and a cheap but very good quality of white wine seem about
the only refreshment obtainable at these little villages.   One asks
in vain for *milch-brod, butter, käse,* or in fact anything acceptable
to the English palate ; the answer to all questions concerning these
things is "*nicht, nicht, nicht.*"—"What have you, then ?" I sometimes ask, the answer to which is almost invariably "*brod und wein.*"
Stone-yards thronged with busy workmen, chipping stone for shipment to cities along the Danube, are a feature of these river-side
villages.   The farther one travels the more frequently gypsies are
encountered on the road.   In almost every band is a maiden, who,
by reason of real or imaginary beauty, occupies the position of pet
of the camp, wears a profusion of beads and trinkets, decorates
herself with wild flowers, and is permitted to do no manner of
drudgery.   Some of these gypsy maidens are really quite beautiful

in spite of their very dark complexions. Their eyes glisten with inborn avarice as I sweep past on my "silver" bicycle, and in their astonishment at my strange appearance and my evidently enormous wealth they almost forget their plaintive wail of "kreuzer! kreuzer!" a cry which readily bespeaks their origin, and is easily recognized as an echo from the land where the cry of "backsheesh" is seldom out of the traveller's hearing.

The roads east of Nezmely are variable, flint-strewn ways predominating ; otherwise the way would be very agreeable, since the gradients are gentle, and the dust not over two inches deep, as against three in most of Austro-Hungary thus far traversed. The weather is broiling hot ; but I worry along perseveringly, through rough and smooth, toward the land of the rising sun. Nearing Budapest the roads become somewhat smoother, but at the same time hillier, the country changing to vine-clad slopes ; and all along the undulating ways I meet wagons laden with huge wine-casks. Reaching Budapest in the afternoon, I seek out Mr. Kosztovitz, of the Budapest Bicycle Club, and consul of the Cyclists' Touring Club, who proves a most agreeable gentleman, and who, besides being an enthusiastic cycler, talks English perfectly. There is more of the sporting spirit in Budapest, perhaps, than in any other city of its size on the Continent, and no sooner is my arrival known than I am taken in hand and practically compelled to remain over at least one day. Svetozar Igali, a noted cycle tourist of the village of Duna Szekesö, now visiting the international exhibition at Budapest, volunteers to accompany me to Belgrade, and perhaps to Constantinople. I am rather surprised at finding so much cycling enthusiasm in the Hungarian capital. Mr. Kosztovitz, who lived some time in England, and was president of a bicycle club there, had the honor of bringing the first wheel into the Austro-Hungarian empire, in the autumn of 1879, and now Budapest alone has three clubs, aggregating nearly a hundred riders, and a still greater number of non-riding members.

Cyclers have far more liberty accorded them in Budapest than in Vienna, being permitted to roam the city almost as untrammelled as in London, this happy condition of affairs being partly the result of Mr. Kosztovitz's diplomacy in presenting a ready drawn-up set of rules and regulations for the government of wheelmen to the police authorities when the first bicycle was introduced, and partly to the police magistrate, being himself an enthusiastic all-

'round sportsman, inclined to patronize anything in the way of athletics. They are even experimenting in the Hungarian army with the view of organizing a bicycle despatch service ; and I am told that they already have a bicycle despatch in successful operation in the Bavarian army. In the evening I am the club's guest at a supper under the shade-trees in the exhibition grounds. Mr. Kosztovitz and another gentleman who can speak English act as interpreters, and here, amid the merry clinking of champagne-glasses, the glare of electric lights, with the ravishing music of an Hungarian gypsy band on our right, and a band of swarthy Servians playing their sweet native melodies on our left, we, among other toasts, drink to the success of my tour. There is a cosmopolitan and exceedingly interesting crowd of visitors at the international exhibition : natives from Bulgaria, Servia, Roumania, and Turkey, in their national costumes ; and mingled among them are Hungarian peasants from various provinces, some of them in a remarkably picturesque dress, that I afterward learn is Croatian.

A noticeable feature of Budapest, besides a predilection for sport among the citizens, is a larger proportion of handsome ladies than one sees in most European cities, and there is, moreover, a certain atmosphere about them that makes them rather agreeable company. If one is travelling around the world with a bicycle, it is not at all inconsistent with Budapest propriety for the wife of the wheelman sitting opposite you to remark that she wishes she were a rose, that you might wear her for a button-hole bouquet on your journey, and to ask whether or not, in that case, you would throw the rose away when it faded. Compliments, pleasant, yet withal as meaningless as the coquettish glances and fan-play that accompany them, are given with a freedom and liberality that put the sterner native of more western countries at his wits' end to return them. But the most delightful thing in all Hungary is its gypsy music. As it is played here beneath its own sunny skies, methinks there is nothing in the wide world to compare with it. The music does not suit the taste of some people, however ; it is too wild and thrilling. Budapest is a place of many languages, one of the waiters in the exhibition café claiming the ability to speak and understand no less than fourteen different languages and dialects.

Nine wheelmen accompany me some distance out of Budapest on Monday morning, and Mr. Philipovitz and two other members continue with Igali and me to Duna Pentele, some seventy-five miles distant; this is our first sleeping-place, the captain making me his guest until our separation and departure in different directions, next morning. During the fierce heat of mid-day we halt for about three hours at Adony, and spend a pleasant after-dinner hour examining the trappings and trophies of a noted sporting gentleman, and witnessing a lively and interesting set-to with fencing foils. There is everything in fire-arms in his cabinet, from an English double-barrelled shot-gun to a tiny air-pistol for shooting flies on the walls of his sitting-room ; he has swords, oars, gymnastic paraphernalia—in fact, everything but boxing gloves.

Arriving at Duna Pentele early in the evening, before supper we swim for an hour in the waters of the Danube. At 9.30 p.m. two of our little company board the up-stream-bound steamer for the return home, and at ten o'clock we are proposing to retire for the night, when lo, in come a half-dozen gentlemen, among them Mr. Ujvärii, whose private wine-cellar is celebrated all the country round, and who now proposes that we postpone going to bed long enough to pay a short visit to his cellar and sample the " finest wine in Hungary." This is an invitation not to be resisted by ordinary mortals, and accordingly we accept, following the gentleman and his friends through the dark streets of the village. Along the dark, cool vault penetrating the hill-side Mr. Ujvärii leads the way between long rows of wine-casks, *heber* * held in arm like a sword at dress parade. The *heber* is first inserted into a cask of red wine, with a perfume and flavor as agreeable as the rose it resembles in color, and carried, full, to the reception end of the vault by the corpulent host with the stately air of a monarch bearing his sceptre. After two rounds of the red wine, two *hebers* of champagne are brought—champagne that plays a fountain of diamond spray three inches above the glass. The following toast is proposed by the host : " The prosperity and welfare of England, America, and Hungary, three countries that are one in their love and appreciation of sport and adventure." The Hungarians have all the Anglo-American love of sport and adventure.

* A glass combination of tube and flask, holding about three pints, with an orifice at each end and the bulb or flask near the upper orifice ; the wine is sucked up into the flask with the breath, and when withdrawn from the cask the index finger is held over the lower orifice, from which the glasses are filled by manipulations of the finger.

From Budapest to Paks, about one hundred and twenty kilometres, the roads are superior to anything I expected to find east
of Germany ; but the thermometer clings around the upper regions,
and everything is covered with dust. Our route leads down the
Danube in an almost directly southern course.

Instead of the poplars of France, and the apples and pears of
Germany, the roads are now fringed with mulberry-trees, both
raw and manufactured silk being a product of this part of Hungary.

My companion is what in England or America would be considered a " character ; " he dresses in the thinnest of racing costumes, through which the broiling sun readily penetrates, wears
racing-shoes, and a small jockey-cap with an enormous poke, beneath which glints a pair of "specs ; " he has rat-trap pedals to his
wheel, and winds a long blue girdle several times around his waist,
consumes raw eggs, wine, milk, a certain Hungarian mineral water,
and otherwise excites the awe and admiration of his sport-admiring
countrymen. Igali's only fault as a road companion is his utter
lack of speed, six or eight kilometres an hour being his natural
pace on average roads, besides footing it up the gentlest of gradients and over all rough stretches. Except for this little drawback,
he is an excellent man to take the lead, for he is a genuine Magyar,
and orders the peasantry about with the authoritative manner of
one born to rule and tyrannize ; sometimes, when the surface is uneven for wheeling, making them drive their clumsy ox-wagons
almost into the road-side ditch in order to avoid any possible chance
of difficulty in getting past. Igali knows four languages : French,
German, Hungarian, and Slavonian, but *Anglaise nicht*, though with
what little French and German I have picked up while crossing
those countries we manage to converse and understand each other
quite readily, especially as I am, from constant practice, getting to
be an accomplished pantomimist, and Igali is also a pantomimist
by nature, and gifted with a versatility that would make a Frenchman envious. Ere we have been five minutes at a *gasthaus* Igali is
usually found surrounded by an admiring circle of leading citizens
—not peasants ; Igali would not suffer them to gather about him
—pouring into their willing ears the account of my journey ; the
words, " San Francisco, Boston, London, Paris, Wien, Pesth, Belgrade, Constantinople, Afghanistan, India, Khiva," etc., which are
repeated in rotation at wonderfully short intervals, being about all

that my linguistic abilities are capable of grasping. The road continues hard, but south of Paks it becomes rather rough ; consequently, halts under the shade of the mulberry-trees for Igali to catch up are of frequent occurrence.

The peasantry, hereabout, seem very kindly disposed and hospitable. Sometimes, while lingering for Igali, they will wonder what I am stopping for, and motion the questions of whether I wish anything to eat or drink ; and this afternoon one of them, whose curiosity to see how I mounted overcomes his patience, offers me a twenty-kreuzer piece to show him. At one village a number of peasants take an old cherry-woman to task for charging me two kreuzers more for some cherries than it appears she ought, and although two kreuzers are but a farthing they make quite a squabble with the poor old woman about it, and will be soothed by neither her voice nor mine until I accept another handful of cherries in lieu of the overcharged two kreuzers.

Szekszard has the reputation, hereabout, of producing the best quality of red wine in all Hungary—no small boast, by the way—and the hotel and wine-gardens here, among them, support an excellent gypsy band of fourteen pieces. Mr. Garäy, the leader of the band, once spent nearly a year in America, and after supper the band plays, with all the thrilling sweetness of the Hungarian muse, " Home, sweet Home," " Yankee Doodle," and " Sweet Violets," for my especial delectation.

A wheelman the fame of whose exploits has preceded him might as well try to wheel through hospitable Hungary without breathing its atmosphere as without drinking its wine ; it isn't possible to taboo it as I tabooed the *vin ordinaire* of France, Hungarians and Frenchmen being two entirely different people.

Notwithstanding music until 11.30 P.M., yesterday, we are on the road before six o'clock this morning—for genuine, unadulterated Hungarian music does not prevent one getting up bright and fresh next day—and about noon we roll into Duna Szekeso, Igali's native town, where we have decided to halt for the remainder of the day to get our clothing washed, one of my shoes repaired, and otherwise prepare for our journey to the Servian capital. Duna Szekeso is a calling-place for the Danube steamers, and this afternoon I have the opportunity of taking observations of a gang of Danubian roustabouts at their noontide meal. They are a swarthy, wild-looking crowd, wearing long hair parted in the middle, or not

parted at all ; to their national costume are added the jaunty trappings affected by river men in all countries. Their food is coarse
black bread and meat, and they take turns in drinking wine from
a wooden tube protruding from a two-gallon watch-shaped cask,
the body of which is composed of a section of hollow log instead of
staves, lifting the cask up and drinking from the tube, as they
would from the bung-hole of a beer-keg. Their black bread would
hardly suit the palate of the Western world ; but there are doubtless a few individuals on both sides of the Atlantic who would willingly be transformed into a Danubian roustabout long enough to
make the acquaintance of yonder rude cask.

After bathing in the river we call on several of Igali's friends,
among them the Greek priest and his motherly-looking wife, Igali
being of the Greek religion. There appears to be the greatest
familiarity between the priests of these Greek churches and their
people, and during our brief visit the priest, languid-eyed, fat, and
jolly, his equally fat and jolly wife, and Igali, caress playfully, and
cut up as many antics as three kittens in a bay window. The farther one travels southward the more amiable and affectionate in
disposition the people seem to become.

Five o'clock next morning finds us wheeling out of Duna Szekeso, and during the forenoon we pass through Baranyavär, a colony of Greek Hovacs, where the women are robed in white drapery
as scant as the statuary which the name of their religion calls to
memory. The roads to-day are variable ; there is little but what is
ridable, but much that is rough and stony enough to compel slow
and careful wheeling. Early in the evening, as we wheel over the
bridge spanning the River Drave, an important tributary of the
Danube, into Eszek, the capital of Slavonia, unmistakable rain-
signs appear above the southern horizon.

CHAPTER VII.

THROUGH SLAVONIA AND SERVIA.

The editor of *Der Drau*, the semi-weekly official organ of the Slavonian capital, and Mr. Freund, being the two citizens of Eszek capable of speaking English, join voices at the supper-table in hoping it will rain enough to compel us to remain over to-morrow, that they may have the pleasure of showing us around Eszek and of inviting us to dinner and supper; and Igali, I am constrained to believe, retires to his couch in full sympathy with them, being possessed of a decided weakness for stopping over and accepting invitations to dine. Their united wish is gratified, for when we rise in the morning it is still raining.

Eszek is a fortified city, and has been in time past an important fortress. It has lost much of its importance since the introduction of modern arms, for it occupies perfectly level ground, and the fortifications consist merely of large trenches that have been excavated and walled, with a view of preventing the city from being taken by storm—not a very overshadowing consideration in these days, when the usual mode of procedure is to stand off and bombard a city into the conviction that further resistance is useless. After dinner the assistant editor of *Der Drau* comes around and pilots us about the city and its pleasant environments. The worthy assistant editor is a sprightly, versatile Slav, and, as together we promenade the parks and avenues, the number and extent of which appear to be the chief glory of Eszek, the ceaseless flow of language and wellnigh continuous interchange of gesticulations between himself and Igali are quite wonderful, and both of them certainly ought to retire to-night far more enlightened individuals than they found themselves this morning.

The Hungarian seems in a particularly happy and gracious mood to-day, as I instinctively felt certain he would be if the fates decreed against a continuation of our journey. When our companion's conversation turns on any particularly interesting sub-

ject I am graciously given the benefit of it to the extent of some
French or German word the meaning of which, Igali has discovered,
I understand. During the afternoon we wander through the intri-
cacies of a yew-shrub maze, where a good-sized area of impenetrably
thick vegetation has been trained and trimmed into a bewildering
net-work of arched walks that almost exclude the light, and Igali
pauses to favor me with the information that this maze is the favorite trysting place of Slavonian nymphs and swains, and furthermore expresses his opinion that the spot must be indeed romantic
and an appropriate place to "come a-wooin'" on nights when the
moonbeams, penetrating through a thousand tiny interspaces, convert the gloomy interior into chambers of dancing light and shadow.
All this information and these comments are embodied in the two
short words, "Amour, luna," accompanied by a few gesticulations,
and is a fair sample of the manner in which conversation is carried
on between us. It is quite astonishing how readily two persons
constantly together will come to understand each other through the
medium of a few words which they know the meaning of in common.

Scores of ladies and gentlemen, the latter chiefly military officers, are enjoying a promenade in the rain-cooled atmosphere, and
there is no mistaking the glances of interest with which many of
them favor—Igali. His pronounced sportsmanlike make-up attracts universal attention and causes everybody to mistake him for
myself—a kindly office which I devoutly wish he would fill until
the whole journey is accomplished. In the Casino garden a dozen
bearded musicians are playing Slavonian airs, and, by request of
the assistant editor, they play and sing the Slavonian national an-
them and a popular air or two besides. The national musical instrument of Slavonia is the "tamborica"—a small steel-stringed
instrument that is twanged with a chip-like piece of wood. Their
singing is excellent in its way, but to the writer's taste there is no
comparison between their tamboricas and the gypsy music of Hungary.

There are no bicycles in all Eszek save ours—though Mr.
Freund, who has lately returned from Paris, has ordered one, with
which he expects to win the admiration of all his countrymen—
and Igali and myself are lionized to our hearts' content ; but this
evening we are quite startled and taken aback by the reappearance
of the assistant editor, excitedly announcing the arrival of a tricycle

in town ! Upon going down, in breathless anticipation of summarily losing the universal admiration of Eszek, we find an itinerant cobbler, who has constructed a machine that would make the rudest bone-shaker of ancient memory seem like the most elegant product of Hartford or Coventry in comparison. The backbone and axletree are roughly hewn sticks of wood, ironed equally rough at the village blacksmith's ; and as, for a twenty-kreuzer piece, the rider mounts and wobbles all over the sidewalk for a short distance, the spectacle would make a stoic roar with laughter, and the good people of the Lower Danubian provinces are anything but stoical.

Six o'clock next morning finds us travelling southward into the interior of Slavonia ; but we are not mounted, for the road presents an unridable surface of mud, stones, and ruts, that causes my companion's favorite ejaculatory expletive to occur with more than its usual frequency. For a portion of the way there is a narrow sidepath that is fairly ridable, but an uninvitingly deep ditch runs unpleasantly near, and no amount of persuasion can induce my companion to attempt wheeling along it. Igali's bump of cautiousness is fully developed, and day by day, as we journey together, I am becoming more and more convinced that he would be an invaluable companion to have accompany one around the world ; true, the journey would occupy a decade, or thereabout, but one would be morally certain of coming out safe and sound in the end.

During our progression southward there has been a perceptible softening in the disposition of the natives, this being more noticeably a marked characteristic of the Slavonians ; the generous southern sun, shining on the great area of Oriental gentleness, casts a softening influence toward the sterner north, imparting to the people amiable and genial dispositions. It takes but comparatively small deeds to win the admiration and applause of the natives of the Lower Danube, with their childlike manners ; and, by slowly meandering along the roadways of Southern Hungary occasionally with his bicycle, Igali has become the pride and admiration of thousands.

For mile after mile we have to trundle our way slowly along the muddy highway as best we can, our road leading through a flat and rather swampy area of broad, waving wheat-fields ; we relieve the tedium of the journey by whistling, alternately, "Yankee Doodle," to which Igali has taken quite a fancy since first hearing it played by the gypsy band in the wine-garden at Szekszard three days ago,

and the Hungarian national air—this latter, of course, falling to
Igali's share of the entertainment. Having been to college in
Paris, Igali is also able to contribute the famous Marseillaise
hymn, and, not to be outdone, I favor him with "God Save the
Queen" and "Britannia Rules the Waves," both of which he thinks
very good tunes—the former seeming to strike his Hungarian ear,
however, as rather solemn. In the middle of the forenoon we
make a brief halt at a rude road-side tavern for some refreshments
—a thick, narrow slice of raw, fat bacon, white with salt, and a
level pint of red wine, satisfying my companion ; but I substitute .
for the bacon a slice of coarse, black bread, much to Igali's wonderment. Here are congregated several Slavonian shepherds, in
their large, ill-fitting, sheepskin garments, with the long wool
turned inward—clothes that apparently serve them alike to keep
out the summer's heat and the winter's cold. One of the peasants, with ideas a trifle befuddled with wine, perhaps, and face all
aglow with admiration for our bicycles, produces a tattered memorandum and begs us to favor him with our autographs, an act that
of itself proves him to be not without a degree of intelligence one
would scarcely look for in a sheepskin-clad shepherd of Slavonia.
Igali gruffly bids the man " begone," and aims a careless kick at the
proffered memorandum ; but seeing no harm in the request, and,
moreover, being perhaps by nature a trifle more considerate of
others, I comply. As he reads aloud, "United States, America," to
his comrades, they one and all lift their hats quite reverently and
place their brown hands over their hearts, for I suppose they
recognize in my ready compliance with the simple request, in comparison with Igali's rude rebuff—which, by the way, no doubt
comes natural enough—the difference between the land of the
prince and peasant, and the land where "liberty, equality, and
fraternity" is not a meaningless motto—a land which I find every
down-trodden peasant of Europe has heard of, and looks upward
to.

Soon after this incident we are passing a prune-orchard, when,
as though for our especial benefit, a couple of peasants working
there begin singing aloud, and with evident enthusiasm, some
national melody, and as they observe not our presence, at my suggestion we crouch behind a convenient clump of bushes and for
several minutes are favored with as fine a duet as I have heard for
many a day ; but the situation becomes too ridiculous for Igali,

The Slavonian Shepherds.

and it finally sends him into a roar of laughter that causes the performance to terminate abruptly, and, rising into full view, we doubtless repay the singers by letting them see us mount and ride into their native village, but a few hundred yards distant.   .

We are to-day passing through villages where a bicycle has never been seen—this being outside the area of Igali's peregrinations—and the whole population invariably turns out *en masse*, clerks, proprietors, and customers in the shops unceremoniously dropping everything and running to the streets ; there is verily a hurrying to and fro of all the citizens ; husbands hastening from magazine to dwelling to inform their wives and families, mothers running to call their children, children their parents, and everybody scampering to call the attention of their sisters, cousins, and aunts, ere we are vanished in the distance, and it be everlastingly too late.

We have been worrying along at some sort of pace, with the exception of the usual noontide halt, since six o'clock this morning, and the busy mosquito is making life interesting for belated wayfarers, when we ride into Sarengrad and put up at the only *gasthaus* in the village.   Our bedroom is situated on the ground floor, the only floor in fact the *gasthaus* boasts, and we are in a fair way of either being lulled to sleep or kept awake, as the case may be, by a howling chorus of wine-bibbers in the public room adjoining ; but here, again, Igali shows up to good advantage by peremptorily ordering the singers to stop, and stop instanter.   The amiably disposed peasants, notwithstanding the wine they have been drinking, cease their singing and become silent and circumspect, in deference to the wishes of the two strangers with the wonderful machines.   We now make a practice of taking our bicycles into our bedroom with us at night, otherwise every right hand in the whole village would busy itself pinching the "gum-elastic" tires and pedal-rubbers, twirling the pedals, feeling spokes, backbone, and forks, and critically examining and commenting upon every visible portion of the mechanism ; and who knows but that the latent cupidity of some easy-conscienced villager might be aroused at the unusual sight of so much "silver" standing around loose (the natives hereabout don't even ask whether the nickelled parts of the bicycle are silver or not ; they take it for granted to be so), and surreptitiously attempt to chisel off enough to purchase an embroidered coat for Sundays ?   From what I can understand of their comments among themselves, it is perfectly consistent with their ideas of the average Englishman that he should bestride a bicycle of solid silver, and if their vocabulary embraced no word corresponding to our "millionnaire," and they desired to use one, they would probably pick upon the word " Englander " as the most appropriate. While we are making our toilets in the morning eager faces are peering inquisitively through the bedroom windows ; a murmur of voices, criticizing us and our strange vehicles, greets our waking moments, and our privacy is often invaded, in spite of Igali's inconsiderate treatment of them whenever they happen to cross his path.

Many of the inhabitants of this part of Slavonia are Croatians —people who are noted for their fondness of finery ; and, as on this sunny Sunday morning we wheel through their villages, the crowds of peasantry who gather about us in all the bravery of their best clothes present, indeed, an appearance gay and picturesque beyond anything hitherto encountered. The garments of the men are covered with braid-work and silk embroidery wherever such ornamentation is thought to be an embellishment, and, to the Croatian mind, that means pretty much everywhere ; and the girls and women are arrayed in the gayest of colors ; those displaying the brightest hues and the greatest contrasts seem to go tripping along conscious of being irresistible. Many of the Croatian peasants are fine, strapping fellows, and very handsome women are observed in the villages—women with great, dreamy eyes, and faces with an expression of languor that bespeaks their owners to be gentleness personified. Igali shows evidence of more susceptibility to female charms than I should naturally have given him credit for, and shows a decided inclination to linger in these beauty-blessed villages longer than is necessary, and as one dark-eyed damsel after another gathers around us, I usually take the initiative in mounting and clearing out.

Were a man to go suddenly flapping his way through the streets of London on the long-anticipated flying-machine, the average Cockney would scarce betray the unfeigned astonishment that is depicted on the countenances of these Croatian villagers as we ride into their midst and dismount.

This afternoon my bicycle causes the first runaway since the trifling affair at Lembach, Austria. A brown-faced peasant woman and a little girl, driving a small, shaggy pony harnessed to a basket-work, four-wheeled vehicle, are approaching; their humble-looking steed betrays no evidence of restiveness until just as I am turning out to pass him, when, without warning, he gives a swift, sudden bound to the right, nearly upsetting the vehicle, and without more ado bolts down a considerable embankment and goes helter-skelter across a field of standing grain.

The old lady pluckily hangs on to the reins, and finally succeeds in bringing the runaway around into the road again without damaging anything save the corn. It might have ended much less satisfactorily, however, and the incident illustrates one possible source of trouble to a 'cycler travelling alone through countries where the people neither understand, nor can be expected to understand, a wheelman's position ; the situation would, of course, be aggravated in a country village where, not speaking the language, one could not make himself understood in his own defence. These people here, if not wise as serpents, are at least harmless as doves ; but, in case of the bicycle frightening a team and causing a runaway with the unpleasant sequel of broken limbs, or injured horse, they would scarce know what to do in the premises, since they would have no precedent to govern them, and, in the absence of any intelligent guidance, might conclude to wreak summary vengeance on the bicycle. In such a case, would a wheelman be justified in using his revolver to defend his bicycle ?

Such is the reverie into which I fall while reclining beneath a spreading mulberry-tree waiting for Igali to catch up ; for he has promised that I shall see the Slavonian national dance sometime to-day, and a village is now visible in the distance. At the Danube-side village of Hamenitz an hour's halt is decided upon to give me the promised opportunity of witnessing the dance in its native land. It is a novel and interesting sight. A round hundred young gallants and maidens are rigged out in finery such as no other people save the Croatian and Slavonian peasants ever wear—the young men braided and embroidered, and the damsels having their hair entwined with a profusion of natural flowers in addition to their costumes of all possible hues. Forming themselves into a large ring, distributed so that the sexes alternate, the young men extend and join their hands in front of the maidens, and the latter join hands behind their partners ; the steel-strung tamboricas strike up a lively twanging air, to which the circle of dancers endeavor to shuffle time with their feet, while at the same time moving around in a circle. Livelier and faster twang the tamboricas, and more and more animated becomes the scene as the dancing, shuffling ring endeavors to keep pace with it. As the fun progresses into the fast and furious stages the youths' hats have a knack of getting into a jaunty position on the side of their heads, and the wearers' faces assume a reckless, flushed appearance, like men half intoxicated, while the maidens' bright eyes and beaming faces betoken unutterable happiness ; finally the music and the shuffling of feet terminate with a rapid flourish, everybody kisses everybody—save, of course, mere luckless onlookers like Igali and myself—and the Slavonian national dance is ended.

To-night we reach the strongly fortified town of Peterwardein, opposite which, just across a pontoon bridge spanning the Danube, is the larger city of Neusatz. At Hamenitz we met Professor Zaubaur, the editor of the *Uj Videk*, who came down the Danube ahead of us by steamboat ; and now, after housing our machines at our *gasthaus* in Peterwardein, he pilots us across the pontoon bridge in the twilight, and into one of those wine-gardens so universal in this part of the world. Here at Neusatz I listen to the genuine Hungarian gypsy music for the last time on the European tour ere bidding the territory of Hungary adieu, for Neusatz is on the Hungarian side of the Danube. The professor has evidently let no grass grow beneath his feet since leaving us scarcely an hour ago at Hamenitz, for he has, in the mean time, ferreted out the only English-speaking person at present in town, the good Frau Schrieber, an Austrian lady, formerly of Vienna, but now at Neusatz with her husband, a well-known advocate. This lady talks English quite fluently. Though not yet twenty-five she is very, very wise, and among other things she informs her admiring friends gathered round about us, listening to the—to them—unintelligible flow of a foreign language, that Englishmen are " very grave beings," a piece of information that wrings from Igali a really sympathetic response—nothing less than the startling announcement that he hasn't seen me smile since we left Budapest together, a week ago ! " Having seen the Slavonian, I ought by all means to see the Hungarian, national dance," Frau Schrieber says ; adding, " It is a nice dance for Englishmen to look at, though it is so very gay that English ladies would neither dance it nor look at it being danced." Ere parting company with this entertaining lady she agrees that, if I will but remain in Hungary permanently, she

11

knows of a very handsome fräulein of sixteen summers, who, having heard of my "wonderful journey," is already predisposed in my favor, and with a little friendly tact and management on her—Frau Schrieber's—part would no doubt be willing to waive the formalities of a long courtship, and yield up hand and heart at my request! . I can scarcely think of breaking in twain my trip around the world even for so tempting a prospect, and I recommend the fair Hungarian to Igali ; but " the fräulein has never heard of Herr Igali, and he will not do."

"Will the fräulein be willing to wait until my journey around the world is completed ? "

"Yes ; she vill vait mit much pleezure ; I vill zee dat she vait ; und I know you vill return, for . an Englishman alvays forgets his promeezes." Henceforth, when Igali and myself enter upon a programme of whistling, " Yankee Doodle" is supplanted by "The girl I left behind me," much to his annoyance, since, not understanding the sentiment responsible for the change, he thinks " Yankee Doodle " a far better tune. So much attached, in fact, has Igali become to the American national air, that he informs the professor and editor of *Uj Videk* of the circumstance of the band playing it at Szekszard. As, after supper, several of us promenade the streets of Neusatz, the professor links his arm in mine, and, taking the cue from Igali, begs me to favor him by whistling it. I try my best to palm this patriotic duty off on Igali, by paying flattering compliments to his style of whistling ; but, after all, the duty falls on me, and I whistle the tune softly, yet merrily, as we walk along, the professor, spectacled and wise-looking, meanwhile exchanging numerous nods of recognition with his fellow-Neusatzers we meet.

The provost-judge of Neusatz shares the honors with Frau Schrieber of knowing more or less English ; but this evening the judge is out of town. The enterprising professor lies in wait for him, however, and at 5.30 on Monday morning, while we are dressing, an invasion of our bed-chamber is made by the professor, the jolly-looking and portly provost-judge, a Slavonian lieutenant of artillery, and a druggist friend of the others. The provost-judge and the lieutenant actually own bicycles and ride them, the only representatives of the wheel in Neusatz and Peterwardein, and the judge is " very angry "—as he expresses it—that Monday is court day, and to-day an unusually busy one, for he would be most happy to wheel with .us to Belgrade.

The lieutenant fetches his wheel and accompanies us to the next village. Peterwardein is a strongly fortified place, and, as a position commanding the Danube so completely, is furnished with thirty guns of large calibre, a battery certainly not to be despised when posted on a position so commanding as the hill on which Peterwardein fortress is built. As the editor and others at Eszek, so here the professor, the judge, and the druggist unite in a friendly protest against my attempt to wheel through Asia, and more especially through China, "for everybody knows it is quite dangerous," they say. These people cannot possibly understand why it is that an Englishman or American, knowing of danger beforehand, will still venture ahead ; and when, in reply to their questions, I modestly announce my intention of going ahead, notwithstanding possible danger and probable difficulties, they each, in turn, shake my hand as though reluctantly resigning me to a reckless determination, and the judge, acting as spokesman, and echoing and interpreting the sentiments of his companions, exclaims, "England and America forever ! it is ze grandest peeples on ze world !" The lieutenant, when questioned on the subject by the judge and the professor, simply shrugs his shoulders and says nothing, as becomes a man whose first duty is to cultivate a supreme contempt for danger in all its forms.

They all accompany us outside the city gates, when, after mutual farewells and assurances of good-will, we mount and wheel away down the Danube, the lieutenant's big mastiff trotting soberly alongside his master, while Igali, sometimes in and sometimes out of sight behind, brings up the rear. After the lieutenant leaves us we have to trundle our weary way up the steep gradients of the Fruskagora Mountains for a number of kilometres. For Igali it is quite an adventurous morning. Ere we had left the shadows of Peterwardein fortress he upset while wheeling beneath some overhanging mulberry-boughs that threatened destruction to his jockey-cap ; soon after parting company with the lieutenant he gets into an altercation with a gang of gypsies about being the cause of their horses breaking loose from their picket-ropes and stampeding, and then making uncivil comments upon the circumstance ; an hour after this he overturns again and breaks a pedal, and when we dismount at Indjia, for our noontide halt, he discovers that his saddle-spring has snapped in the middle. As he ruefully surveys the breakage caused by the roughness of the Fruskagora roads, and sends out to scour the village for a mechanic capable of undertaking the repairs, he eyes my Columbia wistfully, and asks me for the address where one like it can be obtained.   The blacksmith is not prepared to mend the spring, although he makes a good job of the pedal, and it takes a carpenter and his assistant from 1.30 to 4.30 P.M. to manufacture a grooved piece of wood to fit between the spring and backbone so that he can ride with me to Belgrade. It would have been a fifteen-minute task for a Yankee carpenter.

We have been traversing a spur of the Fruskagora Mountains all the morning, and our progress has been slow.   The roads through here are mainly of the natural soil, and correspondingly bad ; but the glorious views of the Danube, with its alternating wealth of green woods and greener cultivated areas, fully recompense for the extra toil.   Prune-orchards, the trees weighed down with fruit yet green, clothe the hill-sides with their luxuriance ; indeed, the whole broad, rich valley of the Danube seems nodding and smiling in the consciousness of overflowing plenty ; for days we have traversed roads leading through vineyards and orchards, and broad areas with promising-looking grain-crops.

It is but thirty kilometres from Indjia to Semlin, on the river-bank opposite Belgrade, and since leaving the Fruskagora Mountains the country has been a level plain, and the roads fairly smooth. But Igali has naturally become doubly cautious since his succession of misadventures this morning, and as, while waiting for him to overtake me, I recline beneath the mulberry-trees near the village of Batainitz and survey the blue mountains of Servia looming up to the southward through the evening haze, he rides up and proposes Batainitz as our halting-place for the night, adding persuasively, "There will be no ferry-boat across to Belgrade to-night, and we can easily catch the first boat in the morning." I reluctantly agree, though advocating going on to Semlin this evening.

While our supper is being prepared we are taken in hand by the leading merchant of the village and " turned loose " in an orchard of small fruits and early pears, and from thence conducted to a large gypsy encampment in the outskirts of the village, where, in acknowledgment of the honor of our visit—and a few kreuzers by way of supplement—the " flower of the camp," a blooming damsel, about the shade of a total eclipse, kisses the backs of our hands, and the men play a strumming monotone with sticks and an inverted wooden trough, while the women dance in a most lively and not ungraceful manner. These gypsy bands are a happy crowd of vagabonds, looking as though they had never a single care in all the world ; the men wear long, flowing hair, and to the ordinary costume of the peasant is added many a gewgaw, worn with a careless, jaunty grace that fails not to carry with it a certain charm in spite of unkempt locks and dirty faces. The women wear a minimum of clothes and a profusion of beads and trinkets, and the children go stark naked or partly dressed.

Unmistakable evidence that one is approaching the Orient appears in the semi-Oriental costumes of the peasantry and roving gypsy bands, as we gradually near the Servian capital. An Oriental costume in Eszek is sufficiently exceptional to be a novelty, and so it is until one gets south of Peterwardein, when the national costumes of Slavonia and Croatia are gradually merged into the tasselled fez, the many-folded waistband, and the loose, flowing pantaloons of Eastern lands. Here at Batainitz the feet are encased in rude raw-hide moccasins, bound on with leathern thongs, and the ankle and calf are bandaged with many folds of heavy red material, also similarly bound. The scene around our *gasthaus*, after our arrival, resembles a popular meeting ; for, although a few of the villagers have been to Belgrade and seen a bicycle, it is only within the last six months that Belgrade itself has boasted one, and the great majority of the Batainitz people have simply heard enough about them to whet their curiosity for a closer acquaintance. Moreover, from the interest taken in my tour at Belgrade on account of the bicycle's recent introduction in that capital, these villagers, but a dozen kilometres away, have heard more of my journey than people in villages farther north, and their curiosity is roused in proportion.

We are astir by five o'clock next morning ; but the same curious crowd is making the stone corridors of the rambling old *gasthaus* impassable, and filling the space in front, gazing curiously at us, and commenting on our appearance whenever we happen to become visible, while waiting with commendable patience to obtain a glimpse of our wonderful machines. They are a motley, and withal a ragged assembly; old women devoutly cross themselves as, after a slight repast of bread and milk, we sally forth with our wheels, prepared to start ; and the spontaneous murmur of admiration which breaks forth as we mount becomes louder and more pronounced as I turn in the saddle and doff my helmet in deference to the homage paid us by hearts which are none the less warm because hidden beneath the rags of honest poverty and semi-civilization. It takes but little to win the hearts of these rude, unsophisticated people. A two hours' ride from Batainitz, over level and reasonably smooth roads, brings us into Semlin, quite an important Slavonian city on the Danube, nearly opposite Belgrade, which is on the same side, but separated from it by a large tributary called the Save. Ferry-boats ply regularly between the two cities, and, after an hour spent in hunting up different officials to gain permission for Igali to cross over into Servian territory without having a regular traveller's passport, we escape from the madding crowds of Semlinites by boarding the ferry-boat, and ten minutes later are exchanging signals with three Servian wheelmen, who have come down to the landing in full uniform to meet and welcome us to Belgrade.

Many readers will doubtless be as surprised as I was to learn that at Belgrade, the capital of the little Kingdom of Servia, independent only since the Treaty of Berlin, a bicycle club was organized in January, 1885, and that now, in June of the same year, they have a promising club of thirty members, twelve of whom are riders owning their own wheels. - Their club is named, in French, *La Société Vélocipédique Serbe ;* in the Servian language it is unpronounceable to an Anglo-Saxon, and printable only with Slav type. The president, Milorade M. Nicolitch Terzibachitch, is the Cyclists' Touring Club Consul for Servia, and is the southeastern picket of that organization, their club being the extreme 'cycle outpost in this direction. Our approach has been announced beforehand, and the club has thoughtfully "seen" the Servian authorities, and so far smoothed the way for our entrance into their country that the officials do not even make a pretence of examining my passport or packages—an almost unprecedented occurrence, I should say, since they are more particular about passports here than perhaps in any other European country, save Russia and Turkey.

Here at Belgrade I am to part company with Igali, who, by the way, has applied for, and just received, his certificate of appointment to the Cyclists' Touring Club Consulship of Duna Szekeső and Mohacs, an honor of which he feels quite proud. True, there is no other 'cycler in his whole district, and hardly likely to be for some time to come ; but I can heartily recommend him to any wandering wheelman happening down the Danube Valley on a tour ; he knows the best wine-cellars in all the country round, and, besides being an agreeable and accommodating road companion, will prove a salutary check upon the headlong career of anyone disposed to over-exertion. I am not yet to be abandoned entirely to my own resources, however ; these hospitable Servian wheelmen couldn't think of such a. thing. I am to remain over as their guest till to-morrow afternoon, when Mr. Douchan Popovitz, the best rider in Belgrade, is delegated to escort me through Servia to the Bulgarian frontier. · When I get there I shall not be much astonished to see a Bulgarian wheelman offer to escort me to Roumelia, and so on clear to Constantinople ; for I certainly never expected to find so jolly and enthusiastic a company of 'cyclers in this corner of the world.

The good fellowship and hospitality of this Servian club know no bounds ; Igali and I are banqueted and driven about in carriages all day.

Belgrade is a strongly fortified city, occupying a commanding hill overlooking the Danube ; it is a rare old town, battle-scarred and rugged ; having been a frontier position of importance in a country that has been debatable ground between Turk and Christian for centuries, it has been a coveted prize to be won and lost on the diplomatic chess-board, or, worse still, the foot-ball of contending armies and wrangling monarchs. Long before the Ottoman Turks first appeared, like a small dark cloud, no bigger than a man's hand, upon the southeastern horizon of Europe, to extend and overwhelm the budding flower of Christianity and civilization in these fairest portions of the continent, Belgrade was an important Roman fortress, and to-day its national museum and antiquarian stores are particularly rich in the treasure-trove of Byzantine antiquities, unearthed from time to time in the fortress itself and the region round about that came under its protection. So plentiful, indeed, are old coins and relics of all sorts at Belgrade, that, as I am standing looking at the collection in the window of an antiquary shop, the proprietor steps out and presents me a small handful of copper coins of Byzantium as a sort of bait that might perchance tempt one to enter and make a closer inspection of his stock.

By the famous Treaty of Berlin the Servians gained their complete independence, and their country, from a principality, paying tribute to the Sultan, changed to an independent kingdom with a Servian on the throne, owing allegiance to nobody, and the people have not yet ceased to show, in a thousand little ways, their thorough

appreciation of the change ; besides filling the picture-galleries of
their museum with portraits of Servian heroes, battle-flags, and
other gentle reminders of their past history, they have, among
other practical methods of manifesting how they feel about the
departure of the dominating crescent from among them, turned
the leading Turkish mosque into a gas-house. One of the most
interesting relics in the Servian capital is an old Roman well,
dug from the brow of the fortress hill to below the level of the
Danube, for furnishing water to the city when cut off from the river
by a besieging army. It is an enormous affair, a tubular brick
wall about forty feet in circumference and two hundred and fifty
feet deep, outside of which a stone stairway, winding round and
round the shaft, leads from top to bottom. Openings through the
wall, six feet high and three wide, occur at regular intervals all the
way down, and, as we follow our ragged guide down, down into
the damp and darkness by the feeble light of a tallow candle in a
broken lantern, I cannot help thinking that these o'erhandy openings leading into the dark, watery depths have, in the tragic history of Belgrade, doubtless been responsible for the mysterious
disappearance of more than one objectionable person. It is not
without certain involuntary misgivings that I take the lantern from
the guide—whose general appearance is, by the way, hardly calculated to be reassuring—and, standing in one of the openings, peer
down into the darksome depths, with him hanging on to my coat
as an act of precaution.

The view from the ramparts of Belgrade fortress is a magnificent panorama, extending over the broad valley of the Danube—
which here winds about as though trying to bestow its favors with
impartiality upon Hungary, Servia, and Slavonia—and of the Save.
The Servian soldiers are camped in small tents in various parts of
the fortress grounds and its environments, or lolling under the shade
of a few scantily verdured trees, for the sun is to-day broiling hot.
With a population not exceeding one and a half million, I am told
that Servia supports a standing army of a hundred thousand men ;
and, when required, every man in Servia becomes a soldier. As one
lands from the ferry-boat and looks about him he needs no interpreter to inform him that he has left the Occident on the other
side of the Save, and to the observant stranger the streets of Belgrade furnish many a novel and interesting sight in the way of
fanciful costumes and phases of Oriental life here encountered for

the first time. In the afternoon we visit the national museum of old coins, arms, and Roman and Servian antiquities.

A banquet in a wine-garden, where Servian national music is dispensed by a band of female musicians, is given us in the evening by the club, and royal quarters are assigned us for the night at the hospitable mansion of Mr. Terzibachitch's father, who is the merchant-prince of Servia, and purveyor to the court. Wednesday morning we take a general ramble over the city, besides visiting the club's head-quarters, where we find a handsome new album has been purchased for receiving our autographs. The Belgrade wheelmen have names painted on their bicycles, as names are painted on steamboats or yachts : "Fairy," "Good Luck," and "Servian Queen," being fair specimens. The cyclers here are sons of leading citizens and business men of Belgrade, and, while they dress and conduct themselves as becomes thorough gentlemen, one fancies detecting a certain wild expression of the eye, as though their civilization were scarcely yet established ; in fact, this peculiar expression is more noticeable at Belgrade, and is apparently more general here than at any other place I visit in Europe. I apprehend it to be a peculiarity that has become hereditary with the citizens, from their city having been so often and for so long the theatre of uncertain fate and distracting political disturbances. It is the half-startled expression of people with the ever-present knowledge of insecurity. But they are a warm-hearted, impulsive set of fellows, and when, while looking through the museum, we happen across Her Britannic Majesty's representative at the Servian court, who is doing the same thing, one of them unhesitatingly approaches that gentleman, cap in hand, and, with considerable enthusiasm of manner, announces that they have with them a countryman of his who is riding around the world on a bicycle. This cooler-blooded and dignified gentleman is not near so demonstrative in his acknowledgment as they doubtless anticipated he would be ; whereat they appear quite puzzled and mystified.

Three carriages with cyclers and their friends accompany us a dozen kilometres out to a wayside *mehana* (the Oriental name hereabouts for hotels, wayside inns, etc.) ; Douchan Popovitz, and Hugo Tichy, the captain of the club, will ride forty-five kilometres with me to Semendria, and at 4 o'clock we mount our wheels and ride away southward into Servia. Arriving at the *mehana*, wine is brought, and then the two Servians accompanying me, and those returning,

kiss each other, after the manner and custom of their country ; then
a general hand-shaking and well-wishes all around, and the carriages turn toward Belgrade, while we wheelmen alternately ride
and trundle over a muddy—for it has rained since noon—and
mountainous road till 7.30, when relatives of Douchan Popovitz, in
the village of Grotzka, kindly offer us the hospitality of their house
till morning, which we hesitate not to avail ourselves of. When
about to part at the *mehana*, the immortal Igali unwinds from
around his waist that long blue girdle, the arranging and rearranging of which has been a familiar feature of the last week's experiences, and presents it to me for a *souvenir* of himself, a courtesy
which I return, by presenting him with several of the Byzantine
coins given to me by the Belgrade antiquary as before mentioned.

Beyond Semendria, where the captain leaves us for the return
journey, we leave the course of the Danube, which I have been following in a general way for over two weeks, and strike due southward up the smaller, but not less beautiful, valley of the Morava
River, where we have the intense satisfaction of finding roads that
are both dry and level, enabling us, in spite of the broiling heat, to
bowl along at a sixteen-kilometre pace to the village, where we
halt for dinner and the usual three hours noontide siesta. Seeing
me jotting down my notes with a short piece of lead-pencil, the
proprietor of the *mehana* at Semendria, where we take a parting
glass of wine with the captain, and who admires America and the
Americans, steps in-doors for a minute, and returns with a telescopic
pencil-case, attached to a silken cord of the Servian national colors,
which he places around my neck, requesting me to wear it around
the world, and, when I arrive at my journey's end, sometimes to
think of Servia.

With Igali's sky-blue girdle encompassing my waist, and the
Servian national colors fondly encircling my neck, I begin to feel
quite a heraldic tremor creeping over me, and actually surprise myself casting wistful glances at the huge antiquated horse pistol
stuck in yonder bull-whacker's ample waistband ; moreover, I really
think that a pair of these Servian moccasins would not be bad
foot-gear for riding the bicycle ! All up the Morava Valley the
roads continue far better than I have expected to find in Servia, and
we wheel merrily along, the Resara Mountains covered with dark
pine forests, skirting the valley on the right, sometimes rising into
peaks of quite respectable proportions. The sun sinks behind

the receding hills, it grows dusk, and finally dark, save the feeble light vouchsafed by the new moon, and our destination still lies several kilometres ahead. But at about nine we roll safely into Jagodina, well-satisfied with the consciousness of having covered one hundred and forty-five kilometres to-day, in spite of delaying our start in the morning until eight o'clock, and the twenty kilometres of indifferent road between Grotzka and Semendria. There has been no reclining under road-side mulberry-trees for my companion to catch up to-day, however ; the Servian wheelman is altogether a speedier man than Igali, and, whether the road is rough or smooth, level or hilly, he is found close behind my rear wheel ; my own shadow follows not more faithfully than does the " best rider in Servia."

We start for Jagodina at 5.30 next morning, finding the roads a little heavy with sand in places, but otherwise all that a wheelman could wish. Crossing a bridge over the Morava River, into Tchupria, we are required not only to foot it across, but to pay a toll for the bicycles, like any other wheeled vehicle. At Tchupria it seems as though the whole town must be depopulated, so great is the throng of citizens that swarm about us. Motley and picturesque even in their rags, one's pen utterly fails to convey a correct idea of their appearance ; besides Servians, Bulgarians, and Turks, and the Greek priests who never fail of being on hand, now appear Roumanians, wearing huge sheep-skin busbies, with the long, ragged edges of the wool dangling about eyes and ears, or, in the case of a more " dudish " person, clipped around smooth at the brim, making the head-gear look like a small, round, thatched roof. Urchins, whose daily duty is to promenade the family goat around the streets, join in the procession, tugging their bearded charges after them ; and a score of dogs, overjoyed beyond measure at the general commotion, romp about, and bark their joyous approval of it all. To have crowds like this following one out of town makes a sensitive person feel uncomfortably like being chased out of a community for borrowing chickens by moonlight, or on account of some irregularity concerning hotel bills. On occasions like this Orientals seemingly have not the slightest sense of dignity ; portly, well-dressed citizens, priests, and military officers press forward among the crowds of peasants and unwashed frequenters of the streets, evidently more delighted with things about them than they have been for many a day before.

At Delegrad we wheel through the battle-field of the same name, where, in 1876, Turks and Servians were arrayed against each other. These battle-scarred hills above Delegrad command a glorious view of the lower Morava Valley, which is hereabouts most beautiful, and just broad enough for its entire beauty to be comprehended. The Servians won the battle of Delegrad, and as I pause to admire the glorious prospect to the southward from the hills, methinks their general showed no little sagacity in opposing the invaders at a spot where the Morava Vale, the jewel of Servia, was spread out like a panorama below his position, to fan with its loveliness the patriotism of his troops—they could not do otherwise than win, with the fairest portion of their well-beloved country spread out before them like a picture. A large cannon, captured from the Turks, is standing on its carriage by the road-side, a mute but eloquent witness of Servian prowess.

A few miles farther on we halt for dinner at Alexinatz, near the old Servian boundary-line, also the scene of one of the greatest battles fought during the Servian struggle for independence. The Turks were victorious this time, and fifteen thousand Servians and three thousand Russian allies yielded up their lives here to superior Turkish generalship, and Alexinatz was burned to ashes. The Russians have erected a granite monument on a hill overlooking the town, in memory of their comrades who perished in this fight.

The roads to-day average even better than yesterday, and at six o'clock we roll into Nisch, one hundred and twenty kilometres from our starting-point this morning, and two hundred and eighty from Belgrade. As we enter the city a gang of convicts working on the fortifications forget their clanking shackles and chains, and the miseries of their state, long enough to greet us with a boisterous howl of approval, and the guards who are standing over them for once, at least, fail to check them, for their attention, too, is wholly engrossed in the same wondrous subject. Nisch appears to be a thoroughly Oriental city, and here I see the first Turkish ladies, with their features hidden behind their white *yashmaks*.

At seven or eight o'clock in the morning, when it is comparatively cool and people are patronizing the market, trafficking and bartering for the day's supply of provisions, the streets present quite an animated appearance ; but during the heat of the day the scene changes to one of squalor and indolence ; respectable citizens are smoking nargilehs (Mark Twain's " hubble-bubble "), or sleeping somewhere out of sight; business is generally suspended, and in every shady nook and corner one sees a swarthy ragamuffin stretched out at full length, perfectly happy and contented if only he is allowed to snooze the hours away in peace.

Human nature is verily the same the world over, and here, in the hotel at Nisch, I meet an individual who recalls a few of the sensible questions that have been asked me from time to time at different places on both continents. This Nisch interrogator is a Hebrew commercial traveller, who has a smattering of English, and who after ascertaining during a short conversation that, when a range of mountains or any other small obstruction is encountered, I get down and push the bicycle up, airs his knowledge of English and of 'cycling to the extent of inquiring whether I don't take a man along to push it up the hills!

Riding out of Nisch this morning we stop just beyond the suburbs to take a curious look at a grim monument of Turkish prowess, in the shape of a square stone structure which the Turks built in 1840, and then faced the whole exterior with grinning rows of Servian skulls partially embedded in mortar. The Servians, naturally objecting to having the skulls of their comrades thus exposed to the gaze of everybody, have since removed and buried them; but the rows of indentations in the thick mortared surface still bear unmistakable evidence of the nature of their former occupants.

An avenue of thrifty prune-trees shades a level road leading out of Nisch for several kilometres, but a heavy thunder-storm during the night has made it rather slavish wheeling, although the surface becomes harder and smoother, also hillier, as we gradually approach the Balkan Mountains, that tower well up toward cloudland immediately ahead. The morning is warm and muggy, indicating rain, and the long, steep trundle, kilometre after kilometre, up the Balkan slopes, is anything but child's play, albeit the scenery is most lovely, one prospect especially reminding me of a view in the Big Horn Mountains of northern Wyoming Territory. On the lower slopes we come to a *mehana*, where, besides plenty of shade-trees, we find springs of most delightfully cool water gushing out of crevices in the rocks, and, throwing our freely perspiring forms beneath the grateful shade and letting the cold water play on our wrists (the best method in the world of cooling one's self when overheated), we both vote that it would be a most agreeable place to spend the heat of the day. But the morning is too young yet to think of thus indulging, and the mountainous prospect ahead warns us that the distance covered to-day will be short enough at the best.

The Balkans are clothed with green foliage to the topmost crags, wild pear-trees being no inconspicuous feature ; charming little valleys wind about between the mountain-spurs, and last night's downpour has imparted a freshness to the whole scene that perhaps it would not be one's good fortune to see every day, even were he here. This region of intermingled vales and forest-clad mountains might be the natural home of brigandage, and those ferocious-looking specimens of humanity with things like long guns in hand, running with scrambling haste down the mountain-side toward our road ahead, look like veritable brigands heading us off with a view to capturing us. But they are peacefully disposed goat-herds, who, alpenstocks in hand, are endeavoring to see "what in the world those queer-looking things are, coming up the road." Their tuneful noise, as they play on some kind of an instrument, greets our ears from a dozen mountain-slopes round about us, as we put our shoulders to the wheel, and gradually approach the summit. Tortoises are occasionally surprised basking in the sun-beams in the middle of the road ; when molested. they hiss quite audibly in protest, but if passed peacefully by they are seen shuffling off into the bushes, as though thankful to escape. Unhappy oxen are toiling patiently upward, literally inch by inch, dragging heavy, creaking wagons, loaded with miscellaneous importations, promi-nent among which I notice square cans of American petroleum.

Men on horseback are encountered, the long guns of the Orient slung at their backs, and knife and pistols in sash, looking altogether ferocious. Not only are these people perfectly harmless, however, but I verily think it would take a good deal of aggravation to make them even think of fighting. The fellow whose horse we frightened down a rocky embankment, at the imminent risk of breaking the neck of both horse and rider, had both gun, knife, and pistols ; yet, though he probably thinks us emissaries of the evil one, he is in no sense a dangerous character, his weapons being merely gewgaws to adorn his person. Finally, the summit of this range is gained, and the long, grateful descent into the valley of the Nissava River begins. The surface during this descent, though averaging very good, is not always of the smoothest ; several dis-mounts are found to be necessary, and many places ridden over

require a quick hand and ready eye to pass. The Servians have
made a capital point in fixing their new boundary-line south of this
mountain-range.

A Belle of the Balkans,

Mountaineers are said to be "always freemen;" one can with
equal truthfulness add that the costumes of mountaineers' wives
and daughters are always more picturesque than those of their sis-

ters in the valleys. In these Balkan Mountains their costumes are a truly wonderful blending of colors, to say nothing of fantastic patterns, apparently a medley of ideas borrowed from Occident and Orient. One woman we have just passed is wearing the loose, flowing pantaloons of the Orient, of a bright-yellow color, a tight-fitting jacket of equally bright blue ; around her waist is folded many times a red and blue striped waistband, while both head and feet are bare. This is no holiday attire ; it is plainly the ordinary everyday costume.

At the foot of the range we halt at a way-side *mehana* for dinner. A daily diligence, with horses four abreast, runs over the Balkans from Nisch to Sophia, Bulgaria, and one of them is halted at the *mehana* for refreshments and a change of horses. Refreshments at these *mehanas* are not always palatable to travellers, who almost invariably carry a supply of provisions along. Of bread nothing but the coarse, black variety common to the country is forthcoming at this *mehana*, and a gentleman, learning from Mr. Popovitz that I have not yet been educated up to black bread, fishes a large roll of excellent *milch-Brod* out of his traps and kindly presents it to us ; and obtaining from the *mehana* some *hune-hen fabrica* and wine we make a very good meal. This *hune-hen fabrica* is nothing more nor less than cooked chicken. Whether *hune-hen fabrica* is genuine Hungarian for cooked chicken, or whether Igali manufactured the term especially for use between us, I cannot quite understand. Be this as it may, before we started from Belgrade, Igali imparted the secret to Mr. Popovitz that I was possessed with a sort of a wild appetite, as it were, for *hune-hen fabrica* and cherries, three times a day, the consequence being that Mr. Popovitz thoughtfully orders those viands whenever we halt. After dinner the mutterings of thunder over the mountains warn us that unless we wish to experience the doubtful luxuries of a road-side *mehana* for the night we had better make all speed to the village of Bela Palanka, twelve kilometres distant over rather hilly roads. In forty minutes we arrive at the Bela Palanka *mehana*, some time before the rain begins. It is but twenty kilometres to Pirot, near the Bulgarian frontier, whither my companion has purposed to accompany me, but we are forced to change this programme and remain at Bela Palanka.

It rains hard all night, converting the unassuming Nissava into a roaring yellow torrent, and the streets of the little Balkan village

Sunday at Bela Palanka.

into mud-holes. It is still raining on Sunday morning, and as Mr. Popovitz is obliged to be back to his duties as foreign correspondent in the Servian National Bank at Belgrade on Tuesday, and the Balkan roads have been rendered impassable for a bicycle, he is compelled to hire a team and wagon to haul him and his wheel back over the mountains to Nisch, while I have to remain over Sunday amid the dirt and squalor and discomforts—to say nothing of a second night among the fleas—of an Oriental village *mehana*. We only made fifty kilometres over the mountains yesterday, but during the three days from Belgrade together the aggregate has been satisfactory, and Mr. Popovitz has proven a most agreeable and interesting companion. When but fourteen years of age he served under the banner of the Red Cross in the war between the Turks and Servians, and is altogether an ardent patriot.

My Sunday in Bela Palanka impresses me with the conviction that an Oriental village is a splendid place not to live in. In dry weather it is disagreeable enough, but to-day it is a disorderly aggregation of miserable-looking villagers, pigs, ducks, geese, chickens, and dogs, paddling around the muddy streets. The Oriental peasant's costume is picturesque or otherwise, according to the fancy of the observer. The red fez or turban, the upper garment, and the ample red sash wound round and round the waist until it is eighteen inches broad, look picturesque enough for anybody ; but when it comes to having the seat of the pantaloons dangling about the calves of the legs, a person imbued with Western ideas naturally thinks that if the line between picturesqueness and a two-bushel gunny-sack is to be drawn anywhere it should most assuredly be drawn here. As I notice how prevalent this ungainly style of nether garment is in the Orient, I find myself getting quite uneasy lest, perchance, anything serious should happen to mine, and I should be compelled to ride the bicycle in a pair of natives, which would, however, be an altogether impossible feat unless it were feasible to gather the surplus area up in a bunch and wear it like a bustle. I cannot think, however, that Fate, cruel as she sometimes is, has anything so outrageous as this in store for me or any other 'cycler.

Although Turkish ladies have almost entirely disappeared from Servia since its severance from Turkey, they have left, in a certain degree, an impress upon the women of the country villages ; although the Bela Palanka maidens, as I notice on the streets in their Sunday clothes to-day, do not wear the regulation *yashmak*, but a head-gear that partially obscures the face, their whole demeanor giving one the impression that their one object in life is to appear the pink of propriety in the eyes of the whole world ; they walk along the streets at a most circumspect gait, looking neither to the right nor left, neither stopping to converse with each other by the way, nor paying any sort of attention to the men. The two proprietors of the *mehana* where I am stopping are subjects for a student of human nature. With their wretched little pigsty of a mehana in this poverty-stricken village, they are gradually accumulating a fortune. Whenever a luckless traveller falls into their clutches they make the incident count for something. They stand expectantly about in their box-like public room ; their whole stock consists of a little diluted wine and mastic, and if a bit of black bread and *smear-käse* is ordered, one is putting it down in the book, while the other is ferreting it out of a little cabinet where they keep a starvation quantity of edibles ; when the one acting as waiter has placed the inexpensive morsel before you, he goes over to the book to make sure that number two has put down enough ; and, although the maximum value of the provisions is perhaps not over twopence, this precious pair will actually put their heads together in consultation over the amount to be chalked down. Ere the shades of Sunday evening have settled down, I have arrived at the conclusion that if these two are average specimens of the Oriental Jew they are financially a totally depraved people.

The rain ceased soon after noon on Sunday, and, although the roads are all but impassable, I pull out southward at five o'clock on Monday morning, trundling up the mountain-roads through mud that frequently compels me to stop and use the scraper. After the summit of the hills between Bela Palanka and Pirot is gained, the road descending into the valley beyond becomes better, enabling me to make quite good time into Pirot, where my passport undergoes an examination, and is favored with a *visé* by the Servian officials preparatory to crossing the Servian and Bulgarian frontier about twenty kilometres to the southward. Pirot is quite a large and important village, and my appearance is the signal for more excitement than the Piroters have experienced for many a day.

While I am partaking of bread and coffee in the hotel, the main street becomes crowded as on some festive occasion, the grown-up people's faces beaming with as much joyous anticipation of what they expect to behold when I emerge from the hotel as the un-

washed countenances of the ragged youngsters around them. Leading citizens who have been to Paris or Vienna, and have learned
something about what sort of road a 'cycler needs, have imparted
the secret to many of their fellow-townsmen, and there is a general
stampede to the highway leading out of town to the southward.
This road is found to be most excellent, and the enterprising people
who have walked, ridden, or driven out there, in order to see me
ride past to the best possible advantage, are rewarded by witnessing what they never saw before—a cycler speeding along past them
at ten miles an hour. This gives such general satisfaction that for
some considerable distance I ride between a double row of lifted
hats and general salutations, and a swelling murmur of applause
runs all along the line.

Two citizens, more enterprising even than the others, have determined to follow me with team and light wagon to a road-side
office ten kilometres ahead, where passports have again to be examined. The road for the whole distance is level and fairly
smooth ; the Servian horses are, like the Indian ponies of the
West, small, but wiry and tough, and although I press forward
quite energetically, the whip is applied without stint, and when
the passport office is reached we pull up alongside it together, but
their ponies' sides are white with lather. The passport officer is
so delighted at the story of the race, as narrated to him by the
others, that he fetches me out a piece of lump sugar and a glass of
water, a common refreshment partaken of in this country.

Yet a third time I am halted by a roadside official and required
to produce my passport, and again at the village of Zaribrod, just
over the Bulgarian frontier, which I reach about ten o'clock. To
the Bulgarian official I present a small stamped card-board check,
which was given me for that purpose at the last Servian examination, but he doesn't seem to understand it, and demands to see the
original passport. When my English passport is produced he examines it, and straightway assures me of the Bulgarian official respect for an Englishman by grasping me warmly by the hand. The
passport office is in the second story of a mud hovel, and is reached
by a dilapidated flight of out-door stairs. My bicycle is left leaning against the building, and during my brief interview with the
officer a noisy crowd of semi-civilized Bulgarians have collected
about, examining it and commenting unreservedly concerning it
and myself. The officer, ashamed of the rudeness of his country-

men and their evidently untutored minds, leans out of the window, and in a chiding voice explains to the crowd that I am a private individual, and not a travelling mountebank going about the country

The Zaribrod Passport Office.

giving exhibitions, and advises them to uphold the dignity of the Bulgarian character by scattering forthwith. But the crowd doesn't scatter to any appreciable extent; they don't care whether I am public or private; they have never seen anything like me and the bicycle before, and the one opportunity of a lifetime is not to be lightly passed over. They are a wild, untamed lot, these Bulgarians here at Zaribrod, little given to self-restraint.

When I emerge, the silence of eager anticipation takes entire possession of the crowd, only to break forth into a spontaneous howl of delight from three hundred bared throats when I mount into the saddle and ride away into—Bulgaria.

My ride through Servia, save over the Balkans, has been most enjoyable, and the roads, I am agreeably surprised to have to record, have averaged as good as any country in Europe, save England and France, though being for the most part unmacadamized ; with wet weather they would scarcely show to such advantage. My impression of the Servian peasantry is most favorable ; they are evidently a warm-hearted, hospitable, and withal a patriotic people, loving their little country and appreciating their independence as only people who have but recently had their dream of self-government realized know how to appreciate it; they even paint the wood-work of their bridges and public buildings with the national colors. I am assured that the Servians have progressed wonderfully since acquiring their full independence ; but as one journeys down the beautiful and fertile valley of the Morava, where improvements would naturally be seen, if anywhere, one falls to wondering where they can possibly have come in.

Some of their methods would, indeed, seem to indicate a most deplorable lack of practicability; one of the most ridiculous, to the writer's mind, is the erection of small, long sheds substantially built of heavy hewn timber supports, and thick, home-made tiles, over ordinary plank fences and gates to protect them from the weather, when a good coating of tar or paint would answer the purpose of preservation much better. These structures give one the impression of a dollar placed over a penny to protect the latter from harm. Every peasant owns a few acres of land, and, if he produces anything above his own wants, he hauls it to market in an ox-wagon with roughly hewn wheels without tires, and whose creaking can plainly be heard a mile away. At present the Servian tills his little freehold with the clumsiest of implements, some his own rude handiwork, and the best imperfectly fashioned and forged on native anvils. His plow is chiefly the forked limb of a tree, pointed with iron sufficiently to enable him to root around in the surface soil. One would think the country might offer a promising field for some enterprising manufacturer of such implements as hoes, scythes, hay-forks, small, strong plows, cultivators, etc.

These people are industrious, especially the women. I have frequently met a Servian peasant woman returning homeward in the evening from her labor in the fields, carrying a fat, heavy baby, a clumsy hoe not much lighter than the youngster, and an earthenware water-pitcher, and, at the same time, industriously spinning wool with a small hand-spindle. And yet some people argue about the impossibility of doing two things at once! Whether these poor women have been hoeing potatoes, carrying the infant, and spinning wool at the same time all day I am unable to say, not having been an eye witness, though I really should not be much astonished if they had.

CHAPTER VIII.

BULGARIA, ROUMELIA, AND INTO TURKEY.

THE road leading into Bulgaria from the Zaribrod custom-house is fairly good for several kilometres, when mountainous and rough ways are encountered ; it is a country of goats and goat-herds. A rain-storm is hovering threateningly over the mountains immediately ahead, but it does not reach the vicinity I am traversing : it passes to the southward, and makes the roads for a number of miles wellnigh impassable. Up in the mountains I meet more than one "Bulgarian national express"—pony pack-trains, carrying merchandise to and fro between Sofia and Nisch. Most of these animals are too heavily laden to think of objecting to the appearance of anything on the road, but some of the outfits are returning from Sofia in "ballast" only ; and one of these, doubtless overjoyed beyond measure at their unaccustomed lissomeness, breaks through all restraint at my approach, and goes stampeding over the rolling hills, the wild-looking teamsters in full tear after them. Whatever of this nature happens in this part of the world the people seem to regard with commendable complacence : instead of wasting time in trying to quarrel about it, they set about gathering up the scattered train, as though a stampede were the most natural thing going.

Bulgaria—at least by the route I am crossing it—is a land of mountains and elevated plateaus, and the inhabitants I should call the "ranchers of the Orient," in their general appearance and demeanor bearing the same relation to the plodding corn-hoer and scythe-swinger of the Morava Valley as the Niobrara cow-boy does to the Nebraska homesteader. On the mountains are encountered herds of goats in charge of men who reck little for civilization, and the upland plains are dotted over with herds of ponies that require constant watching in the interest of scattered fields of grain. For lunch I halt at an unlikely-looking *mehana*, near a cluster of mud hovels, which, I suppose, the Bulgarians consider a village, and am rewarded by the blackest of black bread, in the composition of which sand plays no inconsiderable part, and the remnants of a chicken killed and stewed at some uncertain period of the past. Of all places invented in the world to disgust a hungry, expectant wayfarer, the Bulgarian *mehana* is the most abominable. Black bread and mastic (a composition of gum-mastic and Boston rum, so I am informed) seem to be about the only things habitually kept in stock, and everything about the place plainly shows the proprietor to be ignorant of the crudest notions of cleanliness.

A storm is observed brewing in the mountains I have lately traversed, and, having swallowed my unpalatable lunch, I hasten to mount, and betake myself off toward Sofia, distant thirty kilometres. The road is nothing extra, to say the least, but a howling wind blowing from the region of the gathering storm propels me rapidly, in spite of undulations, ruts, and undesirable road qualities generally. The region is an elevated plateau, of which but a small proportion is cultivated ; on more than one of the neighboring peaks patches of snow are still lingering, and the cool mountain breezes recall memories of the Laramie Plains. Men and women returning homeward on horseback from Sofia are frequently encountered. The women are decked with beads and trinkets and the gewgaws of semi-civilization, as might be the favorite squaws of Squatting Beaver or Sitting Bull, and furthermore imitate their copper-colored sisters of the Far West by bestriding their ponies like men. But in the matter of artistic and profuse decoration of the person the squaw is far behind the peasant woman of Bulgaria. The garments of the men are a combination of sheepskin and a thick, coarse, woollen material, spun by the women, and fashioned after patterns their forefathers brought with them centuries ago when they first invaded Europe. The Bulgarian saddle, like everything else here, is a rudely constructed affair, that answers the double purpose of a pack-saddle or for riding—a home-made, unwieldy thing, that is a fair pony's load of itself.

At 4.30 P.M. I wheel into Sofia, the Bulgarian Capital, having covered one hundred and ten kilometres to-day, in spite of mud, mountains, and roads that have been none of the best. Here again I have to patronize the money-changers, for a few Servian francs which I have are not current in Bulgaria ; and the Israelite, who reserved unto himself a profit of two francs on the pound at Nisch, now seems the spirit of fairness itself along-side a hook-nosed, wizen-faced relative of his here at Sofia, who wants two Servian francs in exchange for each Bulgarian coin of the same intrinsic

value ; and the best I am able to get by going to several different
money-changers is five francs in exchange for seven; yet the
Servian frontier is but sixty kilometres distant, with stages running to it daily ; and the two coins are identical in intrinsic value.
At the Hotel Concordia, in Sofia, in lieu of plates, the meat is served
on round, flat blocks of wood about the circumference of a saucer
—the "trenchers" of the time of Henry VIII.—and two respectable citizens seated opposite me are supping off black bread and
a sliced cucumber, both fishing slices of the cucumber out of a
wooden bowl with their fingers.

Life at the Bulgarian Capital evidently bears its legitimate relative comparison to the life of the country it represents.   One of
Prince Alexander's body-guard, pointed out to me in the bazaar,
looks quite a semi-barbarian, arrayed in a highly ornamented national costume, with immense Oriental pistols in waistband, and
gold-braided turban cocked on one side of his head, and a fierce
mustache. The soldiers here, even the comparatively fortunate ones
standing guard at the entrance to the prince's palace, look as though
they haven't had a new uniform for years and had long since de-
spaired of ever getting one.   A war, and an alliance with some
wealthy nation which would rig them out in respectable uniforms,
would probably not be an unwelcome event to many of them.

While wandering about the bazaar, after supper, I observe that
the streets, the palace grounds, and in fact every place that is lit up
at all, save the minarets of the mosque, which are always illumined
with vegetable oil, are lighted with American petroleum, gas and
coal being unknown in the Bulgarian capital.   There is an evident
want of system in everything these people do.   From my own observations I am inclined to think they pay no heed whatever to
generally accepted divisions of time, but govern their actions entirely by light and darkness.   There is no eight-hour nor ten-hour
system of labor here ; and I verily believe the industrial classes
work the whole time, save when they pause to munch black bread,
and to take three or four hours' sleep in the middle of the night ;
for as I trundle my way through the streets at five o'clock next
morning, the same people I observed at various occupations in the
bazaars are there now, as busily engaged as though they had been
keeping it up all night; as also are workmen building a house ;
they were pegging away at nine o'clock yesterday evening, by the
flickering light of small petroleum lamps, and at five this morning

they scarcely look like men who are just commencing for the day. The Oriental, with his primitive methods and tenacious adherence to the ways of his forefathers, probably enough, has to work these extra long hours in order to make any sort of progress. However this may be, I have throughout the Orient been struck by the industriousness of the real working classes ; but in practicability and inventiveness the Oriental is sadly deficient.

On the way out I pause at the bazaar to drink hot milk and eat a roll of white bread, the former being quite acceptable, for the morning is rather raw and chilly ; the wind is still blowing a gale, and a company of cavalry, out for exercise, are incased in their heavy gray overcoats, as though it were midwinter instead of the twenty-third of June. Rudely clad peasants are encountered on the road, carrying large cans of milk into Sofia from neighboring ranches. I stop several of them with a view of sampling the quality of their milk, but invariably find it unstrained, and the vessels looking as though they had been strangers to scalding for some time. Others are carrying gunny-sacks of *smear-käse* on their shoulders, the whey from which is not infrequently streaming down their backs. Cleanliness is no doubt next to godliness ; but the Bulgarians seem to be several degrees removed from either. They need the civilizing influence of soap quite as much as anything else, and if the missionaries cannot educate them up to Christianity or civilization it might not be a bad scheme to try the experiment of starting a native soap-factory or two in the country.

Savagery lingers in the lap of civilization on the breezy plateaus of Bulgaria, but salvation is coming this way in the shape of an extension of the Roumelian railway from the south, to connect with the Servian line north of the Balkans. For years the freight department of this pioneer railway will have to run opposition against ox-teams, and creaking, groaning wagons ; and since railway stockholders and directors are not usually content with an exclusive diet of black bread, with a wilted cucumber for a change on Sundays, as is the Bulgarian teamster, and since locomotives cannot be turned out to graze free of charge on the hill-sides, the competition will not be so entirely one-sided as might be imagined. Long trains of these ox-teams are met with this morning hauling freight and building-lumber from the railway terminus in Roumelia to Sofia. The teamsters are wearing large gray coats of thick blanketing, with hoods covering the head, a heavy, convenient garment, that keeps

out both rain and cold while on the road, and at night serves for
blanket and mattress; for then the teamster turns his oxen loose
on the adjacent hill-sides to graze, and, after munching a piece of
black bread, he places a small wicker-work wind-break against the
windward side of the wagon, and, curling himself up in his great-
coat, sleeps soundly. Besides the ox-trains, large, straggling trains
of pack-ponies and donkeys occasionally fill the whole roadway;
they are carrying firewood and charcoal from the mountains, or
wine and spirits, in long, slender casks, from Roumelia; while
others are loaded with bales and boxes of miscellaneous merchandise, out of all proportion to their own size.

The road southward from Sofia is abominable, being originally
constructed of earth and large unbroken bowlders; it has not been
repaired for years, and the pack-trains and ox-wagons forever
crawling along have, during the wet weather of many seasons,
tramped the dirt away, and left the surface a wretched waste of
ruts, holes, and thickly protruding stones. It is the worst piece of
road I have encountered in all Europe; and although it is ridable
this morning by a cautious person, one risks and invites disaster
at every turn of the wheel. "Old Boreas" comes howling from the
mountains of the north, and hustles me briskly along over ruts,
holes, and bowlders, however, in a most reckless fashion, furnishing
all the propelling power needful, and leaving me nothing to do but
keep a sharp lookout for breakneck places immediately ahead.

In Servia, the peasants, driving along the road in their wagons,
upon observing me approaching them, being uncertain of the character of my vehicle and the amount of road-space I require, would
ofttimes drive entirely off the road; and sometimes, when they
failed to take this precaution, and their teams would begin to show
signs of restiveness as I drew near, the men would seem to lose
their wits for the moment, and cry out in alarm, as though some
unknown danger were hovering over them. I have seen women
begin to wail quite pitifully, as though they fancied I bestrode an
all-devouring circular saw that was about to whirl into them and
rend team, wagon, and everything asunder. But the Bulgarians
don't seem to care much whether I am going to saw them in twain
or not; they are far less particular about yielding the road, and
both men and women seem to be made of altogether sterner stuff
than the Servians and Slavonians. They seem several degrees less
civilized than their neighbors farther north, judging from their

general appearance and demeanor. They act peaceably and are reasonably civil toward me and the bicycle, however, and personally I rather enjoy their rough, unpolished manners. Although there is a certain element of rudeness and boisterousness about them, compared with anything I have encountered elsewhere in Europe, they seem, on the whole, a good-natured people. We Westerners seldom hear anything of the Bulgarians except in wartimes, and then it is usually in connection with atrocities that furnish excellent sensational material for the illustrated weeklies; consequently I rather expected to have a rough time riding through alone. But, instead of coming out slashed and scarred like a Heidelberg student, I emerge from their territory with nothing more serious than a good healthy shaking up from their ill-conditioned roads and howling winds, and my prejudice against black bread with sand in it partly overcome from having had to eat it or nothing. Bulgaria is a principality under the suzerainty of the Sultan, to whom it is supposed to pay a yearly tribute ; but the suzerainty sits lightly upon the people, since they do pretty much as they please ; and they never worry themselves about the tribute, simply putting it down on the slate whenever it comes due. The Turks might just as well wipe out the account now as at any time, for they will eventually have to whistle for the whole indebtedness.

A smart rain-storm drives me into an uninviting *mehana* near the Roumelian frontier, for two unhappy hours, at noon—a *mehana* where the edible accommodations would wring an " Ugh ! " from an American Indian—and the sole occupants are a blear-eyed Bulgarian, in twenty-year-old sheep-skin clothes, whose appearance plainly indicates an over-fondness for mastic, and an unhappy-looking black kitten. Fearful lest something, perchance, might occur to compel me to spend the night here, I don my gossamers as soon as the rain slacks up a little, and splurge ahead through the mud toward Ichtiman, which, my map informs me, is just on this side of the Kodja Balkans, which rise up in dark wooded ridges at no great distance ahead, to the southward. The mud and rain combine to make things as disagreeable as possible, but before three o'clock I reach Ichtiman, to find that I am in the province of Roumelia, and am again required to produce my passport.

I am now getting well down into territory that quite recently was completely under the dominion of the "unspeakable Turk"— unspeakable, by the way, to the writer in more senses than one—

and is partly so even now, but have as yet seen very little of the "mysterious veiled lady." The Bulgarians are Christian when they are anything, though the great majority of them are nothing religiously. A comparatively comfortable *mehana* is found here at Ichtiman, and the proprietor, being able to talk German, readily comprehends the meaning of *hune-hen fabrica;* but I have to dispense with cherries.

Mud is the principal element of the road leading out of Ichtiman and over the Kodja Balkans this morning. The curious crowd of Ichtimanites that follow me through the mud-holes and filth of their native streets, to see what is going to happen when I get clear of them, are rewarded but poorly for their trouble ; the best I can possibly do being to make a spasmodic run of a hundred yards through the mud, which I do purely out of consideration for their inquisitiveness, since it seems rather disagreeable to disappoint a crowd of villagers who are expectantly following and watching one's every movement, wondering, in their ignorance, why you don't ride instead of walk. It is a long, wearisome trundle up the muddy slopes of the Kodja Balkans, but, after the descent into the Maritza Valley begins, some little ridable surface is encountered, though many loose stones are lying about, and pitch-holes innumerable, make riding somewhat risky, considering that the road frequently leads immediately alongside precipices. Pack-donkeys are met on these mountain-roads, sometimes filling the way, and coming doggedly and indifferently forward, even in places where I have little choice between scrambling up a rock on one side of the road or jumping down a precipice on the other. I can generally manage to pass them, however, by placing the bicycle on one side, and, standing guard over it, push them off one by one as they pass. Some of these Roumelian donkeys are the most diminutive creatures I ever saw ; but they seem capable of toiling up these steep mountain-roads with enormous loads. I met one this morning carrying bales of something far bigger than himself, and a big Roumelian, whose feet actually came in contact with the ground occasionally, perched on his rump ; the man looked quite capable of carrying both the donkey and his load.

The warm and fertile Maritza Valley is reached soon after noon, and I am not sorry to find it traversed by a decent macadamized road ; though, while it has been raining quite heavily up among the mountains, this valley has evidently been favored with a small

Meeting the "Bulgarian Express."

deluge, and frequent stretches are covered with deep mud and sand, washed down from the adjacent hills ; in the cultivated areas of the Bulgarian uplands the grain-fields are yet quite green, but harvesting has already begun in the warmer Maritza Vale, and gangs of Roumelian peasants are in the fields, industriously plying reaping-hooks to save their crops of wheat and rye, which the storm has badly lodged. Ere many miles of this level valley-road are ridden over, a dozen pointed minarets loom up ahead, and at four o'clock I dismount at the confines of the well nigh impassable streets of Tatar Bazardjik, quite a lively little city in the sense that Oriental cities are lively, which means well-stocked bazaars thronged with motley crowds. Here I am delayed for some time by a thunder-storm, and finally wheel away southward in the face of threatening heavens. Several villages of gypsies are camped on the banks of the Maritza, just outside the limits of Tatar Bazardjik ; a crowd of bronzed, half-naked youngsters wantonly favor me with a fusillade of stones as I ride past, and several gaunt, hungry-looking curs follow me for some distance with much threatening clamor. The dogs in the Orient seem to be pretty much all of one breed, genuine mongrel, possessing nothing of the spirit and courage of the animals we are familiar with. Gypsies are more plentiful south of the Save than even in Austria-Hungary, but since leaving Slavonia I have never been importuned by them for alms. Travellers from other countries are seldom met with along the roads here, and I suppose that the wandering Romanies have long since learned the uselessness of asking alms of the natives ; but, since they religiously abstain from anything like work, how they manage to live is something of a mystery.

Ere I am five kilometres from Tatar Bazardjik the rain begins to descend, and there is neither house nor other shelter visible anywhere ahead. The peasants' villages are all on the river, and the road leads for mile after mile through fields of wheat and rye. I forge ahead in a drenching downpour that makes short work of the thin gossamer suit, which on this occasion barely prevents me getting a wet skin ere I descry a thrice-welcome *mehana* ahead and repair thither, prepared to accept, with becoming thankfulness, whatever accommodation the place affords. It proves many degrees superior to the average Bulgarian institution of the same name, the proprietor causing my eyes fairly to bulge out with astonishment by producing a box of French sardines, and bread

13

several shades lighter than I had, in view of previous experience,
expected to find it ; and for a bed provides one of the huge,
thick overcoats before spoken of, which, with the ample hood, envelops the whole figure in a covering that defies both wet and cold.
I am provided with this unsightly but none the less acceptable
garment, and given the happy privilege of occupying the floor of a
small out-building in company with several rough-looking pack-
train teamsters similarly incased ; I pass a not altogether comfortless
night, the pattering of rain against the one small window effectually suppressing such thankless thoughts as have a tendency to
come unbidden whenever the snoring of any of my fellow-lodgers
gets aggravatingly harsh.   In all this company I think I am the
only person who doesn't snore, and when I awake from my rather
fitful slumbers at four o'clock and find the rain no longer pattering
against the window, I arise, and take up my journey toward
Philippopolis, the city I had intended reaching yesterday.

It is after crossing the Kodja Balkans and descending into the
Maritza Valley that one finds among the people a peculiarity that,
until a person becomes used to it, causes no little mystification and
many ludicrous mistakes.   A shake of the head, which with us
means a negative answer, means exactly the reverse with the people
of the Maritza Valley ; and it puzzled me not a little more than once
yesterday afternoon when inquiring whether I was on the right road,
and when patronizing fruit-stalls in Tatar Bazardjik.   One never
feels quite certain about being right when, after inquiring of a native if this is the correct road to Mustapha Pasha or Philippopolis
he replies with a vigorous shake of the head ; and although one
soon gets accustomed to this peculiarity in others, and accepts it
as it is intended, it is not quite so easy to get into the habit yourself.   This queer custom seems to prevail only among the inhabitants of this particular valley, for after leaving it at Adrianople I
see nothing more of it.   Another peculiarity all through Oriental,
and indeed through a good part of Central Europe, is that, instead
of the "whoa" which we use to a horse, the driver hisses like a
goose.

Yesterday evening's downpour has little injured the road between the *mehana* and Philippopolis, the capital of Roumelia, and I
wheel to the confines of that city in something over two hours.
Philippopolis is most beautifully situated, being built on and
around a cluster of several rocky hills ; a situation which, together

with a plenitude of waving trees, imparts a pleasing and pictu-resque effect. With a score of tapering minarets pointing skyward among the green foliage, the scene is thoroughly Oriental; but, like all Eastern cities, "distance lends enchantment to the view." All down the Maritza Valley, and in lesser numbers extending southward and eastward over the undulating plains of Adrianople, are many prehistoric mounds, some twenty-five or thirty feet high, and of about the same diameter. Sometimes in groups, and some-times singly, these mounds occur so frequently that one can often count a dozen at a time. In the vicinity of Philippopolis several have been excavated, and human remains discovered reclining beneath large slabs of coarse pottery set up like an inverted V, thus : Λ, evi-dently intended as a water-shed for the preservation of the bodies. Another feature of the landscape, and one that fails not to strike the observant traveller as a melancholy feature, are the Moham-medan cemeteries. Outside every town and near every village are broad areas of ground thickly studded with slabs of roughly hewn rock set up on end ; cities of the dead vastly more populous than the abodes of life adjacent. A person can stand on one of the Phil-ippopolis heights and behold the hills and vales all around thickly dotted with these rude reminders of our universal fate. It is but as yesterday since the Turk occupied these lands, and was in the habit of making it particularly interesting to any "dog of a Chris-tian" who dared desecrate one of these Mussulman cemeteries with his unholy presence ; but to-day they are unsurrounded by pro-tecting fence or the moral restrictions of dominant Mussulmans, and the sheep, cows, and goats of the "infidel giaour" graze among them ; and oh, shade of Mohammed! hogs also scratch their backs against the tombstones and root around, at their own sweet will, sometimes unearthing skulls and bones, which it is the Turkish custom not to bury at any great depth. The great num-ber and extent of these cemeteries seem to appeal to the unaccus-tomed observer in eloquent evidence against a people whose rule and religion have been of the sword.

While obtaining my breakfast of bread and milk in the Philip-popolis bazaar an Arab ragamuffin rushes in, and, with anxious gesticulations toward the bicycle, which I have from necessity left outside, and cries of "Monsieur, monsieur," plainly announces that there is something going wrong in connection with the machine. Quickly going out I find that, although I left it standing on the narrow

apology for a sidewalk, it is in imminent danger of coming to grief
at the instance of a broadly laden donkey, which, with his load, ver-
itably takes up the whole narrow street, including the sidewalks, as
he slowly picks his way along through mud-holes and protruding
cobble-stones.   And yet Philippopolis has improved wonderfully
since it has nominally changed from a Turkish to a Christian city,
I am told ; the Cross having in Philippopolis not only triumphed
over the Crescent, but its influence is rapidly changing the condition and appearance of the streets.   There is no doubt about the
improvements, but they are at present most conspicuous in the
suburbs, near the English consulate.   It is threatening rain again
as I am picking my way through the crooked streets of Philippopolis toward the Adrianople road ; verily, I seem these days to be
fully occupied in playing hide-and-seek with the elements ; but in
Roumelia at this season it is a question of either rain or insufferable heat, and perhaps, after all, I have reason to be thankful at having the former to contend with rather than the latter.   Two thunder-
storms have to be endured during the forenoon, and for lunch I
reach a *mehana* where, besides eggs roasted in the embers, and
fairly good bread, I am actually offered a napkin that has been
used but a few times—an evidence of civilization that is quite refreshing.

A repetition of the rain-dodging of the forenoon characterizes
the afternoon journey, and while halting at a small village the inhabitants actually take me for a mountebank, and among them collect a handful of diminutive copper coins about the size and thickness of a gold twenty-five-cent piece, and of which it would take at
least twenty to make an American cent, and offer them to me for a
performance.   What with shaking my head for " no " and the villagers naturally mistaking the motion for " yes," according to their
own custom, I have quite an interesting time of it making them understand that I am not a mountebank travelling from one Roumelian
village to another, living on two cents' worth of black sandy bread
per diem, and giving performances for about three cents a time.

For my halting-place to-night I reach the village of Cauheme,
in which I find a *mehana*, where, although the accommodations are
of the crudest nature, the proprietor is a kindly disposed and, withal, a thoroughly honest individual, furnishing me with a reed mat
and a pillow, and making things as comfortable and agreeable as
possible.   Eating raw cucumbers as we eat apples or pears appears

to be universal in Oriental Europe ; frequently, through Bulgaria and Roumelia, I have noticed people, both old and young, gnawing away at a cucumber with the greatest relish, eating it rind and all, without any condiments whatever.

All through Roumelia the gradual decay of the Crescent and the corresponding elevation of the Cross is everywhere evident ; the Christian element is now predominant, and the Turkish authorities play but an unimportant part in the government of internal affairs. Naturally enough, it does not suit the Mussulman to live among people whom his religion and time-honored custom have taught him to regard as inferiors, the consequence being that there has of late years been a general folding of tents and silently stealing away ; and to-day it is no very infrequent occurrence for a whole Mussulman village to pack up, bag and baggage, and move bodily to Asia Minor, where the Sultan gives them tracts of land for settlement. Between the Christian and Mussulman populations of these countries there is naturally a certain amount of the " six of one and half a dozen of the other " principle, and in certain regions, where the Mussulmans have dwindled to a small minority, the Christians are ever prone to bestow upon them the same treatment that the Turks formerly gave them.   There appears to be little conception of what we consider " good manners " among Oriental villagers, and while I am writing out a few notes this evening, the people crowding the *mehana* because of my strange unaccustomed presence stand around watching every motion of my pen, jostling carelessly against the bench, and commenting on things concerning me and the bicycle with a garrulousness that makes it almost impossible for me to write.   The women of these Roumelian villages bang their hair, and wear it in two long braids, or plaited into a streaming white head-dress of some gauzy material, behind ; huge silver clasps, artistically engraved, that are probably heirlooms, fasten a belt around their waists ; and as they walk along barefooted, strings of beads, bangles, and necklaces of silver coins make an incessant jingling.   The sky clears and the moon shines forth resplendently ere I stretch myself on my rude couch to-night, and the sun rising bright next morning would seem to indicate fair weather at last ; an indication that proves illusory, however, before the day is over.

At Khaskhor, some fifteen kilometres from Cauheme, I am able to obtain my favorite breakfast of bread, milk, and fruit, and while

I am in-doors eating it a stalwart Turk considerately mounts guard over the bicycle, resolutely keeping the meddlesome crowd at bay until I get through eating. The roads this morning, though hilly, are fairly smooth, and about eleven o'clock I reach Hermouli, the last town in Roumelia, where, besides being required to produce my passport, I am requested by a pompous lieutenant of *gendarmerie* to produce my permit for carrying a revolver, the first time I have been thus molested in Europe. . Upon explaining, as best I can, that I have no such permit, and that for a *voyageur* permission is not necessary (something about which I am in no way so certain, however, as my words would seem to indicate), I am politely disarmed, and conducted to a guard-room in the police-barracks, and for some twenty minutes am favored with the exclusive society of a uniformed guard and the unhappy reflections of a probable heavy fine, if not imprisonment. I am inclined to think afterward that in arresting and detaining me the officer was simply showing off his authority a little to his fellow-Hermoulites, clustered about me and the bicycle, for, at the expiration of half an hour, my revolver and passport are handed back to me, and without further inquiries or explanations I am allowed to depart in peace.

As though in wilful aggravation of the case, a village of gypsies have their tents pitched and their donkeys grazing in the last Mohammedan cemetery I see ere passing over the Roumelian border into Turkey proper, where, at the very first village, the general aspect of religious affairs changes, as though its proximity to the border should render rigid distinctions desirable. Instead of the crumbling walls and tottering minarets, a group of closely veiled women are observed praying outside a well-preserved mosque, and praying sincerely too, since not even my never-before-seen presence and the attention-commanding bicycle are sufficient to win their attention for a moment from their devotions, albeit those I meet on the road peer curiously enough from between the folds of their muslin *yashmaks*. I am worrying along to day in the face of a most discouraging head-wind, and the roads, though mostly ridable, are none of the best. For much of the way there is a macadamized road that, in the palmy days of the Ottoman dominion, was doubtless a splendid highway, but now weeds and thistles, evidences of decaying traffic and of the proximity of the Roumelian railway, are growing in the centre, and holes and impassable places make cycling a necessarily wide-awake performance.

Mustapha Pasha is the first Turkish town of any importance I come to, and here again my much-required " passaporte " has to be exhibited ; but the police-officers of Mustapha Pasha seem to be exceptionally intelligent and quite agreeable fellows. My revolver is in plain view, in its accustomed place ; but they pay no sort of attention to it, neither do they ask me a whole rigmarole of questions about my linguistic accomplishments, whither I am going, whence I came, etc., but simply glance at my passport, as though its examination were a matter of small consequence anyhow, shake hands, and smilingly request me to let them see me ride.

It begins to rain soon after I leave Mustapha Pasha, forcing me to take refuge in a convenient culvert beneath the road. I have been under this shelter but a few minutes when I am favored with the company of three swarthy Turks, who, riding toward Mustapha Pasha on horseback, have sought the same shelter. These people straightway express their astonishment at finding me and the bicycle under the culvert, by first commenting among themselves ; then they turn a battery of Turkish interrogations upon my devoted head, nearly driving me out of my senses ere I escape. They are, of course, quite unintelligible to me ; for if one of them asks a question a shrug of the shoulders only causes him to repeat the same over and over again, each time a little louder and a little more deliberate. Sometimes they are all three propounding questions and emphasizing them at the same time, until I begin to think that there is a plot to talk me to death and confiscate whatever valuables I have about me. They all three have long knives in their waistbands, and, instead of pointing out the mechanism of the bicycle to each other with the finger, like civilized people, they use these long, wicked-looking knives for the purpose. They may be a coterie of heavy villains for anything I know to the contrary, or am able to judge from their general appearance, and in view of the apparent disadvantage of one against three in such cramped quarters, I avoid their immediate society as much as possible by edging off to one end of the culvert. They are probably honest enough, but as their stock of interrogations seems inexhaustible, at the end of half an hour I conclude to face the elements and take my chances of finding some other shelter farther ahead rather than endure their vociferous onslaughts any longer. They all three come out to see what is going to happen, and I am not ashamed to admit that I stand tinkering around the bicycle in the pelting rain longer than

is necessary before mounting, in order to keep them out in it and
get them wet through, if possible, in revenge for having practically
ousted me from the culvert, and since I have a water-proof, and
they have nothing of the sort, I partially succeed in my plans.

Turkish Amenities.

The road is the same ancient and neglected macadam, but between Mustapha Pasha and Adrianople they either make some pretence of keeping it in repair, or else the traffic is sufficient to keep

down the weeds, and I am able to mount and ride in spite of the down-pour. After riding about two miles I come to another culvert, in which I deem it advisable to take shelter. Here, also, I find myself honored with company, but this time it is a lone cow-herder, who is either too dull and stupid to do anything but stare alternately at me and the bicycle, or else is deaf and dumb, and my recent ex-perience makes me cautious about tempting him to use his tongue. I am forced by the rain to remain cramped up in this last narrow culvert until nearly dark, and then trundle along through an area of stones and water-holes toward Adrianople, which city lies I know not how far to the southeast. While trundling along through the darkness, in the hope of reaching a village or *mehana*, I observe a rocket shoot skyward in the distance ahead, and surmise that it indicates the whereabout of Adrianople ; but it is plainly many a weary mile ahead ; the road cannot be ridden by the uncertain light of a cloud-veiled moon, and I have been forging ahead, over rough ways leading through an undulating country, and most of the day against a strong head-wind, since early dawn. By ten o'clock I happily arrive at a section of country that has not been favored by the afternoon rain, and, no *mehana* making its appearance, I con-clude to sup off the cold, cheerless memories of the black bread and half-ripe pears eaten for dinner at a small village, and crawl beneath some wild prune-bushes for the night.

A few miles wheeling over very fair roads, next morning, brings me into Adrianople, where, at the Hotel Constantinople, I obtain an excellent breakfast of roast lamb, this being the only well-cooked piece of meat I have eaten since leaving Nisch. It has rained every day without exception since it delayed me over Sun-day at Bela Palanka, and this morning it begins while I am eating breakfast, and continues a drenching downpour for over an hour. While waiting to see what the weather is coming to, I wander around the crooked and mystifying streets, watching the animated scenes about the bazaars, and try my best to pick up some knowl-edge of the value of the different coins, for I have had to deal with a bewildering mixture of late, and once again there is a complete change. Medjidis, cheriks, piastres, and paras now take the place of Serb francs, Bulgar francs, and a bewildering list of nickel and copper pieces, down to one that I should think would scarcely purchase a wooden toothpick. The first named is a large silver coin worth four and a half francs ; the cherik might be called a quarter dollar; while piastres and paras are tokens, the former about five cents and the latter requiring about nine to make one cent. There are no copper coins in Turkey proper, the smaller coins being what is called "metallic money," a composition of copper and silver, varying in value from a five-para piece to five piastres.

The Adrianopolitans, drawn to the hotel by the magnetism of the bicycle, are bound to see me ride whether or no, and in their quite natural ignorance of its character, they request me to perform in the small, roughly-paved court-yard of the hotel, and all sorts of impossible places. I shake my head in disapproval and explanation of the impracticability of granting their request, but unfortunately Adrianople is within the circle where a shake of the head is understood to mean "yes, certainly;" and the happy crowd range around a ridiculously small space, and smiling approvingly at what they consider my willingness to oblige, motion for me to come ahead. An explanation seems really out of the question after this, and I conclude that the quickest and simplest way of satisfying everybody is to demonstrate my willingness by mounting and wabbling along, if only for a few paces, which I accordingly do beneath a hack shed, at the imminent risk of knocking my brains out against beams and rafters.

At eleven o'clock I decide to make a start, I and the bicycle being the focus of attraction for a most undignified mob as I trundle through the muddy streets toward the suburbs. Arriving at a street where it is possible to mount and ride for a short distance, I do this in the hope of satisfying the curiosity of the crowd, and being permitted to leave the city in comparative peace and privacy; but the hope proves a vain one, for only the respectable portion of the crowd disperses, leaving me, solitary and alone, among a howling mob of the rag, tag, and bobtail of Adrianople, who follow noisily along, vociferously yelling for me to " bin! bin! " (mount, mount), and " chu! chu! " (ride, ride) along the really unridable streets. This is the worst crowd I have encountered on the entire journey across two continents, and, arriving at a street where the prospect ahead looks comparatively promising, I mount, and wheel forward with a view of outdistancing them if possible; but a ride of over a hundred yards without dismounting would be an exceptional performance in Adrianople after a rain, and I soon find that I have made a mistake in attempting it, for, as I mount, the mob grows fairly wild and riotous with excitement, flinging their red fezes at the wheels, rushing up behind and giving the bicycle smart pushes forward, in their eagerness to see it go faster, and more than one stone comes bounding along the street, wantonly flung by some young savage unable to contain himself. I quickly decide upon allaying the excitement by dismounting, and trundling until the mobs gets tired of following, whatever the distance.

This movement scarcely meets with the approval of the unruly crowd, however, and several come forward and exhibit ten-para pieces as an inducement for me to ride again, while overgrown gamins swarm around me, and, straddling the middle and index fingers of their right hands over their left, to illustrate and emphasize their meaning, they clamorously cry, "*bin! bin! chu! chu! monsieur! chu! chu!*" as well as much other persuasive talk, which, if one could understand, would probably be found to mean in substance, that, although it is the time-honored custom and privilege of Adrianople mobs to fling stones and similar compliments at such unbelievers from the outer world as come among them in a conspicuous manner, they will considerately forego their privileges this time, if I will only "*bin! bin!*" and "*chu! chu!*" The aspect of harmless mischievousness that would characterize a crowd of Occidental youths on a similar occasion is entirely wanting here, their faces wearing the determined expression of people in dead earnest about grasping the only opportunity of a lifetime. Respectable Turks stand on the sidewalk and eye the bicycle curiously, but they regard my evident annoyance at being followed by a mob like this with supreme indifference, as does also a passing *gendarme*, whom I halt, and motion my disapproval of the proceedings. Like the civilians, he pays no sort of attention, but fixes a curious stare on the bicycle, and asks something, the import of which will to me forever remain a mystery.

Once well out of the city the road is quite good for several kilometres, and I am favored with a unanimous outburst of approval from a rough crowd at a suburban *mehana*, because of outdistancing a horseman who rides out from among them to overtake me. At Adrianople my road leaves the Maritza Valley and leads across the undulating uplands of the Adrianople Plains, hilly, and for most of the way of inferior surface. Reaching the village of Hafsa, soon after noon, I am fairly taken possession of by a crowd of turbaned and fezed Hafsaites and soldiers wearing

the coarse blue uniform of the Turkish regulars, and given not
one moment's escape from "*bin! bin!*" until I consent to parade
my modest capabilities with the wheel by going back and forth
along a ridable section of the main street. The population is
delighted. Solid old Turks pat me on the back approvingly, and
the proprietor of the *mehana* fairly hauls me and the bicycle into
his establishment. This person is quite befuddled with *mastic*,
which makes him inclined to be tyrannical and officious; and .
several times within the hour, while I wait for the never-failing
thunder-shower to subside, he peremptorily dismisses both civilians and military out of the *mehana* yard; but the crowd always
filters back again in less than two minutes. Once, while eating
dinner, I look out of the window and find the bicycle has disappeared. Hurrying out, I meet the boozy proprietor and another
individual making their way with alarming unsteadiness up a steep
stairway, carrying the machine between them to an up-stairs room,
where the people will have no possible chance of seeing it. Two
minutes afterward his same whimsical and capricious disposition
impels him to politely remove the eatables from before me, and
with the manners of a showman, he gently leads me away from the
table, and requests me to ride again for the benefit of the very
crowd he had, but two minutes since, arbitrarily denied the privilege
of even looking at the bicycle. Nothing would be more natural
than to refuse to ride under these circumstances; but the crowd
looks so gratified at the proprietor's sudden and unaccountable
change of front, that I deem it advisable, in the interest of being
permitted to finish my meal in peace, to take another short spin;
moreover, it is always best to swallow such little annoyances in
good part.

My route to-day is a continuation of the abandoned macadam
road, the weed-covered stones of which I have frequently found
acceptable in tiding me over places where the ordinary dirt road
was deep with mud. In spite of its long-neglected condition,
occasional ridable stretches are encountered, but every bridge
and culvert has been destroyed, and an honest shepherd, not far
from Hafsa, who from a neighboring knoll observes me wheeling down a long declivity toward one of these uncovered water-
ways, nearly shouts himself hoarse, and gesticulates most frantically in an effort to attract my attention to the danger ahead.
Soon after this I am the innocent cause of two small pack-

mules, heavily laden with merchandise, attempting to bolt from
their driver, who is walking behind.   One of them actually succeeds in escaping, and, although his pack is too heavy to admit of
running at any speed, he goes awkwardly jogging across the rolling
plains, as though uncertain in his own mind of whether he is acting sensibly or not ; but his companion in pack-slavery is less fortunate, since he tumbles into a gully, bringing up flat on his broad
and top-heavy pack with his legs frantically pawing the air.   Stopping to assist the driver in getting the collapsed mule on his feet
again, this individual demands damages for the accident ; so I judge,
at least, from the frequency of the word "medjedie," as he angrily,
yet ruefully, points to the mud-begrimed pack and unhappy, yet
withal laughter-provoking, attitude of the mule ; but I utterly fail
to see any reasonable connection between the uncalled-for scariness
of his mules and the contents of my pocket-book, especially since I
was riding along the Sultan's ancient and deserted macadam, while
he and his mules were patronizing a separate and distinct dirt-road
alongside.   As he seems far more concerned about obtaining a
money satisfaction from me than the rescue of the mule from his
topsy-turvy position, I feel perfectly justified, after several times
indicating my willingness to assist him, in leaving him and proceeding on my way.

    The Adrianople plains are a dreary expanse of undulating graz-
ing-land, traversed by small sloughs and their adjacent cultivated
areas.   Along this route it is without trees, and the villages one
comes to at intervals of eight or ten miles are shapeless clusters of
mud, straw-thatched huts, out of the midst of which, perchance,
rises the tapering minaret of a small mosque, this minaret being,
of course, the first indication of a village in the distance.   Between
Adrianople and Eski Baba, the town I reach for the night, are
three villages, in one of which I approach a Turkish private house
for a drink of water, and surprise the women with faces unveiled.
Upon seeing my countenance peering in the doorway they one
and all give utterance to little screams of dismay, and dart like
frightened fawns into an adjoining room.   When the men appear,
to see what is up, they show no signs of resentment at my abrupt
intrusion, but one of them follows the women into the room, and
loud, angry words seem to indicate that they are being soundly
berated for allowing themselves to be thus caught.   This does not
prevent the women from reappearing the next minute, however,

with their faces veiled behind the orthodox *yashmak*, and through its one permissible opening satisfying their feminine curiosity by critically surveying me and my strange vehicle.

Four men follow me on horseback out of this village, presumably to see what use I make of the machine; at least I cannot otherwise account for the honor of their unpleasantly close attentions—close, inasmuch as they keep their horses' noses almost against my back, in spite of sundry subterfuges to shake them off. When I stop they do likewise, and when I start again they deliberately follow, altogether too near to be comfortable. They are, all four, rough-looking peasants, and their object is quite unaccountable, unless they are doing it for "pure cussedness," or perhaps with some vague idea of provoking me into doing something that would offer them the excuse of attacking and robbing me. The road is sufficiently lonely to invite some such attention. If they are only following me to see what I do with the bicycle, they return but little enlightened, since they see nothing but trundling and an occasional scraping off of mud. At the end of about two miles, whatever their object, they give it up.

Several showers occur during the afternoon, and the distance travelled has been short and unsatisfactory, when just before dark I arrive at Eski Baba, where I am agreeably surprised to find a *mehana*, the proprietor of which is a reasonably mannered individual. Since getting into Turkey proper, reasonably mannered people have seemed wonderfully scarce, the majority seeming to be most boisterous and headstrong. Next to the bicycle the Turks of these interior villages seem to exercise their minds the most concerning whether I have a passport; as I enter Eski Baba; a *gendarme* standing at the police-barrack gates shouts after me to halt and produce "passaporte." Exhibiting my passport at almost every village is getting monotonous, and, as I am going to remain here at least overnight, I ignore the *gendarme's* challenge and wheel on to the *mehana*. Two *gendarmes* are soon on the spot, inquiring if I have a "passaporte;" but, upon learning that I am going no farther to-day, they do not take the trouble to examine it, the average Turkish official religiously believing in never doing anything to-day that can be put off till to-morrow.

The natives of a Turkish interior village are not over-intimate with newspapers, and are in consequence profoundly ignorant, having little conception of anything save what they have been fa-

miliar with and surrounded by all their lives, and the appearance
of the bicycle is indeed a strange visitation, something entirely beyond their comprehension. The *mehana* is crowded by a wildly
gesticulating and loudly commenting and arguing crowd of Turks
and Christians all the evening. Although there seems to be quite
a large proportion of native unbelievers in Eski Baba there is not
a single female visible on the streets this evening ; and from observations next day I judge it to be a conservative Mussulman village,
where the Turkish women, besides keeping themselves veiled with
orthodox strictness, seldom go abroad, and the women who are not
Mohammedan, imbibing something of the retiring spirit of the
dominant race, also keep themselves well in the background.

A round score of dogs, great and small, and in all possible conditions of miserableness, congregate in the main street of Eski Baba
at eventide, waiting with hungry-eyed expectancy for any morsel of
food or offal that may peradventure find its way within their reach.
The Turks, to their credit be it said, never abuse dogs ; but every
male " Christian " in Eski Baba seems to consider himself in duty
bound to kick or throw a stone at one, and scarcely a minute
passes during the whole evening without the yelp of some unfortunate cur. These people seem to enjoy a dog's sufferings ; and one
soulless peasant, who in the course of the evening kicks a half-
starved cur so savagely that the poor animal goes into a fit, and,
after staggering and rolling all over the street, falls down as though
really dead, is the hero of admiring comments from the crowd, who
watch the creature's sufferings with delight. Seeing who can get
the most telling kicks at the dogs seems to be the regular evening's
pastime among the male population of Eski Baba unbelievers, and
everybody seems interested and delighted when some unfortunate
animal comes in for an unusually severe visitation.

A rush mat on the floor of the stable is my bed to-night, with
a dozen unlikely looking natives, to avoid the close companionship
of whom I take up my position in dangerous proximity to a donkey's
hind legs, and not six feet from where the same animal's progeny is
stretched out with all the abandon of extreme youth. Precious little sleep is obtained, for fleas innumerable take liberties with my
person. A flourishing colony of swallows inhabiting the roof keeps
up an incessant twittering, and toward daylight two *muezzins*, one
on the minaret of each of the two mosques near by, begin calling the
faithful to prayer, and howling " *Allah ! Allah !* " with the voices of

men bent on conscientiously doing their duty by making themselves
heard by every Mussulman for at least a mile around, robbing me
of even the short hour of repose that usually follows a sleepless
night.

It is raining heavily again on Sunday morning—in fact, the last
week has been about the rainiest that I ever saw outside of England—and considering the state of the roads south of Eski Baba,
the prospects look favorable for a Sunday's experience in an interior Turkish village.   Men are solemnly squatting around the
benches of the *mehana*, smoking nargilehs and sipping tiny cups of
thick black coffee, and they look on in wonder while I devour a substantial breakfast ; but whether it is the novelty of seeing a 'cycler
feed, or the novelty of seeing *anybody* eat as I am doing, thus early
in the morning, I am unable to say ; for no one else seems to partake
of much solid food until about noontide.   All the morning long,
people swarming around are importuning me with, " *Bin, bin, bin,
monsieur ! " *   The bicycle is locked up in a rear chamber, and thrice
I accommodatingly fetch it out and endeavor to appease their curiosity by riding along a hundred-yard stretch of smooth road in the rear
of the *mehana ;* but their importunities never for a moment cease.
Finally the annoyance becomes so unbearable that the proprietor
takes pity on my harassed head, and, after talking quite angrily to
the crowd, locks me up in the same room with the bicycle.

Iron bars guard the rear windows of the houses at Eski Baba,
and ere I am fairly stretched out on my mat several swarthy faces
appear at the bars, and several voices simultaneously join in the
dread chorus of, " *Bin, bin, bin, monsieur ! bin, bin ! "* compelling
me to close, in the middle of a hot day—the rain having ceased
about ten o'clock—the one small avenue of ventilation in the stuffy
little room.   A moment's privacy is entirely out of the question, for,
even with the window closed, faces are constantly peering in, eager
to catch even the smallest glimpse of either me or the bicycle.   Fate
is also against me to-day, plainly enough, for ere I have been imprisoned in the room an hour the door is unlocked to admit the
mulazim (lieutenant of *gendarmes*), and two of his subordinates,
with long cavalry swords dangling about their legs, after the manner of the Turkish police.

In addition to puzzling their sluggish brains about my passport,
my strange means of locomotion, and my affairs generally, they
have now, it seems, exercised their minds up to the point that they

ought to interfere in the matter of my revolver. But first of all they want to see my wonderful performance of riding a thing that cannot stand alone. After I have favored the *gendarmes* and the assembled crowd by riding once again, they return the compliment by tenderly escorting me down to police headquarters, where, after spending an hour or so in examining my passport, they place that document and my revolver in their strong box, and lackadaisically wave me adieu. Upon returning to the *mehana*, I find a corpulent pasha and a number of particularly influential Turks awaiting my reappearance, with the same diabolical object of asking me to " *bin!* *bin !* " Soon afterward come the two Mohammedan priests, with the same request ; and certainly not less than half a dozen times during the afternoon do I bring out the bicycle and ride, in defer- ence to the insatiable curiosity of the sure enough " unspeakable " Turk ; and every separate time my audience consists not only of the people personally making the request, but of the whole gesticu- lating male population. The proprietor of the *mehana* kindly takes upon himself the office of apprising me when my visitors are people of importance, by going through the pantomime of swelling his features and form up to a size corresponding in proportion relative to their importance, the process of inflation in the case of the pasha being quite a wonderful performance for a man who is not a pro- fessional contortionist.

Once during the afternoon I attempt to write, but I might as well attempt to fly, for the *mehana* is crowded with people who plainly have not the slightest conception of the proprieties. Finally a fez is wantonly flung, by an extra-enterprising youth, at my ink- bottle, knocking it over, and but for its being a handy contrivance, out of which the ink will not spill, it would have made a mess of my notes. Seeing the uselessness of trying to write, I meander forth, and into the leading mosque, and without removing my shoes, tread its sacred floor for several minutes, and stand listening to several devout Mussulmans reciting the Koran aloud, for, be it known, the great fast of Ramadan has begun, and fasting and prayer is now the faithful Mussulman's daily lot for thirty days, his religion forbidding him either eating or drinking from early morn till close · of day. After looking about the interior, I ascend the steep spi- ral stairway up to the minaret balcony whence the *muezzin* calls the faithful to prayer five times a day. As I pop my head out through the little opening leading to the balcony, I am slightly

14

taken aback by finding that small footway already occupied by the *muezzin*, and it is a fair question as to whether the *muezzin's* astonishment at seeing my white helmet appear through the opening is greater, or mine at finding him already in ·possession. However, I brazen it out by joining him, and he, like a sensible

On the Minaret with the Muezzin.

man, goes about his business just the same as if nobody were about. The people down in the streets look curiously up and call one-another's attention to the unaccustomed sight of a white-helmeted 'cycler and a *muezzin* upon the minaret together ; but the fact that I am not interfered with in any way goes far to prove that the Mussulman fanaticism, that we have all heard and read about so often, has wellnigh flickered out in European Turkey ; moreover, I think the Eski Babaus would allow me to do anything, in order to place me under obligations to "*bin ! bin !*" whenever they ask me.

At nine o'clock I begin to grow a trifle uneasy about the fate of my passport and revolver, and, proceeding to the police-barracks, formally demand their return. Nothing has apparently been done concerning either one or the other since they were taken from me, for the *mulazim*, who is lounging on a divan smoking cigarettes, produces them from the same receptacle he consigned them to this afternoon, and lays them before him, clearly as mystified and perplexed as ever about what he ought to do. I explain to him that I wish to depart in the morning, and *gendarmes* are despatched to summon several leading Eski Babans for consultation, in the hope that some of them, or all of them put together, might perchance. arrive at a satisfactory conclusion concerning me. The great trouble appears to be that, while I got the passport *viséd* at Sofia and Philippopolis, I overlooked Adrianople, and the Eski Baba officials, being in the *vilayet* of the latter city, are naturally puzzled to account for this omission ; and, from what I can gather of their conversation, some are advocating sending me back to Adrianople, a suggestion that I straightway announce my disapproval of by again and again calling their attention to the *visé* of the Turkish consul-general in London, and giving them to understand, with much emphasis, that this *visé* answers for every part of Turkey, including the *vilayet* of Adrianople. The question then arises as to whether that has anything to do with my carrying a revolver ; to which I candidly reply that it has not, at the same time pointing out that I have just come through Servia and Bulgaria (countries in which the Turks consider it quite necessary to go armed, though in fact there is quite as much, if not more, necessity for arms in Turkey), and that I have come through both Mustapha Pasha and Adrianople without being molested on account of the revolver ; all of which only seems to mystify them the more, and make them more puzzled than ever about what to do. Finally a brilliant idea occurs to one of them, being nothing less than to shift the weight of the dreadful responsibility upon the authoritative shoulders of a visiting pasha, an important personage who arrived in Eski Baba by carriage about two hours ago, and whose arrival I remember caused quite a flurry of excitement among the natives.

The pasha is found surrounded by a number of bearded Turks, seated cross-legged on a carpet in the open air, smoking nargilehs and cigarettes, and sipping coffee. This pasha is fatter and more unwieldy, if possible, than the one for whose edification I rode the bicycle this afternoon; noticing which, all hopes of being created a pasha upon my arrival at Constantinople naturally vanish, for evidently one of the chief qualifications for a pashalic is obesity, a distinction to which continuous 'cycling, in hot weather is hardly conducive. The pasha seems a good-natured person, after the manner of fat people generally, and straightway bids me be seated on the carpet, and orders coffee and cigarettes to be placed at my disposal while he examines my case. In imitation of those around me I make an effort to sit cross-legged on the mat; but the position is so uncomfortable that I am quickly compelled to change it, and I fancy detecting a merry twinkle in the eye of more than one silent observer at my inability to adapt my posture to the custom of the country. I scarcely think the pasha knows anything more about what sort of a looking document an English passport ought to be, than does the *mulazim* and the leading citizens of Eski Baba; but he goes through the farce of critically examining the *visé* of the Turkish consul-general in London, while another Turk holds his lighted cigarette close to it, and blows from it a feeble glimmer of light. Plainly the pasha cannot make anything more out of it than the others, for many a Turkish pasha is unable to sign his own name intelligibly, using a seal instead; but, probably with a view of favorably impressing those around him, he asks me first if I am an Englishman, and then if I am "a baron," doubtless thinking that an English baron is a person occupying a somewhat similar position in English society to that of a pasha in Turkish: viz., a really despotic sway over the people of his district; for, although there are law and lawyers in Turkey to-day, the pasha, especially in country districts, is still an all-powerful person, practically doing as he pleases.

To the first question I return an affirmative answer; the latter I pretend not to comprehend; but I cannot help smiling at the question and the manner in which it is put—seeing which the pasha and his friends smile in response, and look knowingly at each other, as though thinking, "Ah! he *is* a baron, but don't intend to let us know it." Whether this self-arrived-at decision influences things in my favor I hardly know, but anyhow he tosses me my

' Are you an English Baron ? '

passport, and orders the *mulazim* to return my revolver ; and as I mentally remark the rather jolly expression of the pasha's face, I am inclined to think that, instead of treating the matter with the ridiculous importance attached to it by the *mulazim* and the other people, he regards the whole affair in the light of a few minutes' acceptable diversion. The pasha arrived too late this evening at Eski Baba to see the bicycle : " Will I allow a *gendarme* to go to the *mehana* and bring it for his inspection ? " " I will go and fetch it myself," I explain ; and in ten minutes the fat pasha and his friends are examining the perfect mechanism of an American bicycle by the light of an American kerosene lamp, which has been provided in the meantime. Some of the on-lookers, who have seen me ride to-day, suggested to the pasha that I " *bin ! bin !* " and the pasha smiles approvingly at the suggestion ; but by pantomime I explain to him the impossibility of riding, owing to the nature of the ground and the darkness, and I am really quite surprised at the readiness with which he comprehends and accepts the situation. The pasha is very likely possessed of more intelligence than I have been giving him credit for ; anyhow he has in ten minutes proved himself equal to the situation, which the *mulazim* and several prominent Eski Babans have puzzled their collective brains over for an hour in vain, and, after he has inspected the bicycle, and resumed his cross-legged position on the carpet, I doff my helmet to him and those about him, and return to the *mehana*, well satisfied with the turn affairs have taken.

CHAPTER IX.

THROUGH EUROPEAN TURKEY.

On Monday morning I am again awakened by the *muezzin* calling the Mussulmans to their early morning devotions, and, arising from my mat at five o'clock, I mount and speed away southward from Eski Baba. Not less than a hundred people have collected to see the wonderful performance again.

All pretence of road-making seems to have been abandoned; or, what is more probable, has never been seriously attempted, the visible roadways from village to village being mere ox-wagon and pack-donkey tracks, crossing the wheat-fields and uncultivated tracts in any direction. The soil is a loose, black loam, which the rain converts into mud, through which I have to trundle, wooden scraper in hand; and I not infrequently have to carry the bicycle through the worst places. The morning is sultry, requiring good roads and a breeze-creating pace for agreeable going.

Harvesting and threshing are going forward briskly, but the busy hum of the self-binder and the threshing-machine is not heard; the reaping is done with rude hooks, and the threshing by dragging round and round, with horses or oxen, sleigh-runner shaped, broad boards, roughed with flints or iron points, making the surface resemble a huge rasp. Large gangs of rough-looking Armenians, Arabs, and Africans are harvesting the broad acres of land-owning pashas, the gangs sometimes counting not less than fifty men. Several donkeys are always observed picketed near them, taken, wherever they go, for the purpose of carrying provisions and water. Whenever I happen anywhere near one of these gangs they all come charging across the field, reaping-hooks in hand, racing with each other and good-naturedly howling defiance to competitors. A band of Zulus charging down on a fellow, and brandishing their assegais, could scarcely present a more ferocious front. Many of them wear no covering of any kind on the upper part of the body, no hat, no foot-gear, nothing but a pair of loose,

baggy trousers, while the tidiest man among them would be immediately arrested on general principles in either England or America.
Rough though they are, they appear, for the most part, to be good-
natured fellows, and although they sometimes emphasize their
importunities of "*bin! bin!*" by flourishing their reaping-hooks
threateningly over my head, and one gang actually confiscates the
bicycle, which they lay up on a shock of wheat, and with much
flourishing of reaping-hooks as they return to their labors, warn
me not to take it away, these are simply good-natured pranks,
such as large gangs of laborers are wont to occasionally indulge in
the world over.

Streams have to be forded to-day for the first time in Europe,
several small creeks during the afternoon ; and near sundown I
find my pathway into a village where I propose stopping for the
night, obstructed by a creek swollen bank-full by a heavy thunder-
shower in the hills.   A couple of lads on the opposite bank
volunteer much information concerning the depth of the creek
at different points ; no doubt their evident mystification at not
being understood is equalled only by the amazement at my answers.   Four peasants come down to the creek, and one of them
kindly wades in and shows that it is only waist deep.   Without
more ado I ford it, with the bicycle on my shoulder, and straightway seek the accommodation of the village *mehana*.   This village
is a miserable little cluster of mud hovels, and the best the *mehana*
affords is the coarsest of black-bread and a small salted fish,
about the size of a sardine, which the natives devour without any
pretence of cooking, but which are worse than nothing for me,
since the farther they are away the better I am suited.   Sticking a
flat loaf of black-bread and a dozen of these tiny shapes of salted
nothing in his broad waistband, the Turkish peasant sallies forth
contentedly to toil.

I have accomplished the wonderful distance of forty kilometres to-day, at which I am really quite surprised, considering
everything.   The usual daily weather programme has been faithfully carried out—a heavy mist at morning, that has prevented
any drying up of roads during the night, three hours of oppressive heat—from nine till twelve—during which myraids of
ravenous flies squabble for the honor of drawing your blood, and
then, when the mud begins to dry out sufficient to justify my dispensing with the wooden scraper, thunder-showers begin to be-

stow their unappreciated favor upon the roads, making them well-nigh impassable again. The following morning the climax of vexation is reached when, after wading through the mud for two hours, I discover that I have been dragging, carrying, and trundling my laborious way along in the wrong direction for Tchorlu, which is not over thirty-five kilometres from my starting-point, but it takes me till four o'clock to reach there. A hundred miles on French or English roads would not be so fatiguing, and I wisely take advantage of being in a town where comparatively decent accommodations are obtainable to make up, so far as possible, for this morning's breakfast of black bread and coffee, and my noontide meal of cold, cheerless reflections on the same. The same programme of " bin ! bin ! " from importuning crowds, and police inquisitiveness concerning my "passporte " are endured and survived ; but I spread myself upon my mat to-night thoroughly convinced that a month's cycling among the Turks would worry most people into premature graves.

I am now approaching pretty close to the Sea of Marmora, and next morning I am agreeably surprised to find sandy roads, which the rains have rather improved than otherwise ; and although much is unridably heavy, it is immeasurably superior to yesterday's mud. I pass the country residence of a wealthy pasha, and see the ladies of his harem seated in the meadow hard by, enjoying the fresh morning air. They form a circle, facing inward, and the swarthy eunuch in charge stands keeping watch at a respectful distance. I carry a pocketful of bread with me this morning, and about nine o'clock, upon coming to a ruined mosque and a few deserted buildings, I approach one at which signs of occupation are visible, for some water. This place is simply a deserted Mussulman village, from which the inhabitants probably decamped in a body during the last Russo-Turkish war ; the mosque is in a tumble-down condition, the few dwelling-houses remaining are in the last stages of dilapidation, and the one I call at is temporarily occupied by some shepherds, two of whom are regaling themselves with food of some kind out of an earthenware vessel.

Obtaining the water, I sit down on some projecting boards to eat my frugal lunch, fully conscious of being an object of much furtive speculation on the part of the two occupants of the deserted house ; which, however, fails to strike me as anything extraordinary, since these attentions have long since become an ordinary

every-day affair.   Not even the sulky and rather hang-dog expression of the men, which failed not to escape my observation at my
first approach, awakened any shadow of suspicion in my mind of
their being possibly dangerous characters, although the appearance
of the place itself is really sufficient to make one hesitate about
venturing near ; and upon sober after-thought I am fully satisfied

" And makes a grab for my Revolver."

that this is a resort of a certain class of disreputable characters,
half shepherds, half brigands, who are only kept from turning
full-fledged freebooters by a wholesome fear of retributive justice.
While I am discussing my bread and water one of these worthies
saunters with assumed carelessness up behind me and makes a
grab for my revolver, the butt of which he sees protruding from

the holster. Although I am not exactly anticipating this movement, travelling alone among strange people makes one's faculties of self-preservation almost mechanically on the alert, and my hand reaches the revolver before his does. Springing up, I turn round and confront him and his companion, who is standing in the doorway. A full exposition of their character is plainly stamped on their faces, and for a moment I am almost tempted to use the revolver on them. Whether they become afraid of this or whether they have urgent business of some nature will never be known to me, but they both disappear inside the door ; and, in view of my uncertainty of their future intentions, I consider it advisable to meander on toward the coast.

Ere I get beyond the waste lands adjoining this village I encounter two more of these shepherds, in charge of a small flock ; they are watering their sheep ; and as I go over to the spring, ostensibly to obtain a drink, but really to have a look at them, they both sneak off at my approach, like criminals avoiding one whom they suspect of being a detective. Take it all in all, I am satisfied that this neighborhood is a place that I have been fortunate in coming through in broad daylight ; by moonlight it might have furnished a far more interesting item than the above.

An hour after, I am gratified at obtaining my first glimpse of the Sea of Marmora off to the right, and in another hour I am disporting in the warm clear surf, a luxury that has not been within my reach since leaving Dieppe, and which is a thrice welcome privilege in this land, where the usual ablutions at *mehanas* consist of pouring water on the hands from a tin cup. The beach is composed of sand and tiny shells, the warm surf-waves are clear as crystal, and my first plunge in the Marmora, after a two months' cycle tour across a continent, is the most thoroughly enjoyable bath I ever had ; notwithstanding, I feel it my duty to keep a loose eye on some shepherds perched on a handy knoll, who look as if half inclined to slip down and examine my clothes. The clothes, with, of course, the revolver and every penny I have with me, are almost as near to them as to me, and always, after ducking my head under water, my first care is to take a precautionary glance in their direction. "Cursed is the mind that nurses suspicion," someone has said ; but under the circumstances almost anybody would be suspicious. These shepherds along the Marmora coast favor each other a great deal, and when a person has been the recipient of undesirable attentions from one of them, to look askance at the next one met with comes natural enough.

Over the undulating cliffs and along the sandy beach, my road now leads through the pretty little seaport of Cilivria, toward Constantinople, traversing a most lovely stretch of country, where waving wheat-fields hug the beach and fairly coquet with the waves, and the slopes are green and beautiful with vineyards and fig-gardens, while away beyond the glassy shimmer of the sea I fancy I can trace on the southern horizon the inequalities of the hills of Asia Minor. Greek fishing-boats are plying hither and thither ; one noble sailing-vessel, with all sails set, is slowly ploughing her way down toward the Dardanelles—probably a grain-ship from the Black Sea—and the smoke from a couple of steamers is discernible in the distance. Flourishing Greek fishing-villages and vine-growing communities occupy this beautiful strip of coast, along which the Greeks seem determined to make the Cross as much more conspicuous than the Crescent as possible, by rearing it on every public building under their control, and not infrequently on private ones as well. The people of these Greek villages seem possessed of sunny dispositions, the absence of all reserve among the women being in striking contrast to the demeanor of the Turkish fair sex. These Greek women chatter after me from the windows as I wheel past, and if I stop a minute in the street they gather around by dozens, smiling pleasantly, and plying me with questions, which, of course, I cannot understand. Some of them are quite handsome, and nearly all have perfect white teeth, a fact that I have ample opportunity of knowing, since they seem to be all smiles.

There has been much making of artificial highways leading from Constantinople in this direction in ages past. A road-bed of huge blocks of stone, such as some of the streets of Eastern towns are made impassable with, is traceable for miles, ascending and descending the rolling hills, imperishable witnesses of the wide difference in Eastern and Western ideas of making a road. These are probably the work of the people who occupied this country before the Ottoman Turks, who have also tried their hands at making a macadam, which not infrequently runs close along-side the old block roadway, and sometimes crosses it ; and it is matter of some wonderment that the Turks, instead of hauling material for their road from a distance did not save expense by merely breaking the stones of the old causeway and using the same road-bed. Twice to-day I

have been required to produce my passport, and when toward
evening I pass through a small village, the lone *gendarme* who is
smoking a nargileh in front of the *mehana* where I halt points to
my revolver and demands "passaporte," I wave examination, so
to speak, by arguing the case with him, and by the not always un-
handy plan of pretending not exactly to comprehend his meaning.
"Passaporte! passaporte! *gendarmerie*, me," replies the officer, authoritatively, in answer to my explanation of a *voyageur* being privileged to carry a revolver; while several villagers who have gathered
around us interpose "*Bin! bin! monsieur, bin! bin!*" I have little
notion of yielding up either revolver or passport to this village *gendarme*, for much of their officiousness is simply the disposition to
show off their authority and satisfy their own personal curiosity regarding me, to say nothing of the possibility of coming in for a little
backsheesh. The villagers are worrying me to "*bin! bin!*" at the
same time the *gendarme* is worrying me about the revolver and passport, and knowing from previous experience that the *gendarme*
would never stop me from mounting, being quite as anxious to witness the performance as the villagers, I quickly decide upon killing
two birds with one stone, and accordingly mount, and pick my way
along the rough street out on to the Constantinople road.

The gloaming settles into darkness, and the domes and minarets of Stamboul, which have been visible from the brow of every
hill for several miles back, are still eight or ten miles away, and
rightly judging that the Ottoman Capital is a most bewildering
city for a stranger to penetrate after night, I pillow my head on a
sheaf of oats, within sight of the goal toward which I have been
pedalling for some 2,500 miles since leaving Liverpool. After
surveying with a good deal of satisfaction the twinkling lights that
distinguish every minaret in Constantinople each night during the
fast of Ramadan, I fall asleep, and enjoy, beneath a sky in which
myriads of far-off lamps seem to be twinkling mockingly at the
Ramadan illuminations, the finest night's repose I have had for a
week. Nothing but the prevailing rains have prevented me from
sleeping beneath the starry dome entirely in preference to putting
up at the village *mehanas*.

En route into Stamboul, on the following morning, I meet the
first train of camels I have yet encountered; in the gray of the
morning, with the scenes around so thoroughly Oriental, it seems
like an appropriate introduction to Asiatic life. Eight o'clock

finds me inside the line of earthworks thrown up by Baker Pasha
when the Russians were last knocking at the gates of Constantinople, and ere long·I am trundling through the crooked streets of
the Turkish Capital toward the bridge which connects Stamboul
with Galata and Pera. Even here my ears are assailed with the
eternal importunities to "*bin! bin!*" the officers collecting the
bridge-toll even joining in the request. To accommodate them I
mount, and ride part way across the bridge, and at 9 o'clock on
July 2d, just two calendar months from the start at Liverpool, I
am eating my breakfast in a Constantinople restaurant.

I am not long in finding English-speaking friends, to whom my
journey across the two continents is not unknown, and who kindly
direct me to the Chamber of Commerce Hotel, Rue Omar, Galata,
a home-like establishment, kept by an English lady. I have been
purposing of late to remain in Constantinople during the heated
term of July and August, thinking to shape my course southward
through Asia Minor and down the Euphrates Valley to Bagdad,
and by taking a south-easterly direction as far as circumstances
would permit into India, keep pace with the seasons, thus avoiding
the necessity of remaining over anywhere for the winter. At the
same time I have been reckoning upon meeting Englishmen in
Constantinople who, having travelled extensively in Asia, could
further enlighten me regarding the best route to India. As I
house my bicycle and am shown to my room I take a retrospective
glance across Europe and America, and feel almost as if I have arrived at the half-way house of my journey. The distance from
Liverpool to Constantinople is fully 2,500 miles, which brings the
wheeling distance from San Francisco up to something over 6,000.

So far as the distance wheeled and to be wheeled is concerned,
it is not far from half-way ; but the real difficulties of the journey
are still ahead, although I scarcely anticipate any that time and
perseverance will not overcome. My tour across Europe has been,
on the whole, a delightful journey, and, although my linguistic
shortcomings have made it rather awkward in interior places
where no English-speaking person was to be found, I always managed to make myself understood sufficiently to get along. In the
interior of Turkey a knowledge of French has been considered in-
dispensable to a traveller : but, although a full knowledge of that
language would have made matters much smoother by enabling me
to converse with officials and others, I have nevertheless come

through all right without it; and there have doubtless been occasions when my ignorance has saved me from a certain amount of bother with the *gendarmerie*, who, above all things, dislike to exercise their thinking apparatus. A Turkish official is far less indisposed to act than he is to think; his mental faculties work sluggishly, but his actions are governed largely by the impulse of the moment.

Someone has said that to see Constantinople is to see the entire East; and judging from the different costumes and peoples one meets on the streets and in the bazaars, the saying is certainly not far amiss. From its geographical situation, as well as from its history, Constantinople naturally takes the front rank among the cosmopolitan cities of the world, and the crowds thronging its busy thoroughfares embrace every condition of man between the kid-gloved exquisite without a wrinkle in his clothes and the representative of half-savage Central Asian States incased in sheepskin garments of rudest pattern. The great fast of Ramadan is under full headway, and all true Mussulmans neither eat nor drink a particle of anything throughout the day until the booming of cannon at eight in the evening announces that the fast is ended, when the scene quickly changes into a general rush for eatables and drink. Between eight and nine o'clock in the evening, during Ramadan, certain streets and bazaars present their liveliest appearance, and from the highest-classed restaurant patronized by bey and pasha to the venders of eatables on the streets, all do a rushing business; even the *sujees*. (water-venders), who with leather water-bottles and a couple of tumblers wait on thirsty pedestrians with pure drinking water, at five paras a glass, dodge about among the crowds, announcing themselves with lusty lung, fully alive to the opportunities of the moment.

A few of the coffee-houses provide music of an inferior quality, Constantinople not being a very musical place. A forenoon hour spent in a neighborhood of private residences will repay a stranger for his trouble, since he will during that time see a bewildering assortment of street-venders, from a peregrinating meat-market, with a complete stock dangling from a wooden framework attached to a horse's back, to a grimy individual worrying along beneath a small mountain of charcoal, and each with cries more or less musical. The sidewalks of Constantinople are ridiculously narrow, their only practical use being to keep vehicles from running into the merchandise of the shopkeepers, and to give pedestrians plenty of exercise in jostling each other, and hopping on and off the curbstone to avoid inconveniencing the ladies, who of course are not to be jostled either off the sidewalk or into a sidewalk stock of miscellaneous merchandise. The Constantinople sidewalk is anybody's territory ; the merchant encumbers it with his wares and the coffee-houses with chairs for customers to sit on, the rights of pedestrians being altogether ignored ; the natural consequence is that these latter fill the streets, and the Constantinople Jehu not only has to keep his wits about him to avoid running over men and dogs, but has to use his lungs continually, shouting at them to clear the way. If a seat is taken in one of the coffee-house chairs, a watchful waiter instantly makes his appearance with a tray containing small chunks of a pasty sweetmeat, known in England as " Turkish Delight," one of which you are expected to take and pay half a piastre for, this being a polite way of obtaining payment for the privilege of using the chair. The coffee is served steaming hot in tiny cups holding about two table-spoonfuls, the price varying from ten paras upward, according to the grade of the establishment. A favorite way of passing the evening is to sit in front of one of these establishments, watching the passing throngs, and smoke a nargileh, this latter requiring a good half-hour to do it properly. I undertook to investigate the amount of enjoyment contained in a nargileh one evening, and before smoking it half through concluded that the taste has to be cultivated.

One of the most inconvenient things about Constantinople is the great scarcity of small change. Everybody seems to be short of fractional money save the money-changers—people who are here a genuine necessity, since one often has to patronize them before making the most trifling purchase. Ofttimes the store-keeper will refuse point-blank to sell an article when change is required, solely on account of his inability or unwillingness to supply it. After drinking a cup of coffee, I have had the *kahvajee* refuse to take any payment rather than change a cherik. Inquiring the reason for this scarcity, I am informed that whenever there is any new output of this money the noble army of money-changers, by a liberal and judicious application of backsheesh, manage to get a corner on the lot and compel the general public, for whose benefit it is ostensibly issued, to obtain what they require through them. However this may be, they manage to control its circulation to a great extent ;

for while their glass cases display an overflowing plenitude, even the fruit-vender, whose transactions are mainly of ten and twenty paras, is not infrequently compelled to lose a customer because of his inability to make change. There are not less than twenty money-changers' offices within a hundred yards of the Galata end of the principal bridge spanning the Golden Horn, and certainly not a less number on the Stamboul side.

The money-changer usually occupies a portion of the frontage of a cigarette and tobacco stand ; and on all the business streets one happens at frequent intervals upon these little glass cases full of bowls and heaps of miscellaneous coins, varying in value. Behind sits a business-looking person—usually a Jew—jingling a handful of medjedis, and expectantly eyeing every approaching stranger. The usual percentage charged is, for changing a lira, eighty paras ; thirty paras for a medjedie, and ten for a cherik, the percentage on this latter coin being about five per cent. Some idea of the inconvenience to the public of this state of affairs can be better imagined by the American by reflecting that if this state of affairs existed in Boston he would frequently have to walk around the block and give a money-changer five per cent. for changing a dollar before venturing upon the purchase of a dish of baked beans. If one offers a coin of the larger denominations in payment of an article, even in quite imposing establishments, they look as black over it as though you were trying to palm off a counterfeit, and hand back the change with an ungraciousness and an evident reluctance that makes a sensitive person feel as though he has in some way been unwittingly guilty of a mean action.

Even the principal streets of Constantinople are but indifferently lighted at night, and, save for the feeble glimmer of kerosene lamps in front of stores and coffee-houses, the by-streets are in darkness. Small parties of Turkish women are encountered picking their way along the streets of Galata in charge of a male attendant, who walks a little way behind, if of the better class, or without the attendant in the case of poorer people, carrying small Japanese lanterns. Sometimes a lantern will go out, or doesn't burn satisfactorily, and the whole party halts in the middle of the, perhaps, crowded thoroughfare, and clusters around until the lantern is readjusted. The Turkish lady walks with a slouchy gait, her shroud-like *abbas* adding not a little to the ungracefulness.

Matters are likewise scarcely to be improved by wearing two
15

pairs of shoes, the large, slipper-like overshoes being required by etiquette to be left on the mat upon entering the house she is visiting; and in the case of a strictly orthodox Mussulman lady—and, doubtless, we may also easily imagine in case of a not over-prepossessing countenance—the *yashmak* hides all but the eyes. The eyes of many Turkish ladies are large and beautiful, and peep

Almost persuaded to be a Christian.

from between the white, gauzy folds of the *yashmak* with an effect upon the observant Frank not unlike coquettishly ogling from behind a fan. Handsome young Turkish ladies with a leaning toward Western ideas are no doubt coming to understand this, for many are nowadays met on the streets wearing *yashmaks* that are but a single thickness of transparent gauze that obscures never a feature, at the same time producing the decidedly interesting and taking effect above mentioned. It is readily seen that the wearing of *yashmaks* must be quite a charitable custom in the case of a lady not blessed with a handsome face, since it enables her to appear in public the equal of her more favored sister in commanding whatever homage is to be derived from that mystery which is said to be woman's greatest charm; and if she has but the one redeeming feature of a beautiful pair of eyes, the advantage is obvious. In street-cars, steamboats, and all public conveyances, board or canvas partitions wall off a small compartment for the exclusive use of ladies, where, hidden from the rude gaze of the Frank, the Turkish lady can remove her *yashmak* and smoke cigarettes.

On Sunday, July 12th, in company with an Englishman in the Turkish artillery service, I pay my first visit to Asian soil, taking a *caique* across the Bosphorus to Kadikeui, one of the many delightful seaside resorts within easy distance of Constantinople. Many objects of interest are pointed out, as, propelled by a couple of swarthy, half-naked *caique-jees*, the sharp-prowed *caique* gallantly rides the blue waves of this loveliest of all pieces of land-environed water. More than once I have noticed that a firm belief in the supernatural has an abiding hold upon the average Turkish mind, having frequently during my usual evening promenade through the Galata streets noted the expression of deep and genuine earnestness upon the countenances of fez-crowned citizens giving respectful audience to Arab fortune-tellers, paying twenty-para pieces for the revelations he is favoring them with, and handing over the coins with the business-like air of people satisfied that they are getting its full equivalent. Consequently I am not much astonished when, rounding Seraglio Point, my companion calls my attention to several large sections of whalebone suspended on the wall facing the water, and tells me that they are placed there by the fishermen, who believe them to be a talisman of no small efficacy in keeping the Bosphorus well supplied with fish, they firmly adhering to the story that once, when the bones were removed, the fish nearly all disappeared. The oars used by the *caique-jees* are of quite a peculiar shape, the oar-shaft immediately next the hand-hold swells into a bulbous affair for the next eighteen inches, which is at least four times the circumference of the remainder, and the end of the oar-blade is for some reason made swallow-tailed. The object of the enlarged portion, which of course comes inside the rowlocks, appears to be the double purpose of balancing the weight of the longer portion outside, and also for preventing the oar at all times from escaping into the water. The rowlock is simply a raw-hide loop, kept well greased, and as, toward the end of every stroke, the *caique-jee* leans back to his work, the oar slips several inches, causing a considerable loss of power. The day is warm, the broiling sun shines directly down on the bare heads of the *caique-jees*, and causes the perspiration to roll off their swarthy faces in large beads ; but they lay back to their work manfully, although, from early morning until cannon roar at 8 P.M. neither bite nor sup, not even so much water as to moisten the end of their parched tongues, will pass their lips ; for, although but poor hard-working *caique-jees*, they are true Mussulmans.

Pointing skyward from the summit of the hill back of Seraglio Point are the four tapering minarets of the world-renowned St. Sophia mosque, and a little farther to the left is the Sultana Achmet mosque, the only mosque in all Mohammedanism with six minarets.[1] Near by is the old Seraglio Palace, or rather what is left of it, built by Mohammed II. in 1467, out of materials from the ancient Byzantine palaces, and in a department of which the *sanjiak shereef* (holy standard), *boorda-y shereef* (holy mantle), and other venerated relics of the prophet Mohammed are preserved. To this place, on the 15th of Ramadan, the Sultan and leading dignitaries of the Empire repair to do homage to the holy relics, upon which it would be the highest sacrilege for Christian eyes to gaze. The hem of this holy mantle is reverently kissed by the Sultan and the few leading personages present, after which the spot thus brought in contact with human lips is carefully wiped with an embroidered napkin dipped in a golden basin of water ; the water used in this ceremony is then supposed to be of priceless value as a purifier of sin, and is carefully preserved, and, corked up in tiny phials, is distributed among the sultanas, grand dignitaries, and prominent people of the realm, who in return make valuable presents to the lucky messengers and Mussulman ecclesiastics employed in its distribution. This precious liquid is doled out drop by drop, as though it were nectar of eternal life received direct from heaven, and, mixed with other water, is drunk immediately upon breaking fast each evening during the remaining fifteen days of Ramadan.

Arriving at Kadikeui, the opportunity presents of observing something of the high-handed manner in which Turkish pashas are wont to expect from inferiors their every whim obeyed. We meet a friend of my companion, a pasha, who for the remainder of the afternoon makes one of our company. Unfortunately for a few other persons the pasha is in a whimsical mood to-day and inclined to display for our benefit rather arbitrary authority toward others.

The first individual coming under his immediate notice is a young man torturing a harp. Summoning the musician, the pasha summarily orders him to play " Yankee Doodle." The musician

---

[1] The writer arrived in Constantinople with the full impression that it was the mosque of St. Sophia that has the famous six minarets, having, I am quite sure, seen it thus quite frequently accredited in print, and I mention this especially, in order that readers who may have been similarly misinformed may know that the above account is the correct one.

does not know it, and humbly begs the pasha to name something more familiar. "Yankee Doodle!" replies the pasha peremptorily. The poor man looks as though he would willingly relinquish all hopes of the future if only some present avenue of escape would offer itself; but nothing of the kind seems at all likely. The musician appeals to my Turkish-speaking friend, and begs him to request me to favor him with the tune. I am of course only too glad to help him stem the rising tide of the pasha's wrath by whistling the tune for him; and after a certain amount of preliminary twanging he strikes up and manages to blunder through "Yankee Doodle." The pasha, after ascertaining from me that the performance is creditable, considering the circumstances, forthwith hands him more money than he would collect among the poorer patrons of the place in two hours. Soon a company of five strolling acrobats and conjurers happens along, and these likewise are summoned into the "presence" and ordered to proceed. Many of the conjurer's tricks are quite creditable performances; but the pasha occasionally interferes in the proceedings just in the nick of time to prevent the prestidigitator finishing his manipulations, much to the pasha's delight. Once, however, he cleverly manages to hoodwink the pasha, and executes his trick in spite of the latter's interference, which so amuses the pasha that he straightway gives him a medjedie. Our return boat to Galata starts at seven o'clock, and it is a ten minutes' drive down to the landing. At fifteen minutes to seven the pasha calls for a public carriage to take us down to the steamer.

"There are no carriages, Pasha Effendi. Those three are all engaged by ladies and gentlemen in the garden," exclaims the waiter, respectfully.

"Engaged or not engaged, I want that open carriage yonder," replies the pasha authoritatively, and already beginning to show signs of impatience." Boxhanna!" (hi, you, there!) "drive around here," addressing the driver.

The driver enters a plea of being already engaged. The pasha's temper rises to the point of threatening to throw carriage, horses, and driver into the Bosphorus if his demands are not instantly complied with. Finally the driver and everybody else interested collapse completely, and, entering the carriage, we are driven to our destination without another murmur. Subsequently I learned that a government officer, whether a pasha or of lower rank, has the

"Play 'Yankee Doodle,'" said the Pasha.

power of taking arbitrary possession of a public conveyance over the head of a civilian, so that our pasha was, after all, only sticking up for the rights of himself and my friend of the artillery, who likewise wears the mark by which a military man is in Turkey always distinguishable from a civilian—a longer string to the tassel of his fez.

This is the last day of Ramadan, and the following Monday ushers in the three days' feast of Biàram, which is in substance a kind of a general carousal to compensate for the rigid self-denial of the thirty days' fasting and prayer just ended. The government offices and works are all closed, everybody is wearing new clothes, and holiday-making engrosses the public attention. A friend proposes a trip on a Bosphorus steamer up as far as the entrance to the Black Sea. The steamers are profusely decorated with gay-colored flags, and at certain hours all war-ships anchored in the Bosphorus, as well as the forts and arsenals, fire salutes, the roar and rattle of the great guns echoing among the hills of Europe and Asia, that here confront each other, with but a thousand yards of dancing blue waters between them. All along either lovely shore villages and splendid country-seats of wealthy pashas and Constantinople merchants dot the verdure-clad slopes. Two white marble kiosks of the Sultan are pointed out. The old castles of Europe and Asia face each other on opposite sides of the narrow channel. They were famous fortresses in their day, but, save as interesting relics of a bygone age, they are no longer of any use.

At Therapia are the summer residences of the different ambassadors, the English and French the most conspicuous. The extensive grounds of the former are most beautifully terraced, and evidently fit for the residence of royalty itself. Happy indeed is the Constantinopolitan whose income commands a summer villa in Therapia, or at any of the many desirable locations in plain view within this earthly paradise of blue waves and sunny slopes, and a yacht in which to wing his flight whenever and wherever fancy bids him go. In the glitter and glare of the mid-day sun the scene along the Bosphorus is lovely, yet its loveliness is plainly of the earth ; but as we return cityward in the eventide the dusky shadows of the gloaming settle over everything. As we gradually approach, the city seems half hidden behind a vaporous veil, as though, in imitation of thousands of its fair occupants, it were hiding its comeliness behind the *yashmak ;* the scores of tapering minarets, and the towers, and the masts of the crowded shipping of all nations rise above the mist, and line with delicate tracery the western sky, already painted in richest colors by the setting sun.

On Saturday morning, July 18th, the sound of martial music announces the arrival of the soldiers from Stamboul, to guard the streets through which the Sultan will pass on his way to a certain mosque to perform some ceremony in connection with the feast just over. At the designated place I find the streets already lined with Circassian cavalry and Ethiopian zouaves ; the latter in red and blue zouave costumes and immense turbans. Mounted *gendarmes* are driving civilians about, first in one direction and then in another, to try and get the streets cleared, occasionally fetching some unlucky wight in the threadbare shirt of the Galata plebe a stinging cut across the shoulders with short raw-hide whips—a glaring injustice that elicits not the slightest adverse criticism from the spectators, and nothing but silent contortions of face and body from the individual receiving the attention. I finally obtain a good place, where nothing but an open plank fence and a narrow plot of ground thinly set with shrubbery intervenes between me and the street leading from the palace. In a few minutes the approach of the Sultan is announced by the appearance of half a dozen Circassian outriders, who dash wildly down the streets, one behind the other, mounted on splendid dapple-gray chargers ; then come four close carriages, containing the Sultan's mother and leading ladies of the imperial harem, and a minute later appears a mounted guard, two abreast, keen-eyed fellows, riding slowly, and critically eyeing everybody and everything as they proceed ; behind them comes a gorgeously arrayed individual in a perfect blaze of gold braid and decorations, and close behind him follows the Sultan's carriage, surrounded by a small crowd of pedestrians and horsemen, who buzz around the imperial carriage like bees near a hive, the pedestrians especially dodging about hither and thither, hopping nimbly over fences, crossing gardens, etc., keeping pace with the carriage meanwhile, as though determined upon ferreting out and destroying anything in the shape of danger that may possibly be lurking along the route. My object of seeing the Sultan's face is gained ; but it is only a momentary glimpse, for besides the horsemen flitting around the carriage, an officer suddenly appears in front of my position and unrolls a broad scroll of paper with something printed on it, which he holds up. Whatever the scroll is, or the object of its display may be, the Sultan bows his acknowledgments, either to the scroll or to the officer holding it up.

Ere I am in the Ottoman capital a week, I have the opportunity of witnessing a fire, and the workings of the Constantinople Fire Department. While walking along Tramway Street, a hue and cry of "*yangoon var! yangoon var!*" (there is fire! there is fire!) is raised, and three barefooted men, dressed in the scantiest linen clothes, come charging pell-mell through the crowded streets, flourishing long brass hose-nozzles to clear the way; behind them comes a crowd of

Constantinople Fire Laddies.

about twenty others, similarly dressed, four of whom are bearing on their shoulders a primitive wooden pump, while others are carrying leathern water-buckets. They are trotting along at quite a lively pace, shouting and making much unnecessary commotion, and lastly comes their chief on horseback, cantering close at their heels, as though to keep the men well up to their pace. The crowds of pedestrians, who refrain from following after the firemen, and who scurried for the sidewalks at their approach, now resume their place in the middle of the street; but again the wild cry of "*yangoon var!*" resounds along the narrow street, and the same scene of citizens scuttling to the sidewalks, and a hurrying fire brigade followed by a noisy crowd of *gamins*, is enacted over again, as another and yet another of these primitive organizations go scooting swiftly past. It is said that these nimble-footed firemen do almost miraculous work, considering the material they have at command—an assertion which I think is not at all unlikely; but the wonder is that destructive fires are not much more frequent, when the fire department is evidently so inefficient. In addition to the regular police force and fire department, there is a system of night watchmen, called *bekjees*, who walk their respective beats throughout the night, carrying staves heavily shod with iron, with which they pound the flagstones with a resounding "thwack!"

Owing to the hilliness of the city and the roughness of the streets, much of the carrying business of the city is done by *hamals*, a class of sturdy-limbed men, who, I am told, are mostly Armenians. They wear a sort of pack-saddle, and carry loads the mere sight of which makes the average Westerner groan. For carrying such trifles as crates and hogsheads of crockery and glass-ware, and puncheons of rum, four *hamals* join strength at the ends of two stout poles. Scarcely less marvellous than the weights they carry is the apparent ease with which they balance tremendous loads, piled high up above them, it being no infrequent sight to see a stalwart *hamal* with a veritable Saratoga trunk, for size, on his back, with several smaller trunks and valises piled above it, making his way down Step Street, which is as much as many pedestrians can do to descend without carrying anything. One of these *hamals*, meandering along the street with six or seven hundred pounds of merchandise on his back, has the legal right—to say nothing of the evident moral right—to knock over any unloaded citizen who too tardily yields the way. From observations made on the spot, one cannot help thinking that there is no law in any country to be compared to this one, for simon-pure justice between man and man. These are most assuredly the strongest-backed and hardest working men I have seen anywhere. They are remarkably trustworthy and sure-footed, and their chief ambition, I am told, is to save sufficient money to return to the mountains and valleys of their native Armenia, where most of them have wives patiently awaiting their coming, and purchase a piece of land upon which to spend their declining years in ease and independence.

Far different is the daily lot of another *habitué* of the streets of this busy capital—large, pugnacious-looking rams, that occupy pretty much the same position in Turkish sporting circles that thoroughbred bull-dogs do in England, being kept by young Turks solely on account of their combative propensities and the facilities thereby afforded for gambling on the prowess of their favorite animals. At all hours of the day and evening the Constantinople sport may be met on the streets leading his woolly pet tenderly with a string, often carrying something in his hand to coax the ram along. The wool of these animals is frequently clipped to give them a fanciful aspect, the favorite clip being to produce a lion-like appearance, and they are always carefully guarded against the fell influence of the "evil eye" by a circlet of blue beads and pendent charms suspended from the neck. This latter precautionary measure is not confined to these hard-headed contestants for the championship of Galata, Pera, and Stamboul, however, but grace the necks of a goodly proportion of all animals met on the streets, notably the saddle-ponies, whose services are offered on certain street-corners to the public.

Occasionally one notices among the busy throngs a person wearing a turban of dark green; this distinguishing mark being the sole privilege of persons who have made the pilgrimage to Mecca. All true Mussulmans are supposed to make this pilgrimage some time during their lives, either in person or by employing a substitute to go in their stead, wealthy pashas sometimes paying quite large sums to some *imam* or other holy person to go as their proxy, for the holier the substitute the greater is supposed to be the benefit to the person sending him. Other persons are seen with turbans of a lighter shade of green than the returned Mecca pilgrims. These are people related in some way to the reigning sovereign.

Constantinople has its peculiar attractions as the great centre of the Mohammedan world as represented in the person of the Sultan, and during the five hundred years of the Ottoman dominion here, almost every Sultan and great personage has left behind him some interesting reminder of the times in which he lived and the wonderful possibilities of unlimited wealth and power. A stranger will scarcely show himself upon the streets ere he is discovered and accosted by a guide. From long experience these men can readily distinguish a new arrival, and they seldom make a mistake regarding his nationality. Their usual mode of self-introduction is to ap-

proach him, and ask if he is looking for the American consulate, or
the English post-office, as the case may be, and if the stranger
replies in the affirmative, to offer to show the way. Nothing is
mentioned about charges, and the uninitiated new arrival naturally
wonders what kind of a place he has got into, when, upon offering
what his experience in Western countries has taught him to consider a most liberal recompense, the guide shrugs his shoulders,
and tells you that he guided a gentleman the same distance yesterday and the gentleman gave—usually about double what you are
offering, no matter whether it be one cherik or half a dozen.

An afternoon ramble with a guide through Stamboul embraces
the Museum of Antiquities, the St. Sophia Mosque, the Costume
Museum, the thousand and one columns, the Tomb of Sultan Mahmoud, the world-renowned Stamboul Bazaar, the Pigeon Mosque,
the Saraka Tower, and the Tomb of Sultan Suliman I. Passing
over the Museum of Antiquities, which to the average observer is
very similar to a dozen other institutions of the kind, the visitor
very naturally approaches the portals of the St. Sophia Mosque
with expectations enlivened by having already read wondrous accounts of its magnificence and unapproachable grandeur. But, let
one's fancy riot as it will, there is small fear of being disappointed
in the "finest mosque in Constantinople." At the door one either
has to take off his shoes and go inside in stocking-feet, or, in addition to the entrance fee of two cheriks, "backsheesh" the attendant
for the use of a pair of overslippers. People with holes in their
socks and young men wearing boots three sizes too small are the
legitimate prey of the slipper-man, since the average human would
yield up almost his last piastre rather than promenade around in St.
Sophia with his big toe protruding through his foot-gear like a
mud-turtle's head, or run the risk of having to be hauled barefooted to his hotel in a hack, from the impossibility of putting his
boots on again. Devout Mussulmans are bowing their foreheads
down to the mat-covered floor in a dozen different parts of the
mosque as we enter; tired-looking pilgrims from a distance are
curled up in cool corners, happy in the privilege of peacefully
slumbering in the holy atmosphere of the great edifice they have,
perhaps, travelled hundreds of miles to see; a dozen half-naked
youngsters are clambering about the railings and otherwise disporting themselves after the manner of unrestrained juveniles everywhere—free to gambol about to their hearts' content, providing

they abstain from making a noise that would interfere with devotions.

Upon the marvellous mosaic ceiling of the great dome is a figure of the Virgin Mary, which the Turks have frequently tried to cover up by painting it over ; but paint as often as they will, the figure will not be concealed. On one of the upper galleries are the "Gate of Heaven" and "Gate of Hell," the former of which the Turks once tried their best to destroy; but every arm that ventured to raise a tool against it instantly became paralyzed, when the would-be destroyers naturally gave up the job. In giving the readers these facts I earnestly request them not to credit them to my personal account ; for, although earnestly believed in by a certain class of Christian natives here, I would prefer the responsibility for their truthfulness to rest on the broad shoulders of tradition rather than on mine.

The Turks never call the attention of visitors to these reminders of the religion of the infidels who built the structure, at such an enormous outlay of money and labor, little dreaming that it would become one of the chief glories of the Mohammedan world. But the door-keeper who follows visitors around never neglects to point out the shape of a human hand on the wall, too high up to be closely examined, and volunteer the intelligence that it is the imprint of the hand of the first Sultan who visited the mosque after the occupation of Constantinople by the Osmanlis. Perhaps, however, the Mussulman, in thus discriminating between the traditions of the Greek residents and the alleged hand-mark of the first Sultan, is actuated by a laudable desire to be truthful so far as possible ; for there is nothing improbable about the story of the hand-mark, inasmuch as a hole chipped in the masonry, an application of cement, and a pressure of the Sultan's hand against it before it hardened, give at once something for visitors to look at through future centuries and shake their heads incredulously about.

Not the least of the attractions are two monster wax candles, which, notwithstanding their lighting up at innumerable fasts and feasts, for the guide does not know how many years past, are still eight feet long by four in circumference ; but more wonderful than the monster wax candles, the brass tomb of Constantine's daughter, set in the wall over one of the massive doors, the Sultan's hand-mark, the figure of the Virgin Mary, and the green columns brought from Baalbec ; above everything else is the wonderful mosaic-work. The mighty dome and the whole vast ceiling are mosaic-work in which tiny squares of blue, green, and gold crystal are made to work out patterns. The squares used are tiny particles having not over a quarter-inch surface ; and the amount of labor and the expense in covering the vast ceiling of this tremendous structure with incomputable myriads of these small particles fairly stagger any attempt at comprehension.

An interesting hour can next be spent in the Costume Museum, where life-size figures represent the varied and most decidedly picturesque costumes of the different officials of the Ottoman capital in previous ages, the janizaries, and natives of the different provinces. Some of the head-gear in vogue at Constantinople before the fez were tremendous affairs, but the fez is certainly a step too far in the opposite direction, being several degrees more uncomfortable than nothing in the broiling sun ; the fez makes no pretence of shading the eyes, and excludes every particle of air from the scalp. The thousand and one columns are in an ancient Greek reservoir that formerly supplied all Stamboul with water. The columns number but three hundred and thirty-four in reality, but each column is in three parts, and by stretching the point we have the fanciful "thousand-and-one." The reservoir is reached by descending a flight of stone steps ; it is filled in with earth up to the upper half of the second tier of columns, so that the lower tier is buried altogether. This filling up was done in the days of the janizaries, as it was found that those frisky warriors were carrying their well-known theory of "right being might and the Devil take the weakest" to the extent of robbing unprotected people who ventured to pass this vicinity after dark, and then consigning them to the dark depths of the deserted reservoir. The reservoir is now occupied during the day by a number of Jewish silk-weavers, who work here on account of the dampness and coolness being beneficial to the silk.

The tomb of Mahmoud is next visited on the way to the Bazaar. The several coffins of the Sultan Mahmoud and his Sultana and princesses are surrounded by massive railings of pure silver ; monster wax candles are standing at the head and foot of each coffin, in curiously wrought candlesticks of solid silver that must weigh a hundred pounds each at least ; ranged around the room are silver caskets, inlaid with mother-of-pearl, in which rare illumined copies of the Koran are carefully kept, the attendant who opened one for my inspection using a silk pocket-handkerchief to turn the leaves.

The Stamboul Bazaar well deserves its renown, since there is
nothing else of its kind in the whole world to compare with it. Its
labyrinth of little stalls and shops if joined together in one straight
line would extend for miles ; and a whole day might be spent quite
profitably in wandering around, watching the busy scenes of bar-
gaining and manufacturing. Here, in this bewildering maze of
buying and selling, the peculiar life of the Orient can be seen to
perfection ; the "mysterious veiled lady" of the East is seen
thronging the narrow traffic-ways and seated in every stall ; water-
venders and venders of *carpooses* (water-melons) and a score of different eatables are meandering through. Here, if your guide be an
honest fellow, he can pilot you into stuffy little holes full of antique articles of every description, where genuine bargains can be
picked up; or, if he be dishonest, and in league with equally dishonest tricksters, whose places are antiquaries only in name, he can
lead you where everything is basest imitation. In the former case,
if anything is purchased he comes in for a small and not undeserved commission from the shopkeeper, and in the latter for perhaps as much as thirty per cent. I am told that one of these
guides, when escorting a party of tourists with plenty of money
to spend and no knowledge whatever of the real value or genuine-
ness of antique articles, often makes as much as ten or fifteen pounds
sterling a day commission.

On the way from the Bazaar we call at the Pigeon Mosque, so
called on account of being the resort of thousands of pigeons, that
have become quite tame from being constantly fed by visitors and
surrounded by human beings. A woman has charge of a store of
seeds and grain, and visitors purchase a handful for ten paras and
throw to the pigeons, who flock around fearlessly in the general
scramble for the food. At any hour of the day Mussulman ladies
may be seen here feeding the pigeons for the amusement of their
children. From the Pigeon Mosque we ascend the Saraka Tower,
the great watch-tower of Stamboul, from the summit of which the
news of a fire in any part of the city is signalled, by suspending
huge frame-work balls covered with canvas from the ends of projecting poles in the day, and lights at night. . Constant watch and
ward is kept over the city below by men snugly housed in quarters
near the summit, who, in addition to their duties as watchmen,
turn an honest cherik occasionally by supplying cups of coffee to
visitors.

No fairer site ever greeted human vision than the prospect from the Tower of Saraka. Stamboul, Galata, Pera, and Scutari, with every suburban village and resort for many a mile around, can be seen to perfection from the commanding height of Saraka Tower. The guide can here point out every building of interest in Stamboul—the broad area of roof beneath which the busy scenes of Stamboul Bazaar are enacted from day to day, the great Persian khan, the different mosques, the Sultan's palaces at Pera, the Imperial kiosks up the Bosphorus, the old Grecian aqueduct, along which the water for supplying the great reservoir of the thousand and one columns used to be conducted, the old city walls, and scores of other interesting objects too numerous to mention here. On the opposite hill, across the Golden Horn, Galata Watch-tower points skyward above the mosques and houses of Galata and Pera. The two bridges connecting Stamboul and Galata are seen thronged with busy traffic ; a forest of masts and spars is ranged all along the Golden Horn ; steamboats are plying hither and thither across the Bosphorus ; the American cruiser Quinnebaug rides at anchor opposite the Imperial water-side palace ; the blue waters of the Sea of Marmora and the Gulf of Ismidt are dotted here and there with snowy sails or lined with the smoke of steamships ; all combined to make the most lovely panorama imaginable, and to which the coastwise hills and more lofty mountains of Asia Minor in the distance form a most appropriate background.

From this vantage-point the guide will not neglect whetting the curiosity of his charge for more sight-seeing by pointing out everything that he imagines would be interesting ; he points out a hill above Scutari, whence, he says, a splendid view can be had of "all Asia Minor," and "we could walk there and back in half a day, or go quicker with horses or donkeys ;" he reminds you that to-morrow is the day for the howling dervishes in Scutari, and tells you that by starting at one we can walk out to the English cemetery, and return to Scutari in time for the howling dervishes at four o'clock, and manages altogether to get his employer interested in a programme, which, if carried out, would guarantee him employment for the next week. On the way back to Galata we visit the tomb of Sulieman I., the most magnificent tomb in Stamboul. Here, before the coffins of Sulieman I., Sulieman II., and his brother Ahmed, are monster wax candles, that have stood sentry here for three hundred and fifty years ; and the mosaic dome of the beautiful edifice is studded with what are popularly believed to be genuine diamonds, that twinkle down on the curiously gazing visitor like stars from a miniature heaven. The attendant tells the guide, in answer to an inquiry from me, that no one living knows whether they are genuine diamonds or not, for never, since the day it was finished, over three centuries and a half ago, has anyone been permitted to go up and examine them. The edifice was so perfectly and solidly built in the beginning, that no repairs of any kind have ever been necessary ; and it looks almost like a new building to-day.

Not being able to spare the time for visiting all the objects of interest enumerated by the guide, I elect to see the howling dervishes as the most interesting among them. Accordingly we take the ferry-boat across to Scutari on Thursday afternoon in time to visit the English cemetery before the dervishes begin their peculiar services. We pass through one of the largest Mussulman cemeteries of Constantinople, a bewildering area of tombstones beneath a grove of dark cypresses, so crowded and disorderly that the oldest gravestones seem to have been pushed down, or on one side, to make room for others of a later generation, and these again for still others. In happy comparison to the disordered area of crowded tombstones in the Mohammedan graveyard is the English cemetery, where the soldiers who died at the Scutari hospital during the Crimean war were buried, and the English residents of Constantinople now bury their dead. The situation of the English cemetery is a charming spot, on a sloping bluff, washed by the waters of the Bosphorus, where the requiem of the murmuring waves is perpetually sung for the brave fellows interred there. An Englishman has charge ; and after being in Turkey a month it is really quite refreshing to visit this cemetery, and note the scrupulous neatness of the grounds. The keeper must be industry personified, for he scarcely permits a dead leaf to escape his notice ; and the four angels beaming down upon the grounds from the national monument erected by England, in memory of the Crimean heroes, were they real visitors from the better land, could doubtless give a good account of his stewardship.

The howling dervishes have already begun to howl as we open the portals leading into their place of worship by the influence of a cherik placed in the open palm of a sable eunuch at the door ; but it is only the overture, for it is half an hour later when the inter-

16

esting part of the programme begins. The first hour seems to be devoted to preliminary meditations and comparatively quiet ceremonies; but the cruel-looking instruments of self-flagellation hanging on the wall, and a choice and complete assortment of drums and other noise-producing but unmelodious instruments, remind the visitor that he is in the presence of a peculiar people. Sheepskin mats almost cover the floor of the room, which is kept scrupulously clean, presumably to guard against the worshippers soiling their lips whenever they kiss the floor, a ceremony which they perform quite frequently during the first hour'; and everyone who presumes to tread within that holy precinct removes his over-shoes, if he is wearing any, otherwise he enters in his stockings.

At five o'clock the excitement begins; thirty or forty men are ranged around one end of the room, bowing themselves about most violently, and keeping time to the movements of their bodies with shouts of "*Allah! Allah!*" and then branching off into a howling chorus of Mussulman supplications, that, unintelligible as they are to the infidel ear, are not altogether devoid of melody in the expression, the Turkish language abounding in words in which there is a world of mellifluousness. A dancing dervish, who has been patiently awaiting at the inner gate, now receives a nod of permission from the priest, and, after laying aside an outer garment, waltzes nimbly into the room, and straightway begins spinning round like a ballet-dancer in Italian opera, his arms extended, his long skirt forming a complete circle around him as he revolves, and his eyes fixed with a determined gaze into vacancy. Among the howlers is a negro, who is six feet three at least, not in his socks, but in the finest pair of under-shoes in the room, and whether it be in the ceremony of kissing the floor, knocking foreheads against the same, kissing the hand of the priest, or in the howling and bodily contortions, this towering son of Ham performs his part with a grace that brings him conspicuously to the fore in this respect. But as the contortions gradually become more violent, and the cry of "*Allah akbar! Allah hai!*" degenerates into violent grunts of "h-o-o-o-o-a-hoo-hoo," the half-exhausted devotees fling aside everything but a white shroud, and the perspiration fairly streams off them, from such violent exercise in the hot weather and close atmosphere of the small room. The exercises make rapid inroads upon the tall negro's powers of endurance, and he steps to one side and takes a breathing-spell of five minutes, after which he resumes his place again,

and, in spite of the ever-increasing violence of both lung and muscular exercise, and the extra exertion imposed by his great height,
he keeps it up heroically to the end.

For twenty-five minutes by my watch, the one lone dancing
dervish—who appears to be a visitor merely, but is accorded the
brotherly privilege of whirling round in silence while the others
howl—spins round and round like a tireless top, making not the
slightest sound, spinning in a long, persevering, continuous whirl,
as though determined to prove himself holier than the howlers, by
spinning longer than they can keep up their howling—a fair test
of fanatical endurance, so to speak.   One cannot help admiring the
religious fervor and determination of purpose that impel this lone
figure silently around on his axis for twenty-five minutes, at a speed
that would upset the equilibrium of anybody but a dancing dervish
in thirty seconds ; and there is something really heroic in the
manner in which he at last suddenly stops, and, without uttering a
sound or betraying any sense of dizziness whatever from the exercise, puts on his coat again and departs in silence, conscious, no
doubt, of being a holier person than all the howlers put together,
even though they are still keeping it up.   As unmistakable signals
of distress are involuntarily hoisted by the violently exercising
devotees, and the weaker ones quietly fall out of line, and the military precision of the twists of body and bobbing and jerking of
head begins to lose something of its regularity, the six " encoura-
gers," ranged on sheep-skins before the line of howling men, like
non-commissioned officers before a squad of new recruits, increase
their encouraging cries of " Allah ! Allah akbar ! " as though fearful
that the din might subside, on account of the several already exhausted organs of articulation, unless they chimed in more lustily
and helped to swell the volume.

Little children now come trooping in, seeking with eager anticipation the happy privilege of being ranged along the floor like
sardines in a tin box, and having the priest walk along their bodies, stepping from one to the other along the row, and returning
the same way, while two assistants steady him by holding his hands.
In the case of the smaller children, the priest considerately steps
on their thighs, to avoid throwing their internal apparatus out of
gear ; but if the recipient of his holy attentions is, in his estimation,
strong enough to run the risk, he steps square on their backs.
The little things jump up as sprightly as may be, kiss the priest's

hand fervently, and go trooping out of the door, apparently well pleased with the novel performance. Finally human nature can endure it no longer, and the performance terminates in a long, despairing wail of "*Allah! Allah! Allah!*" The exhausted devotees, soaked wet with perspiration, step forward, and receive what I take to be rather an inadequate reward for what they have been subjecting themselves to—viz., the privilege of kissing the priest's already much-kissed hand, and at 5.45 P.M. the performance is over. I take my departure in time to catch the six o'clock boat for Galata, well satisfied with the finest show I ever saw for a cherik.

I have already made mention of there being many beautiful sea-side places to which Constantinopolitans resort on Sundays and holidays, and among them all there is no lovelier spot than the island of Prinkipo, one of the Prince's Islands group, situated some twelve miles from Constantinople, down the Gulf of Ismidt. Shelton Bey (Colonel Shelton), an English gentleman, who superintends the Sultan's cannon-foundry at Tophana, and the well-known author of Shelton's "Mechanic's Guide," owns the finest steam-yacht on the Bosphorus, and three Sundays out of the five I remain here, this gentleman and his excellent lady kindly invite me to visit Prinkipo with them for the day.

On the way over we usually race with the regular passenger steamer, and as the Bey's yacht is no plaything for size and speed, we generally manage to keep close enough to amuse ourselves with the comments on the beauty and speed of our little craft from the crowded deck of the other boat. Sometimes a very distinguished person or two is aboard the yacht with our little company, personages known to the Bey, who having arrived on the passenger-boat, accept invitations for a cruise around the island, or to dine aboard the yacht as she rides at anchor before the town. But the advent of the "Americanish Velocipediste" and his glistening machine, a wonderful thing that Prinkipo never saw the like of before, creates a genuine sensation, and becomes the subject of a nine-days' wonder. Prinkipo is a delightful gossipy island, occupied during the summer by the families of wealthy Constantinopolitans and leading business men, who go to and fro daily between the little island and the city on the passenger-boats regularly plying between them, and is visited every Sunday by crowds in search of the health and pleasure afforded by a day's outing.

While here at Constantinople I received by mail from America a

Prinkipo the Beautiful.

Butcher spoke cyclometer, and on the second visit to Prinkipo I measured the road which has been made around half the island ; the distance is four English miles and a fraction. The road was built by refugees employed by the Sultan during the last Russo-Turkish war, and is a very good one ; for part of the distance it leads between splendid villas, on the verandas of which are seen groups of the wealth and beauty of the Osmanli capital, Armenians, Greeks, and Turks—the latter ladies sometimes take the privilege of dispensing with the *yashmak* during their visits to the comparative seclusion of Prinkipo villas—with quite a sprinkling of English and Europeans. The sort of impression made upon the imaginations of Prinkipo young ladies by the bicycle is apparent from the following comment made by a bevy of them confidentially to Shelton Bey, and kindly written out by him, together with the English interpretation thereof. The Prinkipo ladies' compliment to the first bicycle rider visiting their beautiful island is : " *O Bizdan kayáore ghyurulduzug em nezaketli sadi bir dakika utchum ghyuriorus nazaman bir dah backiorus O bittum gitmush.*" (He glides noiselessly and gracefully past ; we see him only for a moment ; when we look again he is quite gone) The men are of course less poetical, their ideas running more to the practical side of the possibilities of the new arrival, and they comment as follows : " *Onum beyghir hich-bir-shèy yemiore hich-bir-shèy ichmiore hich yorumliore ma sheitan gibi ghitiore.*" (His horse, he eats nothing, drinks nothing, never gets tired, and goes like the very devil.) It is but fair to add, however, that any bold Occidental contemplating making a descent on Prinkipo with a " sociable " with a view to delightful moonlight rides with the fair authors of the above poetic contribution will find himself " all at sea " upon his arrival, unless he brings a three-seated machine, so that the mamma can be accommodated with a seat behind, since the daughters of Prinkipo society never wander forth by moonlight, or any other light, unless thus accompanied, or by some equally staid and solicitous relative.

For the Asiatic tour I have invented a " bicycle tent "—a handy contrivance by which the bicycle is made to answer the place of tent poles. The material used is fine, strong sheeting, that will roll up into a small space, and to make it thoroughly water-proof, I have dressed it with boiled linseed oil. My footgear henceforth will be Circassian moccasins, with the pointed toes sticking up like the prow of a Venetian galley. I have had a pair made to order by a native shoemaker in Galata, and, for either walking or pedalling, they are ahead of any foot-gear I ever wore; they are as easy as a three-year-old glove, and last indefinitely, and for fancifulness in appearance, the shoes of civilization are nowhere.

Three days before starting out I receive friendly warnings from both the English and American consul that Turkey in Asia is infested with brigands, the former going the length of saying that if he had the power he would refuse me permission to meander forth upon so risky an undertaking. I have every confidence, however, that the bicycle will prove an effectual safeguard against any undue familiarity on the part of these frisky citizens. Since reaching Constantinople the papers here have published accounts of recent exploits accomplished by brigands near Eski Baba. I have little doubt but that more than one brigand was among my highly interested audiences there on that memorable Sunday.

The Turkish authorities. seem to have made themselves quite familiar with my intentions, and upon making application for a *teskéré* (Turkish passport) they required me to specify, as far as possible, the precise route

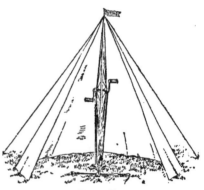

Bicycle Tent.

I intend traversing from Scutari to Ismidt, Angora, Erzeroum, and beyond, to the Persian frontier. An English gentleman who has lately travelled through Persia and the Caucasus tells me that the Persians are quite agreeable people, their only fault being the one common failing of the East: a disposition to charge whatever they think it possible to obtain for anything. The Circassians seem to be the great bugbear in Asiatic Turkey. I am told that once I get beyond the country that these people range over—who are regarded as a sort of natural and half-privileged freebooters—I shall be reasonably safe from molestation. It is a common thing in Constantinople when two men are quarrelling for one to threaten to give a Circassian a couple of medjedis to kill the other. The Circassian is to Turkey what the mythical "bogie" is to England; mothers threaten undutiful daughters, fathers unruly sons, and everybody their enemies generally, with the Circassian, who, however, unlike the " bogie " of the English household, is a real material presence, popularly understood to be ready for any devilment a person may hire him to do.

The bull-dog revolver, under the protecting presence of which I have travelled thus far, has to be abandoned here at Constantinople, having proved itself quite a wayward weapon since it came from the gunsmith's hands in Vienna, who seemed to have upset the internal mechanism in some mysterious manner while boring out the chambers a trifle to accommodate European cartridges. My experience thus far is that a revolver has been more ornamental than useful ; but I am now about penetrating far different countries to any I have yet traversed. Plenty of excellently finished German imitations of the Smith & Wesson revolver are found in the magazines of Constantinople ; but, apart from it being the duty of every Englishman or American to discourage, as far as his power goes, the unscrupulousness of German manufacturers in placing upon foreign markets what are, as far as outward appearance goes, the exact counterparts of our own goods, for half the money, a genuine American revolver is a different weapon from its would-be imitators, and I hesitate not to pay the price for the genuine article. Remembering the narrow escape on several occasions of having the bull-dog confiscated by the Turkish *gendarmerie,* and having heard, moreover, in Constantinople, that the same class of officials in Turkey in Asia will most assuredly want to confiscate the Smith & Wesson as a matter of private speculation and enterprise, I obtain through the British consul a *teskéré* giving me special permission to carry a revolver. Subsequent events, however, proved this precaution to be unnecessary, for a more courteous, obliging, and gentlemanly set of fellows, according to their enlightenment, I never met anywhere, than the government officials of Asiatic Turkey.

Were I to make the simple statement that I am starting into Asia with a pair of knee-breeches that are worth fourteen English pounds (about sixty-eight dollars) and offer no further explanation, I should, in all probability, be accused of a high order of prevarication. Nevertheless, such is the fact ; for among other subterfuges to outwit possible brigands, and kindred citizens, I have made cloth-covered buttons out of Turkish liras (eighteen shillings English), and sewed them on in place of ordinary buttons. Pantaloon buttons at $54 a dozen are a luxury that my wildest dreams

never soared to before, and I am afraid many a thrifty person will
condemn me for extravagance ; but the " splendor " of the Orient
demands it ; and the extreme handiness of being able to cut off a
button, and with it buy provisions enough to load down a mule,
would be all the better appreciated if one had just been released
from the hands of the Philistines with nothing but his clothes—

A Notice of my Journey in the Sultan's Official Organ.

and buttons—and the bicycle. With these things left to him, one
could afford to regard the whole matter as a joke, expensive, perhaps, but nevertheless a joke compared with what might have been.

The Constantinople papers have advertised me to start on Monday, August 10th, " direct from Scutari." I have received friendly
warnings from several Constantinople gentlemen, that a band of
brigands, under the leadership of an enterprising chief named

Mahmoud Pehlivan, operating about thirty miles out of Scutari, have beyond a doubt received intelligence of this fact from spies here in the city, and, to avoid running direct into the lion's mouth, I decide to make the start from Ismidt, about twenty-five miles beyond their rendezvous. A Greek gentleman, who is a British subject, a Mr. J. T. Corpi, whom I have met here, fell into the hands of this same gang, and being known to them as a wealthy gentleman, had to fork over £3,000 ransom; and he says I would be in great danger of molestation in venturing from Scutari to Ismidt after my intention to do so has been published.

CHAPTER X.

THE START THROUGH ASIA.

In addition to a cycler's ordinary outfit and the before-mentioned small wedge tent I provide myself with a few extra spokes, a cake of tire cement, and an extra tire for the rear wheel. This latter, together with twenty yards of small, stout rope, I wrap snugly around the front axle ; the tent and spare underclothing, a box of revolver cartridges, and a small bottle of sewing-machine oil are consigned to a luggage-carrier behind ; while my writing materials, a few medicines and small sundries find a repository in my White-house sole-leather case on a Lamson carrier, which also accommodates a suit of gossamer rubber.

The result of my study of the various routes through Asia is a determination to push on to Teheran, the capital of Persia, and there spend the approaching winter, completing my journey to the Pacific next season.

Accordingly nine o'clock on Monday morning, August 10th, finds me aboard the little Turkish steamer that plies semi-weekly between Ismidt and the Ottoman capital, my bicycle, as usual, the centre of a crowd of wondering Orientals. This Ismidt steamer, with its motley crowd of passengers, presents a scene that upholds with more eloquence than words Constantinople's claim of being the most cosmopolitan city in the world ; and a casual observer, judging only from the evidence aboard the boat, would pronounce it also the most democratic. There appears to be no first, second, or third class ; everybody pays the same fare, and everybody wanders at his own sweet will into every nook and corner of the upper deck, perches himself on top of the paddle-boxes, loafs on the pilot's bridge, or reclines among the miscellaneous assortment of freight piled up in a confused heap on the fore-deck; in short, everybody seems perfectly free to follow the bent of his inclinations, except to penetrate behind the scenes of the aftmost deck, where, carefully hidden from the rude gaze of the male passengers by a canvas partition, the Moslem ladies have their little world of gossip and coffee, and fragrant cigarettes. Every public conveyance in the Orient has this walled-off retreat, in which Osmanli fair ones can remove their *yashmaks*, smoke cigarettes, and comport themselves with as much freedom as though in the seclusion of their apartments at home.

Greek and Armenian ladies mingle with the main-deck passengers, however, the picturesque costumes of the former contributing not a little to the general Oriental effect of the scene. The dress of the Armenian ladies differs but little from Western costumes, and their deportment would wreathe the benign countenance of the Lord Chamberlain with a serene smile of approval; but the minds and inclinations of the gentle Hellenic dames seem to run in rather a contrary channel. Singly, in twos, or in cosey, confidential coteries, arm in arm, they promenade here and there, saying little to each other or to anybody else. By the picturesqueness of their apparel and their seemingly bold demeanor they attract to themselves more than their just share of attention; but with well-feigned ignorance of this they divide most of their time and attention between rolling cigarettes and smoking them. Their heads are bound with jaunty silk handkerchiefs; they wear rakish-looking short jackets, down the back of which their luxuriant black hair dangles in two tresses; but the crowning masterpiece of their costume is that wonderful garment which is neither petticoat nor pantaloons, and which can be most properly described as "indescribable," which tends to give the wearer rather an unfeminine appearance, and is not to be compared with the really sensible and not unpicturesque nether garment of a Turkish lady.

The male companions of these Greek women are not a bit behind them in the matter of gay colors and startling surprises of the Levantine clothier's art, for they likewise are in all the bravery of holiday attire. There is quite a number of them aboard, and they now appear at their best, for they are going to take part in wedding festivities at one of the little Greek villages that nestle amid the vine-clad slopes along the coast—white-painted villages, that from the deck of the moving steamer look as though they have been placed here and there by nature's artistic hand for the sole purpose of embellishing the lovely green frame-work that surrounds the blue waters of the Ismidt Gulf. Several of these merry-makers enliven the passing hours with music and dancing, to the delight of a numerous audience, while a second ever-changing but never-dispersing audience is gathered around the bicycle.

The verbal comments and Solomon-like opinions, given in expressive pantomime, of this latter garrulous gathering concerning the machine and myself, I can of course but partly understand; but occasionally some wiseacre suddenly becomes inflated with the idea that he has succeeded in unravelling the knotty problem, and forthwith proceeds to explain, for the edification of his fellow-passengers, the *modus operandi* of riding it, supplementing his words by the most extraordinary gestures. The audience is usually very attentive and highly interested in these explanations, and may be considerably enlightened by their self-constituted tutors, whose sole advantage over their auditors, so far as bicycles are concerned, consists simply in a belief in the superiority of their own particular powers of penetration. But to the only person aboard the steamer who really does know anything at all about the subject, the chief end of their exposition seems to be gained when they have duly impressed upon the minds of their hearers that the bicycle is to ride on, and that it goes at a rate of speed quite beyond the comprehension of their—the auditors'—minds; "*Bin, bin, bin! Chu, chu, chu! Haidi, haidi, haidi!*" being repeated with a vehemence that is intended to impress upon them little less than flying-Dutchman speed.

The deck of a Constantinople steamer affords splendid opportunity for character study, and the Ismidt packet is no exception. Nearly every person aboard has some characteristic, peculiar and distinct from any of the others. At intervals of about fifteen minutes a couple of Armenians, bare-footed, bare-legged, and ragged, clamber with much difficulty and scraping of shins over a large pile of empty chicken-crates to visit one particular crate. Their collective baggage consists of a thin, half-grown chicken tied by both feet to a small bag of barley, which is to prepare it for the useful but inglorious end of all chickendom. They have imprisoned their unhappy charge in a crate that is most difficult to get at. Why they didn't put it in one of the nearer crates, what their object is in climbing up to visit it so frequently, and why they always go together, are problems of the knottiest kind.

A far less difficult riddle is the case of a middle-aged man, whose costume and avocation explain nothing, save that he is not an Osmanli. He is a passenger homeward bound to one of the coast vil-

lages, and he constantly circulates among the crowd with a basket
of water-melons, which he has brought aboard "on spec," to vend
among his fellow-passengers, hoping thereby to gain sufficient to
defray the cost of his passage.   Seated on whatever they can find to
perch upon, near the canvas partition, all unmoved by the gay and
stirring scenes before them, is a group of Mussulman pilgrims from
some interior town, returning from a pilgrimage to Stamboul—
fine-looking Osmanli graybeards, whose haughty reserve not even
the bicycle is able to completely overcome, although it proves more
efficacious in subduing it and waking them out of their habitual

contemplative attitude than anything else aboard.   Two of these
men are of magnificent physique ; their black eyes, rather full lips,
and swarthy skins betraying Arab blood.   In addition to the long
daggers and antiquated pistols so universally worn in the Orient,
they are armed with fine, large, pearl-handled revolvers, and they
sit cross-legged, smoking cigarette after cigarette in silent meditation, paying no heed even to the merry music and the dancing of the
Greeks.

    At Jelova, the first village the steamer halts at, a coupleof
zaptiehs come aboard with two prisoners whom they are convey-

ing to Ismidt. These men are lower-class criminals, and their wretched appearance betrays the utter absence of hygienic considerations on the part of the Turkish prison authorities ; they evidently have had no cause to complain of any harsh measures for the enforcement of personal cleanliness. Their foot-gear consists of pieces of rawhide, fastened on with odds and ends of string ; and pieces of coarse sacking tacked on to what were once clothes barely suffice to cover their nakedness ; bare-headed—their bushy hair has not for months felt the smoothing influence of a comb, and their hands and faces look as if they had just endured a seven-years' famine of soap and water. This latter feature is a sure sign that they are not Turks, for prisoners are most likely allowed full liberty to keep themselves clean, and a Turk would at least have come out into the world with a clean face.

The *zaptiehs* squat down together and smoke cigarettes, and allow their charges full liberty to roam wheresoever they will while on board, and the two prisoners, to all appearances perfectly oblivious of their rags, filth, and the degradation of their position, mingle freely with the passengers ; and, as they move about, asking and answering questions, I look in vain among the latter for any sign of the spirit of social Pharisaism that in a Western crowd would have kept them at a distance. Both these men have every appearance of being the lowest of criminals—men capable of any deed in the calendar within their mental and physical capacities ; they may even be members of the very gang I am taking this steamer to avoid ; but nobody seems to either pity or condemn them ; everybody acts toward them precisely as they act toward each other. Perhaps in no other country in the world does this social and moral apathy obtain among the masses to such a degree as in Turkey.

While we lie to for a few minutes to disembark passengers at the village where the before-mentioned wedding festivities are in progress, four of the seven imperturbable Osmanlis actually arise from the one position they have occupied unmoved since coming aboard, and follow me to the foredeck, in order to be present while I explain the workings and mechanism of the bicycle to some Armenian students of Roberts College, who can speak a certain amount of English. Having listened to my explanations without understanding a word, and, without condescending to question the Armenians, they survey the machine some minutes in silence and then return to their former positions, their cigarettes, and their meditations, paying not the slightest heed to several *caique* loads of Greek merry-makers who have rowed out to meet the new arrivals, and are paddling around the steamer, filling the air with music. Finding that there is someone aboard that can converse with me, the Greeks, desirous of seeing the bicycle in action, and of introducing a novelty into the festivities of the evening, ask me to come ashore and be their guest until the arrival of the next Ismidt boat—a matter of three days. Offer declined with thanks, but not without reluctance, for these Greek merry-makings are well worth seeing.

The Ismidt packet, like everything else in Turkey, moves at a snail's pace, and although we got under way in something less than an hour after the advertised starting-time, which, for Turkey, is quite commendable promptness, and the distance is but fifty-five miles, we call at a number of villages *en route*, and it is 6 P.M. when we tie up at the Ismidt wharf.

" Five piastres, Effendi, " says the ticket-collector, as, after waiting till the crowd has passed the gang-plank, I follow with the bicycle and hand him my ticket.

" What are the five piastres for ? " I ask.   For answer, he points to my wheel.

" Baggage, " I explain.

" Baggage yoke, cargo, " he replies ; and I have to pay it. The fact is, that, never having seen a bicycle before, he don't know whether it is cargo or baggage ; but whenever a Turkish official has no precedent to follow, he takes care to be on the right side in case there is any money to be collected ; otherwise he is not apt to be so particular.   This is, however, rather a matter of private concern than of zealousness in the performance of his official duties ; the possibilities of peculation are ever before him.

While satisfying the claim of the ticket-collector a deck-hand comes forward and, pointing to the bicycle, blandly asks me for backsheesh. He asks, not because he has put a finger to the machine, or been asked to do so, but, being a thoughtful, far-sighted youth, he is looking out for the future. The bicycle is something he never saw on his boat before ; but the idea that these things may now become common among the passengers wanders through his mind, and that obtaining backsheesh on this particular occasion will establish a precedent that may be very handy hereafter ; so he makes a most respectful salaam, calls me " Bey Effendi, " and smilingly requests two

piastres backsheesh. After him comes the passport officer, who, besides the *teskeri* for myself, demands a special passport for the machine. He likewise is in a puzzle (it don't take much, by the by, to
puzzle the brains of a Turkish official), because the bicycle is something he has had no previous dealings with ; but as this is a matter
in which finances play no legitimate part—though probably his demand for a passport is made for no other purpose than that of getting backsheesh—a vigorous protest, backed up by the unanimous,
and most certainly vociferous, support of a crowd of wharf-loafers,
and my fellow-passengers, who, having disembarked, are waiting
patiently for me to come and ride down the street, either overrules
or overawes the officer and secures my relief.

Impatient at consuming a whole day in reaching Ismidt, I have
been thinking of taking to the road immediately upon landing,
and continuing till dark, taking my chances of reaching some suitable stopping-place for the night. But the good people of Ismidt
raise their voices in protest against what they professedly regard as
a rash and dangerous proposition. As I evince a disposition to override their well-meant interference and pull out, they hurriedly send
for a Frenchman, who can speak sufficient English to make himself
intelligible. Speaking for himself, and acting as interpreter in
echoing the words and sentiments of the others, the Frenchman
straightway warns me not to start into the interior so late in the day,
and run the risk of getting benighted in the brush ; for " Much very
bad people, very bad people! are between Ismidt and Angora ;
Circassians plenty, " he says, adding that the worst characters are
near Ismidt, and that the nearer I get to Angora the better I shall
find the people. As by this time the sun is already setting behind
the hills, I conclude that an early start in the morning will, after all,
be the most sensible course.

During the last Russo-Turkish war thousands of Circassian refugees migrated to this part of Asia Minor. Having a restless, roving disposition, that unfits them for the laborious and uneventful
life of a husbandman, many of them remain even to the present day
loafers about the villages, maintaining themselves nobody seems to
know how. The belief appears to be unanimous, however, that
they are capable of any deviltry under the sun, and that, while
their great specialty and favorite occupation is stealing horses, if
this becomes slack or unprofitable, or even for the sake of a little
pleasant variety, these freebooters from the Caucasus have no hes-

17

itation about turning highwaymen whenever a tempting occasion offers. All sorts of advice about the best way to avoid being robbed is volunteered by the people of Ismidt. My watch-chain, L. A. W. badge, and everything that appears of any value, they tell me, must be kept strictly out of sight, so as not to excite the latent cupidity of such Circassians as I meet on the road or in the villages. Some advocate the plan of adorning my coat with Turkish official buttons, shoulder-straps, and trappings, to make myself look like a government officer ; others think it would be best to rig myself up as a full-blown *zaptieh*, with whom, of course, neither Circassian nor any other guilty person would attempt to interfere.

To these latter suggestions I point out that, while they are very good, especially the *zaptieh* idea, so far as warding off Circassians is concerned, my adoption of a uniform would most certainly get me into hot water with the military authorities of every town and village, owing to my ignorance of the vernacular, and cause me no end of vexatious delay. To this the quick-witted Frenchman replies by at once offering to go with me to the resident pasha, explain the matter to him, and get a letter permitting me to wear the uniform ; which offer I gently but firmly decline, being secretly of the opinion that these excessive precautions are all unnecessary. From the time I left Hungary I have been warned so persistently of danger ahead, and have so far met nothing really dangerous, that I am getting sceptical about there being anything like the risk people seem to think. Without being blind to the fact that there is a certain amount of danger in travelling alone through a country where it is the universal custom either to travel in company or to take a guard, I feel quite confident that the extreme novelty of my conveyance will make so profound an impression on the Asiatic mind that, even did they know that my buttons are gold coins of the realm, they would hesitate seriously to molest me. From past observations among people seeing the bicycle ridden for the first time, I believe that with a hundred yards of smooth road it is quite possible for a cycler to ride his way into the good graces of the worst gang of freebooters in Asia.

Having decided to remain here over-night, I seek the accommodation of a rudely comfortable hotel, kept by an Armenian, where, at the supper-table, I am first made acquainted with the Asiatic dish called "*pillau*," that is destined to form no inconsiderable part of my daily bill of fare for several weeks. *Pillau* is a dish that is met with in one disguise or another all over Asia. With a foundation of boiled rice, it receives a variety of other compounds, the nature of which will appear as they enter into my daily experiences. In deference to the limited knowledge of each other's language possessed by myself and the proprietor, I am invited into the cookhouse and permitted to take a peep at the contents of several dif-

My Bill of Fare.

ferent pots and kettles simmering over a slow fire in a sort of brick trench, to point out to the waiter such dishes as I think I shall like. Failing to find among the assortment any familiar acquaintances, I try the *pillau*, and find it quite palatable, preferring it to anything else the house affords.

Our friend the Frenchman is quite delighted at the advent of a bicycle in Ismidt, for in his younger days, he tells me with much enthusiasm, he used to be somewhat partial to whirling wheels himself ; and when he first came here from France, some eighteen years ago, he actually brought with him a bone-shaker, with which, for the first summer, he was wont to surprise the natives. This relic of by-gone days has been stowed away among a lot of old traps ever since, all but forgotten ; but the appearance of a mounted wheelman recalls it to memory, and this evening, in honor of my visit, it is brought once more to light, its past history explained by its owner, and its merits and demerits as a vehicle in comparison with my bicycle duly discussed. The bone-shaker has wheels heavy enough for a dog-cart ; the saddle is nearly all gnawed away by mice, and it presents altogether so antiquated an appearance that it seems a relic rather of a past century than of a past decade. Its owner assays to take a ride on it ; but the best he can do is to wabble around a vacant space in front of the hotel, the awkward motions of the old bone-shaker affording intense amusement to the crowd. After supper this chatty and entertaining gentleman brings his wife, a rotund, motherly-looking person, to see the bicycle ; she is a Levantine Greek, and besides her own *lingua franca*, her husband has improved her education to the extent of a smattering of rather misleading English. Desiring to be complimentary in return for my riding back and forth a few times for her special benefit, the lady comes forward as I dismount and, smiling complacently upon me, remarks, "How very *grateful* you ride, monsieur ! " and her husband and tutor, desiring also to say something complimentary, echoes, " Much *grateful*—very."

The Greeks seem to be the life and poetry of these sea-coast places on the Ismidt gulf. My hotel faces the water ; and for hours after dark a half-dozen *caique*-loads of serenaders are paddling about in front of the town, making quite an entertaining concert in the silence of the night, the pleasing effect being heightened by the well-known softening influence of the water, and not a little enhanced by a display of rockets and Roman candles.

Earlier in the evening, while taking a look at Ismidt and the surrounding scenery, in company with a few sociable natives, who point out beauty-spots in the surrounding landscape with no little enthusiasm, I am impressed with the extreme loveliness of the situation. The town itself, now a place of thirteen thousand inhabitants, is the Nicomedia of the ancients. It is built in the form of a

Greeks Enjoying Themselves.

crescent, facing the sea ; the houses, many of them painted white,
are terraced upon the slopes of the green hills, whose sides and
summits are clothed with verdure, and whose bases are laved by
the blue waves of the gulf, which here, at the upper extremity, narrows to about a mile and a half in width ; white villages dot the
green mountain-slopes on the opposite shore, prominent among
them being the Armenian town of Bahgjadjik, where for a number
of years has been established an American missionary-school, a
branch, I think, of Roberts College.   Every mile of visible country,
whether gently sloping or more rugged and imposing, is green
with luxuriant vegetation, and the waters of the gulf are of that
deep-blue color peculiar to mountain-locked inlets ; the bright
green hills, the dancing blue waters, and the white painted villages
combine to make a scene so lovely in the chastened light of early·
eventide that, after the Bosporus, I think I never saw a place more
beautiful !   Besides the loveliness of the situation, the little moun-
tain-sheltered inlet makes an excellent anchorage for shipping ; and
during the late war, at the well-remembered crisis when the Russian
armies were bearing down on Constantinople and the British fleet
received the famous order to pass through the Dardanelles with
or without the Sultan's permission, the head-waters of the Ismidt
gulf became, for several months, the rendezvous of the ships.

CHAPTER XI.

EARLY dawn on Tuesday morning finds me already astir and groping about the hotel in search of some of the slumbering employees to let me out. Pocketing a cold lunch in lieu of eating breakfast, I mount and wheel down the long street leading out of the eastern end of town. On the way out I pass a party of caravan-teamsters who have just arrived with a cargo of mohair from Angora ; their pack-mules are fairly festooned with strings of bells of all sizes, from a tiny sleigh-bell to a solemn-voiced sheet-iron affair the size of a two-gallon jar. These bells make an awful din ; the men are unpacking the weary animals, shouting both at the mules and at each other, as if their chief object were to create as much noise as possible ; but as I wheel noiselessly past, they cease their unpacking and their shouting, as if by common consent, and greet me with that silent stare of wonder that men might be supposed to accord to an apparition from another world. For some few miles a rough macadam road affords a somewhat choppy but nevertheless ridable surface, and further inland it develops into a fairly good roadway, where a dismount is unnecessary for several miles.

The road leads along a depression between a continuation of the mountain-chains that inclose the Ismidt gulf, which now run parallel with my road on either hand at the distance of a couple of miles, some of the spurs on the south range rising to quite an imposing height. For four miles out of Ismidt the country is flat and swampy ; beyond that it changes to higher ground ; and the swampy flat, the higher ground, and the mountain-slopes are all covered with timber and a dense growth of underbrush, in which wild-fig shrubs and the homely but beautiful ferns of the English commons, the Missouri Valley woods, and the California foot-hills, mingle their respective charms, and hob-nob with scrub-oak, chestnut, walnut, and scores of others. The whole face of the country is covered with this dense thicket, and the first little hamlet I pass on the road is nearly hidden in it, the roofs of the houses being barely visible above the green sea of vegetation. Orchards and little patches of ground that have been cleared and cultivated are hidden entirely, and one cannot help thinking that if this interminable forest of brushwood were once to get fairly ablaze, nothing could prevent it from destroying everything these villagers possess.

A foretaste of what awaits me farther in the interior is obtained even within the first few hours of the morning, when a couple of horsemen canter at my heels for miles; they seem delighted beyond measure, and their solicitude for my health and general welfare is quite affecting. When I halt to pluck some blackberries, they solemnly pat their stomachs and shake their heads in chorus, to make me understand that blackberries are not good things to eat; and by gestures they notify me of bad places in the road which are yet out of sight ahead. Rude *mehanas*, now called *khans*, occupy little clearings by the

A Circassian Refugee.

roadside, at intervals of a few miles; and among the *habitués* congregated there I notice several of the Circassian refugees on whose account friends at Ismidt and Constantinople have shown themselves so concerned for my safety.

They are dressed in the long Cossack coats of dark cloth peculiar to the inhabitants of the Caucasus ; two rows of bone or metal cartridge-cases adorn their breast, being fitted into flutes or pockets made for them ; they wear either top boots or top boot-legs, and the counterpart of my own moccasins ; and their head-dress is a tall black lamb's-wool turban, similar to the national head-gear of the Persians. They are by far the best-dressed and most respectable-looking men one sees among the groups ; for while the majority of the natives are both ragged and barefooted, I don't re-member ever seeing Circassians either. To all outward appear-ances they are the most trustworthy men of them all ; but there is really more deviltry concealed beneath the smiling exterior of one of these homeless mountaineers from Circassia than in a whole village of the less likely-looking natives here, whose general cut-throat appearance—an effect produced, more than anything else, by the universal custom of wearing all the old swords, knives, and pistols they can get hold of—really counts for nothing. In pict-uresqueness of attire some of these khan loafers leave nothing to be desired ; and although I am this morning wearing Igali's ceru-lean scarf as a sash, the tri-colored pencil string of Servia around my neck, and a handsome pair of Circassian moccasins, I am abso-lutely nowhere by the side of many a native here whose entire wardrobe wouldn't fetch half a medjedie in a Galata auction-room.

The great light of Central Asian hospitality casts a glimmer even up into this out-of-the-way northwestern corner of the conti-nent, though it seems to partake more of the Nevada interpretation of the word than farther in the interior. Thrice during the fore-noon I am accosted with the invitation "mastic ? cogniac ? coffee ? " by road-side khan-jees or their customers who wish me to stop and let them satisfy their consuming curiosity at my novel bagar (horse), as many of them jokingly allude to it. Beyond these three beverages and the inevitable nargileh, these wayside khans provide nothing ; vishner syrup (a pleasant extract of the vishner cherry ; a spoonful in a tumbler of water makes a most agreeable and re-freshing sherbet), which is my favorite beverage on the road, being an inoffensive, non-intoxicating drink, is not in sufficient demand among the patrons of the khans to justify keeping it in stock.

An ancient bowlder causeway traverses the route I am following, but the blocks of stone composing it have long since become mis-placed and scattered about in confusion, making it impassable for

wheeled vehicles ; and the natural dirt-road alongside it is covered
with several inches of dust which is continually being churned up
by mule-caravans bringing mohair from Angora and miscellaneous
merchandise from Ismidt. Camel-caravans make smooth tracks,
but they seldom venture to Ismidt at this time of the year, I am
told, on account of the bellicose character of the mosquitoes that
inhabit this particular region ; their special mode of attack being
to invade the camels' sensitive nostrils, which drives these patient
beasts of burden to the last verge of distraction, sometimes even
worrying them to death. Stopping for dinner at the village of Sabanja, the scenes familiar in connection with a halt for refreshments in the Balkan Peninsula are enacted ; though for bland and
childlike assurance there is no comparison between the European
Turk and his brother in Asia Minor. More than one villager approaches me during the few minutes I am engaged in eating dinner, and blandly asks me to quit eating and let him see me ride ;
one of them, with a view of putting it out of my power to refuse,
supplements his request with a few green apples which no European could eat without bringing on an attack of cholera morbus,
but which Asiatics consume with impunity. After dinner I request
the proprietor to save me from the madding crowd long enough to
round up a few notes, which he attempts to do by locking me in
a room over the stable. In less than ten minutes the door is unlocked, and in walks the headman of the village, making a most
solemn and profound salaam as he enters. He has searched out a
man who fought with the English in the Crimea, according to his
—the man's—own explanation, and who knows a few words of
Frank language and has brought him along to interpret. Without
the slightest hesitation he asks me to leave off writing and come
down and ride, in order that he may see the performance, and—
he continues, artfully—that he may judge of the comparative merits
of a horse and a bicycle.

This peculiar trait of the Asiatic character is further illustrated
during the afternoon in the case of a caravan leader whom I meet
on an unridable stretch of road. " *Bin! bin!* " says this person,
as soon as his mental faculties grasp the idea that the bicycle is
something to ride on. " *Mimkin, deyil ; fenna yole ; duz yole lazim* "
(impossible ; bad road ; good road necessary), I reply, airing my
limited stock of Turkish. Nothing daunted by this answer, the
man blandly requests me to turn about and follow his caravan until

Sabanjans Worrying Me to Ride.

ridable road is reached—a good mile —in order that he may be enlightened. It is, perhaps, superfluous to add that, so far as I know, this particular individual's ideas of 'cycling are as hazy and undefined to-day as they ever were.

The principal occupation of the Sabanjans seems to be killing time; or perhaps waiting for something to turn up. Apple and pear-orchards are scattered about among the brush, looking utterly neglected; they are old trees mostly, and were planted by the more enterprising ancestors of the present owners, who would appear to be altogether unworthy of their sires, since they evidently do nothing in the way of trimming and pruning, but merely accept such blessings as unaided nature vouchsafes to bestow upon them. Moss-grown gravestones are visible here and there amid the thickets; the graveyards are neither protected by fence nor shorn of brush; in short, this aggressive undergrowth appears to be altogether too much for the energies of the Sabanjans; it seems to be encroaching upon them from every direction, ruthlessly pursuing them even to their very door-sills; like Banquo's ghost, it will not down, and the people have evidently retired discouraged from the contest. Higher up on the mountain-slopes the underbrush gives place to heavier timber, and small clearings abound, around which the unsubdued forest stands like a solid wall of green, the scene reminding one quite forcibly of backwoods clearings in Ohio; and were it not for the ancient appearance of the Sabanja minarets, the old bowlder causeway, and other evidences of declining years, one might easily imagine himself in a new country instead of the cradle of our race.

At Sabanja the wagon-road terminates, and my way becomes execrable beyond anything I ever encountered; it leads over a low mountain-pass, following the track of the ancient roadway, that on the acclivity of the mountain has been torn up and washed about, and the stone blocks scattered here and piled up there by the torrents of centuries, until it would seem to have been the sport and plaything of a hundred Kansas cyclones. Round about and among this disorganized mass, caravans have picked their way over the pass from the first dawn of commercial intercourse; following the same trail year after year, the stepping-places have come to resemble the steps of a rude stairway. From the summit of the pass is obtained a comprehensive view of the verdure-clad valley; here and there white minarets are seen protruding above the verdant area,

like lighthouses from a green sea; villages dot the lower slopes of
the mountains, while a lake, covering half the width of the valley
for a dozen miles, glimmers in the mid-day sun, making altogether
a scene that in some countries would long since have been immor-
talized on canvas or in verse. The descent is even rougher, if
anything, than the western side, but it leads down into a tiny valley that, if situated near a large city, would resound with the voices
of merry-makers the whole summer long. The undergrowth of
this morning's observations has entirely disappeared ; wide-spreading chestnut and grand old sycamore trees shade a circumscribed
area of velvety greensward and isolated rocks ; a tiny stream, a
tributary of the Sackaria, meanders along its rocky bed, and forest-
clad mountains tower almost perpendicularly around the charming
little vale save one narrow outlet to the east. There is not a human
being in sight, nor a sound to break the silence save the murmuring
of the brook, as I fairly clamber down into this little sylvan retreat;
but a wreath of smoke curling above the trees some distance from
the road betrays the presence of man. The whole scene vividly
calls to mind one of those marvellous mountain-retreats in which
writers of banditti stories are wont to pitch their heroes' silken
tent—no more appropriate rendezvous for a band of story-book
free-booters could well be imagined.

Short stretches of ridable mule-paths are found along this valley as I follow the course of the little stream eastward ; they are by
no means continuous, by reason of the eccentric wanderings of the
rivulet ; but after climbing the rough pass one feels thankful for
even small favors, and I plod along, now riding, now walking, occasionally passing little clusters of mud huts and meeting with
pack animals en route to Ismidt with the season's shearing of mohair.
"Alla Franga!" is the greeting I am now favored with, instead of
the "Ah, l'Anglais !" of Europe, as I pass people on the road;
and the bicycle is referred to as an araba, the name the natives
give their rude carts, and a name which they seem to think is quite
appropriate for anything with wheels.

Following the course of the little tributary for several miles,
crossing and recrossing it a number of times, I finally emerge with
it into the valley of Sackaria. There are some very good roads
down this valley, which is narrow, and in places contracts to but
little more than a mere neck between the mountains. At one of the
narrowest points the mountains present an almost perpendicular

face of rock, and here are the remnants of an ancient stone wall
reputed to have been built by the Greeks, somewhere about the

Down the Sakaria.

twelfth century, in anticipation of an invasion of the Turks from
the south. The wall stretches across the valley from mountain to

river, and is quite a massive affair ; an archway has been cut through it for the passage of caravans. Soon after passing through this opening I am favored with the company of a horseman, who follows me for three or four miles, and thoughtfully takes upon himself the office of telling me when to *bin* and when not to *bin*, according as he thinks the road suitable for 'cycling or not, until he discovers that his gratuitous advice produces no visible effect on my movements, when he desists and follows along behind in silence like a sensible fellow. About five o'clock in the afternoon I cross the Sackaria on an old stone bridge, and half an hour later roll into Geiveh, a large village situated in the middle of a triangular valley about seven miles in width. My cyclometer shows a trifle over forty miles from Ismidt; it has been a variable forty miles; I shall never forget the pass over the old causeway, the view of the Sabanja Valley from the summit, nor the lovely little retreat on the eastern side.

Trundling through the town in quest of a *khan*, I am soon surrounded by a clamorous crowd ; and passing the house or office of the *mudir* or headman of the place, that person sallies forth, and, after ascertaining the cause of the commotion, begs me to favor the crowd and himself by riding round a vacant piece of ground hard by. After this performance, a respectable-looking man beckons me to follow him, and he takes me—not to his own house to be his guest, for Geiveh is too near Europe for this sort of thing—to a *khan* kept by a Greek with a mote in one eye, where a "shake down" on the floor, a cup of coffee or a glass of vishner is obtainable, and opposite which another Greek keeps an eating-house. There is no separate kitchen in this latter establishment as in the one at Ismidt ; one room answers for cooking, eating, nargileh-smoking, coffee-sipping, and gossiping ; and while I am eating, a curious crowd watches my every movement with intense interest. Here, as at Ismidt, I am requested to examine for myself the contents of several pots. Most of them contain a greasy mixture of chopped meat and tomatoes stewed together, with no visible difference between them save in the sizes of the pieces of meat ; but one vessel contains *pillau*, and of this and some inferior red wine I make my supper. Prices for eatables are ridiculously low; I hand him a cherik for the supper ; he beckons me out of the back door, and there, with none save ourselves to witness the transaction, he counts me out two piastres change, which left him ten cents for the supper. He has probably been guilty of the awful crime of charging me about three farthings over the regular price, and was afraid to venture upon so iniquitous a proceeding in the public room lest the Turks should perchance detect him in cheating an Englishman, and revenge the wrong by making him feed me for nothing.

It rains quite heavily during the night, and while waiting for it to dry up a little in the morning, the Geivehites voluntarily tender me much advice concerning the state of the road ahead, being governed in their ideas according to their knowledge of a 'cycler's mountain-climbing ability. By a round dozen of men, who penetrate into my room in a body ere I am fairly dressed, and who, after solemnly salaaming in chorus, commence delivering themselves of expressive pantomime and gesticulations, I am led to understand that the road from Geiveh to Tereklu is something fearful for a bicycle. One fat old Turk, undertaking to explain it more fully, after the others have exhausted their knowledge of sign language, swells himself up like an inflated toad and imitates the labored respiration of a broken-winded horse in order to duly impress upon my mind the physical exertion I may expect to put forth in "riding"—he also paws the air with his right foot—over the mountain-range that looms up like an impassable barrier three miles east of the town. The Turks as a nation have the reputation of being solemn-visaged, imperturbable people, yet one occasionally finds them quite animated and "Frenchy" in their behavior—the bicycle may, however, be in a measure responsible for this.

The soil around Geiveh is a red clay that, after a shower, clings to the rubber tires of the bicycle as though the mere resemblance in color tended to establish a bond of sympathy between them that nothing could overcome. I pass the time until ten o'clock in avoiding the crowd that has swarmed the *khan* since early dawn, and has been awaiting with Asiatic patience ever since. At ten o'clock I win the gratitude of a thousand hearts by deciding to start, the happy crowd deserting half-smoked nargilehs, rapidly swallowing tiny cups of scalding-hot coffee in their anxiety lest I vault into the saddle at the door of the *khan* and whisk out of their sight in a moment—an idea that is flitting through the imaginative mind of more than one Turk present, as a natural result of the stories his wife has heard from his neighbor's wife, whose sister, from the roof of her house, saw me ride around the vacant space at the *mudir's* request yesterday. The Oriental imagination of scores of wonder-

18

ing villagers has been drawn upon to magnify that modest performance into a feat that fills the hundreds who didn't see it with the
liveliest anticipations, and a murmuring undercurrent of excitement
thrills the crowd as the word goes round that I am about to start.
A minority of the people learned yesterday that I wouldn't ride
across the stones, water-ditches, and mud-holes of the village
streets, and these at once lead the way, taking upon themselves the
office of conducting me to the road leading to the Kara Su Pass ;
while the less enlightened majority press on behind, the more restless spirits worrying me to ride, those of more patient disposition
maintaining a respectful silence, but wondering why on earth I am
walking.

The road they conduct me to is another of those ancient stone
causeways that traverse this section of Asia Minor in all directions. This one and several others I happen to come across are
but about three feet wide, and were evidently built for military
purposes by the more enterprising people who occupied Constantinople and the adjacent country before the Turks—narrow stone
pathways built to facilitate the marching of armies during the rainy
season when the natural ground hereabout is all but impassable.
These stone roads were probably built during the Byzantine occupation. Fairly smooth mule-paths lead along-side this relic of departed greatness and energy, and the warm sun having dried the
surface, I mount and speed away from the wondering crowd, and
in four miles reach the foot of the Kara Su Pass. From this spot I
can observe a small caravan, slowly picking its way down the mountain ; the animals are sometimes entirely hidden behind rocks, as
they follow the windings and twistings of the trail down the rugged slope which the old Turk this morning thought would make me
puff to climb.

A little stream called the Kara Su, or black water, comes dancing out of a rocky avenue near by ; and while I am removing my
foot-gear to ford it, I am joined by several herdsmen who are tending flocks of the celebrated Angora goats and the peculiar fat-tailed
sheep of the East, which are grazing on neighboring knolls. These
gentle shepherds are not overburdened with clothing, their nakedness being but barely covered ; but they wear long sword-knives
and old flint-lock, bell-mouthed horse-pistols—weapons that give
them a ferocious appearance that seems strangely at variance with
their peaceful occupation. They gather about me with a familiarity

that impresses me anything but favorably toward them ; they critically examine my clothing from helmet to moccasins, eying my various belongings wistfully, tapping my leather case, and pinching the rear package to try and ascertain the nature of its contents. I gather from their remarks about "*para*" (a term used in a general sense for money, as well as for the small coin of that name), as they regard the leather case with a covetous eye, that they are inclined to the opinion that it contains money ; and there is no telling the fabulous wealth their untutored minds are associating with the supposed treasure-chest of a Frank who rides a silver "*araba*."

Evidently these fellows have never heard of the tenth commandment ; or, having heard of it, they have failed to read, mark, learn, and inwardly digest it for the improvement of their moral natures ; for covetousness beams forth from every lineament of their faces and every motion of their hands. Seeing this, I endeavor to win them from the moral shackles of their own gloomy minds by pointing out the beautiful mechanism of my machine ; I twirl the pedals and show them how perfect are the bearings of the rear wheel ; I pinch the rubber tire to show them that it is neither iron nor wood, and call their attention to the brake, fully expecting in this usually winsome manner to fill them with gratitude and admiration, and make them forget all about my baggage and clothes. But these fellows seem to differ from those of their countrymen I left but a short time ago ; my other effects interest them far more than the wheel does, and one of them, after wistfully eying my moccasins, a handsomer pair, perhaps, than he ever saw before, points ruefully down to his own rude sandals of thong-bound raw-hide, and casts a look upon his comrades that says far more eloquently than words, "What a shame that such lovely moccasins should grace the feet of a Frank and an unbeliever—ashes on his head— while a true follower of the Prophet like myself should go about almost barefooted !" There is no mistaking the natural bent of these gentle shepherds' inclinations, and as, in the absence of a rusty sword and a seventeenth-century horse pistol, they doubtless think I am unarmed, my impression from their bearing is that they would, at least, have tried to frighten me into making them a present of my moccasins and perhaps a few other things. In the innocence of their unsophisticated natures, they wist not of the compact little weapon reposing beneath my coat that is as superior to their entire armament as is a modern gunboat to the wooden walls of the last century.   Whatever their intentions may be, however, they are doomed never to be carried out, for their attention is now attracted by the caravan, whose approach is heralded by the jingle of a thousand bells.

The next two hours find me engaged in the laborious task of climbing a mere bridle-path up the rugged mountain slope, along which no wheeled vehicle has certainly ever been before.   There is in some places barely room for pack animals to pass between the masses of rocks, and at others, but a narrow ledge between a perpendicular rock and a sheer precipice.   The steepest portions are worn into rude stone stairways by the feet of pack animals that toiled over this pass just as they toiled before America was discovered and have been toiling ever since ; and for hundreds of yards at a stretch I am compelled to push the bicycle ahead, rear wheel aloft, in the well-known manner of going up-stairs.   While climbing up a rather awkward place, I meet a lone Arab youth, leading his horse by the bridle, and come near causing a serious accident. It was at the turning of a sharp corner that I met this swarthy-faced youth face to face, and the sudden appearance of what both he and the horse thought was a being from a far more distant sphere than the western half of our own so frightened them both that I expected every minute to see them go toppling over the precipice.   Reassuring the boy by speaking a word or two of Turkish, and seeing the impossibility of either passing him or of his horse being able to turn around, I turn about and retreat a short distance, to where there is more room.   He is not quite assured of my terrestrial character even yet ; he is too frightened to speak, and he trembles visibly as he goes past, greeting me with a leer of mingled fear and suspicion ; at the same time making a brave but very sickly effort to ward off any evil designs I might be meditating against him by a pitiful propitiatory smile which will haunt my memory for weeks ; though I hope by plenty of exercise to escape an attack of the nightmare.

This is the worst mountain climbing I have done with a bicycle ; all the way across the Rockies there is nothing approaching this pass for steepness ; although on foot or horseback it would of course not appear so formidable.   When part way up, a bank of low hanging clouds come rolling down to meet me, enveloping the mountain in fog, and bringing on a disagreeable drizzle which scarcely improves the situation.

Five miles from the bottom of the pass and three hours from Geiveh I reach a small *postaya-khan*, occupied by one *zaptieh* and the station-keeper, where I halt for a half hour and get the *zaptieh* to brew me a cup of coffee, feeling the need of a little refreshment after the stiff tugging of the last two hours. Coffee is the only refreshment obtainable here, and, though the weather looks anything but propitious, I push ahead toward a regular roadside *khan*, which I am told I shall come to at the distance of another hour—the natives of Asia Minor know nothing of miles or kilometres, but reckon the distance from point to point by the number of hours it usually takes to go on horseback. Reaching this *khan* at three o'clock, I call for something to satisfy the cravings of hunger, and am forthwith confronted with a loaf of black bread, villanously heavy, and given a preliminary peep into a large jar of a crumbly white substance as villanously odoriferous as the bread is heavy, and which I think the proprietor expects me to look upon as cheese. This native product seems to be valued by the people here in proportion as it is rancid, being regarded by them with more than affection when it has reached a degree of rancidness and odoriferousness that would drive a European—barring perhaps, a Limburger—out of the house. These two delicacies, and the inevitable tiny cups of black bitter coffee make up all the edibles the *khan* affords ; so seeing the absence of any alternative, I order bread and coffee, prepared to make the most of circumstances. The proprietor being a kindly individual, and thinking perhaps that limited means forbid my indulgence in such luxuries as the substance in the earthenware jar, in the kindness of his heart toward a lone stranger, scoops out a small portion with his unwashed hand, puts it in a bowl of water and stirs it about a little by way of washing it, drains the water off through his fingers, and places it before me.

While engaged in the discussion of this delectable meal, a caravan of mules arrives in charge of seven rough-looking Turks, who halt to procure a feed of barley for their animals, the supplying of which appears to be the chief business of the *khan-jee*. No sooner have these men alighted and ascertained the use of the bicycle, than I am assailed with the usual importunities to ride for their further edification. It would be quite as reasonable to ask a man to fly as to ride a bicycle anywhere near the *khan;* but in the innocence of their hearts and the dulness of their Oriental understandings they think differently. They regard my objections as the result of a perverse and contrary disposition, and my explanation of "*mimkin deyil*" as but a groundless excuse born of my unwillingness to oblige. One old gray-beard, after examining the bicycle, eyes me meditatively for a moment, and then comes forward with a humorous twinkle in his eye, and pokes me playfully in the ribs, and makes a peculiar noise with the mouth: "q-u-e-e-k," in an effort to tickle me into good-humor and compliance with their wishes; in addition to which, the artful old dodger, thinking thus to work on my vanity, calls me "Pasha Effendi." Finding that toward their entreaties I give but the same reply, one of the younger men coolly advocates the use of force to coerce me into giving them an exhibition of my skill on the *araba*. As far as I am able to interpret, this bold visionary's argument is: "Behold, we are seven; Effendi is only one; we are good Mussulmans—peace be with us—he is but a Frank—ashes on his head—let us make him *bin*."

CHAPTER XII.

THE other members of the caravan company, while equally anxious to see the performance, and no doubt thinking me quite an unreasonable person, disapprove of the young man's proposition ; and the *khan-jee* severely reprimands him for talking about resorting to force, and turning to the others, he lays his forefingers together and says something about Franks, Mussulmans, Turks, and Ingilis ; meaning that even if we are Franks and Mussulmans, we are not prevented from being at the same time allies and brothers.

From the *khan* the ascent is more gradual, though in places muddy and disagreeable from the drizzling rain which still falls, and about 4 P.M. I arrive at the summit. The descent is smoother, and shorter than the western slope, but is even more abrupt ; the composition is a slaty, blue clay, in which the caravans have worn trails so deep in places that a mule is hidden completely from view. There is no room for animals to pass each other in these deep trench-like trails, and were any to meet, the only possible plan is for the ascending animals to be backed down until a wider place is reached. There is little danger of the larger caravans being thus caught in these "traps for the unwary," since each can hear the other's approach and take precautions; but single horsemen and small parties must sometimes find themselves obliged to either give or take, in the depths of these queer highways of commerce. It is quite an awkward task to descend with the bicycle, as for much of the way the trail is not even wide enough to admit of trundling in the ordinary manner, and I have to adopt the same tactics in going down as in coming up the mountain, with the difference, that on the eastern slope I have to pull back quite as stoutly as I had to push forward on the western. In going down I meet a man with three donkeys, but fortunately I am able to scramble up the bank sufficiently to let him pass. His donkeys are loaded with half-ripe grapes, which he is perhaps taking all the way to

Constantinople in this slow and laborious manner, and he offers me some as an inducement for me to ride for his benefit. Some wheelmen, being possessed of a sensitive nature, would undoubtedly think they had a right to feel aggrieved or insulted if offered a bunch of unripe grapes as an inducement to go ahead and break their necks ; but these people here in Asia Minor are but simple-hearted, overgrown children ; they will go straight to heaven when they die, every one of them.

At six o'clock I roll into Tereklu, having found ridable road a mile or so before reaching town. After looking at the cyclometer I begin figuring up the number of days it is likely to take me to reach Teheran, if yesterday and to-day have been expository of the country ahead ; forty and one-third miles yesterday and nineteen and a half to-day, thirty miles a day—rather slow progress for a wheelman, I mentally conclude ; but, although I would rather ride from "Land's End to John O'Groat's" for a task, than bicycle over the ground I have traversed between here and Ismidt, I find the tough work interlarded with a sufficiency of novel and interesting phases to make the occupation congenial. Upon dismounting at Tereklu, I find myself but little fatigued with the day's exertions, and with a view to obtaining a little peace and freedom from importunities to ride after supper, I gratify Asiatic curiosity several times before undertaking to allay the pangs of hunger—a piece of self-denial quite commendable, even if taken in connection with the idea of self-protection, when one reflects that I had spent the day in severe exercise, and had eaten since morning only a piece of bread.

Not long after my arrival at Tereklu I am introduced to another peculiar and not unknown phase of the character of these people, one that I have sometimes read of, but was scarcely prepared to encounter before being on Asian soil three days. From some of them having received medical favors from the medicine chest of travellers and missionaries, the Asiatics have come to regard every Frank who passes through their country as a skilful physician, capable of all sorts of wonderful things in the way of curing their ailments ; and immediately after supper I am waited upon by my first patient, the *mulazim* of the Tereklu *zaptiehs*. He is a tall, pleasant-faced fellow, whom I remember as having been wonderfully courteous and considerate while I was riding for the people before supper, and he is suffering with neuralgia in his lower jaw. He comes and seats himself beside me, rolls a cigarette in silence, lights it, and hands it to me, and then, with the confident assurance of a child approaching its mother to be soothed and cured of some ailment, he requests me to cure his aching jaw, seemingly having not the slightest doubt of my ability to afford him instant relief. I ask him why he don't apply to the *hakim* (doctor) of his native town. He rolls another cigarette, makes me throw the half-consumed one away, and having thus ingratiated himself a trifle deeper into my affections, he tells me that the Tereklu *hakim* is "*fenna ;*" in other words, no good, adding that there is a *duz hakim* at Gieveh, but Gieveh is over the Kara Su *dagh*. At this juncture he seems to arrive at the conclusion that perhaps I require a good deal of coaxing and good treatment, and, taking me by the hand, he leads me in that affectionate, brotherly manner down the street and into a coffee-*khan*, and spends the next hour in pressing upon me coffee and cigarettes, and referring occasionally to his aching jaw. The poor fellow tries so hard to make himself agreeable and awaken my sympathies, that I really begin to feel myself quite an ingrate in not being able to afford him any relief, and slightly embarrassed by my inability to convince him that my failure to cure him is not the result of indifference to his sufferings.

Casting about for some way of escape without sacrificing his good-will, and having in mind a box of pills I have brought along, I give him to understand that I am at the top of the medical profession as a stomach-ache *hakim*, but as for the jaw-ache I am, unfortunately, even worse than his compatriot over the way. Had I attempted to persuade him that I was not a doctor at all, he would not have believed me ; his mind being unable to grasp the idea of a Frank totally unacquainted with the noble Æsculapian art; but he seems quite aware of the existence of specialists in the profession, and notwithstanding my inability to deal with his particular affliction, my modest confession of being unexcelled in another branch of medicine seems to satisfy him. My profound knowledge of stomachic disorders and their treatment excuses my ignorance of neuralgic remedies.

There seems to be a larger proportion of superior dwelling-houses in Tereklu than in Gieveh, although, to the misguided mind of an unbeliever from the West, they have cast a sort of a funereal shadow over this otherwise desirable feature of their town by building their principal residences around a populous cemetery, which plays the part of a large central square. The houses are mostly two-story frame buildings, and the omnipresent balconies and all the windows are faced with close lattice work, so that the Osmanli ladies can enjoy the luxury of gazing contemplatively out on the area of disorderly grave-stones without being subjected to the prying eyes of passers-by. In the matter of veiling their faces the women of these interior towns place no such liberal—not to say coquettish—interpretation upon the office of the *yashmak* as do their sisters of the same religion in and about Constantinople. The ladies of Tereklu, seemingly, have a holy horror of displaying any of their facial charms ; the only possible opportunity offered of seeing anything, is to obtain an occasional glimpse of the one black eye with which they timidly survey you through a small opening in the folds of their shroud-like outer garment, that encases them from head to foot ; and even this peeping window of their souls is frequently hidden behind the impenetrable *yashmak.*

Mussulman women are the most gossipy and inquisitive creatures imaginable ; a very natural result, I suppose, of having had their feminine rights divine under constant restraint and suppression by the peculiar social position women occupy in Mohammedan countries. When I have arrived in town and am surrounded and hidden from outside view by a solid wall of men, it is really quite painful to see the women standing in small groups at a distance trying to make out what all the excitement is about. Nobody seems to have a particle of sympathy for their very natural inquisitiveness, or even to take any notice of their presence. It is quite surprising to see how rapidly the arrival of the Frank with the wonderful *araba* becomes known among these women from one end of town to another ; in an incredibly short space of time, groups of shrouded forms begin to appear on the housetops and other vantage-points, craning their necks to obtain a glimpse of whatever is going on.

In the innocence of an unsophisticated nature, and a feeling of genuine sympathy for their position, I propose collecting these scattered groups of neglected females together and giving an exhibition for their especial benefit, but the men evidently regard the idea of going to any trouble out of consideration for them as quite ridiculous ; indeed, I am inclined to think they regard it as evidence that I am nothing less than a gay Lothario, who is betraying alto-

gether too much interest in their women; for the old school Osmanli encompasses those hapless mortals about with a green wall of
jealousy, and regards with disapproval, even so much as a glance in
their direction. While riding on one occasion, this evening, I noticed
one over-inquisitive female become so absorbed in the proceedings
as to quite forget herself, and approach nearer to the crowd than
the Tereklu idea of propriety would seem to justify. In her absent-
mindedness, while watching me ride slowly up and dismount, she
allowed her *yashmak* to become disarranged and reveal her features.
This awful indiscretion is instantly detected by an old Blue-beard
standing by, who eyes the offender severely, but says nothing; if
she is one of his own wives, or the wife of an intimate friend, the
poor lady has perhaps earned for herself a chastisement with a
stick later in the evening.

Human nature is pretty much the same in the Orient as anywhere else; the degradation of woman to a position beneath her
proper level has borne its legitimate fruits; the average Turkish
woman is said to be as coarse and unchaste in her conversation as
the lowest outcasts of Occidental society, and is given to assailing
her lord and master, when angry, with language anything but
choice.

It is hardly six o'clock when I issue forth next morning, but
there are at least fifty women congregated in the cemetery, alongside which my route leads. During the night they seem to have
made up their minds to grasp the only opportunity of "seeing the
elephant" by witnessing my departure; and as, "when a woman
will she will," etc., applies to Turkish ladies as well as to any others,
in their laudable determination not to be disappointed they have
been patiently squatting among the gray tombstones since early
dawn. The roadway is anything but smooth, nevertheless one
could scarce be so dead to all feelings of commiseration as to remain unmoved by the sight of that patiently waiting crowd of
shrouded females; accordingly I mount and pick my way along the
street and out of town. Modest as is this performance, it is the
most marvellous thing they have seen for many a day; not a
sound escapes them as I wheel by, they remain as silent as though
they were the ghostly population of the graveyard they occupy, for
which, indeed, shrouded as they are in white from head to foot,
they might easily be mistaken by the superstitious.

My road leads over an undulating depression between the higher hills, a region of small streams, wheat-fields, and irrigating ditches, among which several trails, leading from Tereklu to numerous villages scattered among the mountains and neighboring small valleys, make it quite difficult to keep the proper road. Once I wander off my proper course for several miles; finding out my mistake I determine upon regaining the Torbali trail by a short cut across the stubble-fields and uncultivated knolls of scrub oak. This brings me into an acquaintanceship with the shepherds and husbandmen, and the ways of their savage dogs, that proves more lively than agreeable. Here and there I find primitive threshing-floors; they are simply spots of level ground selected in a central position and made smooth and hard by the combined labors of the several owners of the adjoining fields, who use them in common. Rain in harvest is very unusual; therefore the trouble and expense of covering them is considered unnecessary. At each of these threshing-centres I find a merry gathering of villagers, some threshing out the grain, others winnowing it by tossing it aloft with wooden, flat-pronged forks; the wind blows the lighter chaff aside, while the grain falls back into the heap. When the soil is sandy, the grain is washed in a neighboring stream to take out most of the grit, and then spread out on sheets in the sun to dry before being finally stored away in the granaries. The threshing is done chiefly by the boys and women, who ride on the same kind of broad sleigh-runner-shaped boards described in European Turkey.

The sight of my approaching figure is, of course, the signal for a general suspension of operations, and a wondering as to what sort of being I am. If I am riding along some well-worn by-trail, the women and younger people invariably betray their apprehensions of my unusual appearance, and seldom fail to exhibit a disposition to flee at my approach, but the conduct of their dogs causes me not a little annoyance. They have a noble breed of canines throughout the Angora goat country—fine animals, as large as Newfoundlands, with a good deal the appearance of the mastiff; and they display their hostility to my intrusion by making straight at me, evidently considering me fair game. These dogs are invaluable friends, but as enemies and assailants they are not exactly calculated to win a 'cycler's esteem. In my unusual appearance they see a strange, undefinable enemy bearing down toward their friends and owners, and, like good, faithful dogs, they hesitate not to commence the attack; sometimes there is a man among the threshers and winnowers who retains presence of mind enough to notice the dogs sallying forth to attack me, and to think of calling them back; but oftener I have to defend myself as best I can, while the gaping crowd, too dumfounded and overcome at my unaccountable appearance to think of anything else, simply stare as though expecting to see me sail up into space out of harm's way, or perform some other miraculous feat. My general tactics are to

Lively Times.

dismount if riding, and manœuvre the machine so as to keep it between myself and my savage assailant if there be but one; and if more than one, make feints with it at them alternately, not forgetting to caress them with a handy stone whenever occasion offers. There is a certain amount of cowardice about these animals notwithstanding their size and fierceness; they are afraid and suspicious of the bicycle as of some dreaded supernatural object; and although I am sometimes fairly at my wit's end to keep them at bay, I manage to avoid the necessity of shooting any of them. I have learned that to kill one of these dogs, no matter how great the provocation, would certainly get me into serious trouble with the natives, who value them very highly and consider the wilful killing of one little short of murder ; hence my forbearance.

When I arrive at a threshing-floor, and it is discovered that I am actually a human being and do not immediately encompass the destruction of those whose courage has been equal to awaiting my arrival, the women and children who have edged off to some distance now approach, quite timidly though, as if not quite certain of the prudence of trusting their eyesight as to the peaceful nature of my mission ; and the men vie with each other in their eagerness to give me all desired information about my course ; sometimes accompanying me a considerable distance to make sure of guiding me aright. But their contumacious canine friends seem anything but reassured of my character or willing to suspend hostilities ; in spite of the friendly attitude of their masters and the peacefulness of the occasion generally, they make furtive dashes through the ranks of the spectators at me as I wheel round the small circular threshing-floor, and savagely snap at the revolving wheels. Sometimes, after being held in check until I am out of sight beyond a knoll, these vindictive and determined assailants will sneak around through the fields, and, overtaking me unseen, make stealthy onslaughts upon me from the brush ; my only safety is in unremitting vigilance. Like the dogs of most semi-civilized peoples, they are but imperfectly trained to obey ; and the natives dislike checking them in their attacks upon anybody, arguing that so doing interferes with the courage and ferocity of their attack when called upon for a legitimate occasion.

It is very questionable, to say the least, if inoffensive wayfarers should be expected to quietly submit to the unprovoked attack of ferocious animals large enough to tear down a man, merely in view of possibly checking their ferocity at some other time. When capering wildly about in an unequal contest with three or four of these animals, while conscious of having the means at hand to give them all their quietus, one feels as though he were at that particular moment doing as the Romans do, with a vengeance ; nevertheless, it has to be borne, and I manage to come through with nothing worse than a rent in the leg of my riding trousers.

Finally, after fording several small streams, giving half a dozen threshing-floor exhibitions, and running the gauntlet of no end of warlike canines, I reach the lost Torbali trail, and, find it running parallel with a range of hills, intersecting numberless small streams, across which are sometimes found precarious foot-bridges consisting of a tree-trunk felled across it from bank to bank, the work of some enterprising peasant for his own particular benefit rather than the outcome of public spirit. Occasionally I bowl merrily along stretches of road which nature and the caravans together have made smooth enough even to justify a spurt ; but like a fleeting dream, this favorable locality passes to the rearward, and is followed by another mountain-slope whose steep grade and rough surface reads " trundle only."

They seem the most timid people hereabout I ever saw. . Few of them but show unmistakable signs of being frightened at my approach, even when I am trundling—the nickel-plate glistening in the sunlight, I think, inspires them with awe even at a distance— and while climbing this hill I am the innocent cause of the ignominious flight of a youth riding a donkey. While yet two hundred yards away, he reins up and remains transfixed for one transitory moment, as if making sure that his eyes are not deceiving him, or that he is really awake, and then hastily turns tail and bolts across the country, belaboring his long-eared charger into quite a lively gallop in his wild anxiety to escape from my awe-inspiring presence ; and as he vanishes across a field, he looks back anxiously to reassure himself that I am not giving chase. Ere kind friends and thoughtful well-wishers, with all their warnings of danger, are three days' journey behind, I find myself among people who run away at my approach. Shortly afterward I observe this bold donkey-rider half a mile to the left, trying to pass me and gain my rear unobserved. Others whom I meet this forenoon are more courageous ; instead of resorting to flight, they keep boldly on their general course, simply edging off to a respectful distance from my road ; some even venture to keep the road, taking care to give me a sufficiently large margin over and above my share of the way to insure against any possibility of giving offence ; while others will even greet me with a feeble effort to smile, and a timid, hesitating look, as if undecided whether they are not venturing too far. Sometimes I stop and ask these lion-hearted specimens whether I am on the right road, when they give a hurried reply and immediately take themselves off, as if startled at their own temerity.

These, of course, are lone individuals, with no companions to bolster up their courage or witness their cowardice ; the conduct of a party is often quite the reverse. Sometimes they seem determined not to let me proceed without riding for them, whether rocky ridge, sandy depression, or mountain-slope characterizes our meeting place, and it requires no small stock of forbearance and tact to get away from them without bringing on a serious quarrel. They take hold of the machine whenever I attempt to leave them, and give me to understand that nothing but a compliance with their wishes will secure my release ; I have known them even try the effect of a little warlike demonstration, having vague ideas of gaining their object by intimidation ; and this sort of thing is kept up until their own stock of patience is exhausted, or until some more reasonable member of the company becomes at last convinced that it really must be "*mimkin deyil,*" after all ; whereupon they let me go, ending the whole annoying, and yet really amusing, performance by giving me the most minute particulars of the route ahead, and parting in the best of humor. To lose one's temper on these occasions, or to attempt to forcibly break away, is quickly discovered to be the height of folly ; they themselves are brimful of good humor, and from beginning to end their countenances are wreathed in smiles ; although they fairly detain me prisoner the while, they would never think of attempting any real injury to either myself or the bicycle. Some of the more enterprising even express their determination of trying to ride the machine themselves ; but I always make a firm stand against any such liberties as this ; and, rough, half-civilized fellows though they often are, armed, and fully understanding the advantage of numbers, they invariably yield this point when they find me seriously determined not to allow it.

Descending into a narrow valley, I reach a road-side *khan,* adjoining a thrifty-looking melon-garden—this latter a welcome sight, since the day is warm and sultry ; and a few minutes' quiet, soulful communion with a good ripe water-melon, I think to myself, will be just about the proper caper to indulge in after being worried with dogs, people, small streams, and unridable hills since six o'clock.

" *Carpoose ?* " I inquire, addressing the proprietor of the *khan,* who issues forth from the stable.

" *Peeki, effendi,*" he answers, and goes off to the garden for the melon. Smiling sweetly at vacancy, in joyous anticipation of the coming feast and the soothing influence I feel sure of its exerting upon my feelings, somewhat ruffled by the many annoyances of the morning, I seek a quiet, shady corner, thoughtfully loosening my revolver-belt a couple of notches ere sitting down. In a minute the *khan-jee* returns, and hands me a "cucumber" about the size of a man's forearm.

"That isn't a *carpoose;* I want a *carpoose—a su carpoose!*" I explain.

"*Su carpoose, yoke!*" he replies ; and as I have not yet reached that reckless disregard of possible consequences to which I afterward attain, I shrink from tempting Providence by trying conclusions with the overgrown and untrustworthy cucumber ; so bidding the *khan-jee* adieu, I wheel off down the valley. I find a fair proportion of good road along this valley ; the land is rich, and though but rudely tilled, it produces wonderfully heavy crops of grain when irrigated. Small villages, surrounded by neglected-looking orchards and vineyards, abound at frequent intervals. Wherever one finds an orchard, vineyard, or melon-patch, there is also almost certain to be seen a human being evidently doing nothing but sauntering about, or perhaps eating an unripe melon.

This naturally creates an unfavorable impression upon a traveller's mind ; it means either that the kleptomaniac tendencies of the people necessitate standing guard over all portable property, or that the Asiatic follows the practice of hovering around all summer, watching and waiting for nature to bestow her blessings upon his undeserving head. Along this valley I meet a Turk and his wife bestriding the same diminutive donkey, the woman riding in front and steering their long-eared craft by the terror of her tongue in lieu of a bridle. The fearless lady halts her steed as I approach, trundling my wheel, the ground being such that riding is possible but undesirable. "What is that for, effendi?" inquires the man, who seems to be the more inquisitive of the two. "Why, to *bin*, of course ! don't you see the saddle?" says the woman, without a moment's hesitation ; and she bestows a glance of reproach upon her worse half for thus betraying his ignorance, twisting her neck round in order to send the glance straight at his unoffending head. This woman, I mentally conclude, is an extraordinary specimen of her race ; I never saw a quicker-witted person anywhere ; and I am not at all surprised to find her proving herself a phenomenon in other things. When a Turkish female meets a stranger on the road, and more especially a . Frank, her first thought and most natural impulse is to make sure

19

that no part of her features is visible—about other parts of her person she is less particular.   This remarkable woman, however, flings custom to the winds, and instead of drawing the ample folds of her *abbas* about her, uncovers her face entirely, in order to obtain a better view ; and, being unaware of my limited understanding, she begins discussing bicycle in quite a chatty manner.   I fancy her poor husband looks a trifle shocked at this outrageous conduct of the partner of his joys and sorrows ; but he remains quietly and discreetly in the background ; whereupon I register a silent vow never more to be surprised at anything, for that long-suffering and submissive being, the hen-pecked husband, is evidently not unknown even in Asiatic Turkey.

Another mountain-pass now has to be climbed ; it is only a short distance—perhaps two miles—but all the way up I am subjected to the disagreeable experience of having my footsteps dogged by two armed villagers.   There is nothing significant or exceptional about their being armed, it is true ; but what their object is in stepping almost on my heels for the whole distance up the acclivity is beyond my comprehension.   Uncertain whether their intentions are honest or not, it is anything but reassuring to have them following within sword's reach of one's back, especially when trundling a bicycle up a lonely mountain-trail.   I have no right to order them back or forward, neither do I care to have them think I entertain suspicions of their intentions, for in all probability they are but honest villagers, satisfying their curiosity in their own peculiar manner, and doubtless deriving additional pleasure from seeing one of their fellow-mortals laboriously engaged while they leisurely follow.   We all know how soul-satisfying it is for some people to sit around and watch their fellow-man saw wood.   Whenever I halt for a breathing-spell they do likewise ; when I continue on, they promptly take up their line of march, following as before in silence ; and when the summit is reached, they seat themselves on a rock and watch my progress down the opposite slope.

A couple of miles down grade brings me to Torbali, a place of several thousand inhabitants with a small covered bazaar and every appearance of a thriving interior town, as thrift goes in Asia Minor.   It is high noon, and I immediately set about finding the wherewithal to make a substantial meal.   I find that upon arriving at one of these towns, the best possible disposition to make of the bicycle is to deliver it into the hands of some respectable Turk,

request him to preserve it from the meddlesome crowd, and then pay
no further attention to it until ready to start.   Attempting to keep
watch over it oneself is sure to result in a dismal failure, whereas
an Osmanli gray-beard becomes an ever-willing custodian, regards

A Faithful Guardian.

its safe-keeping as appealing to his honor, and will stand guard over
it for hours if necessary, keeping the noisy and curious crowds of
his townspeople at a respectful distance by brandishing a thick
stick at anyone who ventures to approach too near.   These men
will never accept payment for this highly appreciated service, it
seems to appeal to the Osmanli's spirit of hospitality ; they seem

happy as clams at high tide while gratuitously protecting my property, and I have known them to unhesitatingly incur the displeasure
of their own neighbors by officiously carrying the bicycle off into an
inner room, not even granting the assembled people the harmless
privilege of looking at it from a distance—for there might be some
among the crowd possessed of the *fenna ghuz* (evil eye), and rather
than have them fix their baleful gaze upon the important piece of
property left under his charge by a stranger, he chivalrously braves
the displeasure of his own people ; smiling complacently at their
shouts of disapproval, he triumphantly bears it out of their sight
and from the fell influence of the possible *fenna ghuz*. Another
strange and seemingly paradoxical phase of these occasions is that
when the crowd is shouting out its noisiest protests against the
withdrawal of the machine from popular inspection, any of the
protestors will eagerly volunteer to help carry the machine inside,
should the self-important personage having it in custody condescend
to make the slightest intimation that such service would be acceptable.

Handing over the bicycle, then, to the safe-keeping of a respectable *kahvay-jee* (coffee-*khan* employee) I sally forth in quest of eatables. The *kahvay-jee* has it immediately carried inside and set up on
one of the divans, in which elevated position he graciously permits
it to be gazed upon by the people, who swarm into his *khan* in such
numbers as to make it impossible for him to transact any business.
Under the guidance of another volunteer, who, besides acting the
part of guide, takes particular care that I get lumping weight, etc.,
I proceed to the *ett-jees* and procure some very good mutton-chops,
and from there to the *ekmek-jees* for bread. This latter person
straightway volunteers to cook my chops. Sending to his residence
for a tin dish, some chopped onions and butter, he puts them in
his oven, and in a few minutes sets them before me, browned and
buttered. Meanwhile, he has despatched a youth somewhere on
another errand, who now returns and supplements the savory chops
with a small dish of honey in the comb and some green figs. Seated
on the generous-hearted *ekmek-jee's* dough-board, I make a dinner good enough for anybody.

While discussing these acceptable viands, I am somewhat
startled at hearing one of the worst " cuss-words " in the English
language repeated several times by one of the two Turks engaged
in the self-imposed duty of keeping people out of the place while

I am eating—a kindly piece of courtesy that wins for them my warmest esteem. The old fellow proves to be a Crimean veteran, and, besides a much-prized medal he brought back with him, he somehow managed to acquire this discreditable, perhaps, but nevertheless unmistakable, memento of having at some time or other campaigned it with "Tommy Atkins." I try to engage him in conversation, but find that he doesn't know another solitary word of English. He simply repeats the profane expression alluded to in a parrot-like manner without knowing anything of its meaning; has, in fact, forgotten whether it is English, French, or Italian. He only knows it as a "Frank" expression, and in that he is perfectly right: it is a frank expression, a very frank expression indeed. As if determined to do something agreeable in return for the gratifying interest I seem to be taking in him on account of this profanity, he now disappears, and shortly returns with a young man, who turns out to be a Greek, and the only representative of Christendom in Torbali. The old Turk introduces him as a "Ka-ris-ti-ahn" (Christian) and then, in reply to questioners, explains to the interested on-lookers that, although an Englishman, and, unlike the Greeks, friendly to the Turks, I also am a "Ka-ris-ti-ahn;" one of those queer specimens of humanity whose perverse nature prevents them from embracing the religion of the Prophet, and thereby gaining an entrance into the promised land of the kara ghuz kiz (black-eyed houris). During this profound exposition of my merits and demerits, the wondering people stare at me with an expression on their faces that plainly betrays their inability to comprehend so queer an individual; they look as if they think me the oddest specimen they have ever met, and taking into due consideration my novel mode of conveyance, and that many Torbali people never before saw an Englishman, this is probably not far from a correct interpretation of their thoughts.

Unfortunately, the streets and environments of Torbali are in a most wretched condition; to escape sprained ankles it is necessary to walk with a great deal of caution, and the idea of bicycling through them is simply absurd. Nevertheless the populace turns out in high glee, and their expectations run riot as I relieve the kahvay-jee of his faithful vigil and bring forth my wheel. They want me to bin in their stuffy little bazaar, crowded with people and donkeys; mere alley-ways with scarcely a twenty yard stretch from one angle to another; the surface is a disorganized mass of holes and stones over which the wary and hesitative donkey picks his way with the greatest care ; and yet the popular clamor is " *Bin, bin ; bazaar, bazaar !* " The people who have been showing me how courteously and considerately it is possible for Turks to treat a stranger, now seem to have become filled with a determination not to be convinced by anything I say to the contrary ; and one of the most importunate and headstrong among them sticks his bearded face almost up against my own placid countenance (I have already learned to wear an unruffled, martyr-like expression on these howling occasions) and fairly shrieks out, "*Bin ! bin !*" as though determined to hoist me into the saddle, whether or no, by sheer force of his own desire to see me there. This person ought to know better, for he wears the green turban of holiness, proving him to have made a pilgrimage to Mecca, but the universal desire to see the bicycle ridden seems to level all distinctions.

All this tumult, it must not be forgotten, is carried on in perfect good humor ; but it is, nevertheless, very annoying to have it seem that I am too boorish to repay their kindness by letting them see me ride ; even walking out of town to avoid gratifying them, as some of them doubtless think. These little embarrassments are some of the penalties of not knowing enough of the language to be able to enter into explanations. Learning that there is a piece of wagon-road immediately outside the town, I succeed in silencing the clamor to some extent by promising to ride when the *araba yole* is reached ; whereupon hundreds come flocking out of town, following expectantly at my heels. Consoling myself with the thought that perhaps I will be able to mount and shake the clamorous multitude off by a spurt, the promised *araba ¯yole* is announced ; but the fates are plainly against me to-day, for I find this road leading up a mountain slope from the very beginning. The people cluster expectantly around, while I endeavor to explain that they are doomed to disappointment—that to be disappointed in their expectations to see the *araba* ridden is plainly their *kismet*, for the hill is too steep to be ridden. They laugh knowingly and give me to understand that they are not quite such simpletons as to think that an *araba* cannot be ridden along an *araba yole*. "This is an *araba yole*," they argue, "you are riding an *araba ;* we have seen even our own clumsily-made *arabas* go up here time and again, therefore it is evident that you are not sincere," and they gather closer around and spend another ten minutes in coaxing. It is a ridiculous position to be in ; these people use the most endearing terms imaginable ; some of them kiss the bicycle and would get down and kiss my dust-begrimed moccasins if I would permit it ; at coaxing they are the most persevering people I ever saw. To convince them of the impossibility of riding up the hill I allow a muscular young Turk to climb into the saddle and try to propel himself forward while I hold him up. This has the desired effect, and they accompany me farther up the slope to where they fancy it to be somewhat less steep, a score of all too-willing hands being extended to assist in trundling the machine. Here again I am subjected to another interval of coaxing ; and this same annoying programme is carried out several times before I obtain my release. They are the most headstrong, persistent people I have yet encountered ; the natural pig-headed disposition of the " unspeakable Turk" seems to fairly run riot in this little valley, which at the point where Torbali is situated contracts to a mere ravine between rugged heights.

For a full mile up the mountain road, and with a patient insistence quite commendable in itself, they persist in their aggravating attentions ; aggravating, notwithstanding that they remain in the best of humor, and treat me with the greatest consideration in every other respect, promptly and severely checking any unruly conduct among the youngsters, which once or twice reveals itself in the shape of a stone pitched into the wheel, or some other pleasantry peculiar to the immature Turkish mind. At length one enterprising young man, with wild visions of a flying wheelman descending the mountain road with lightning-like velocity, comes prominently to the fore, and unblushingly announces that they have been bringing me along the wrong road ; and, with something akin to exultation in his gestures, motions for me to turn about and ride back. Had the others seconded this brilliant idea there was nothing to prevent me from being misled by the statement ; but his conduct is at once condemned ; for though pig-headed, they are honest of heart, and have no idea of resorting to trickery to gain their object. It now occurs to me that perhaps if I turn round and ride down hill a short distance they will see that my trundling up hill is really a matter of necessity instead of choice, and thus rid me of their undesirable presence.

Hitherto the slope has been too abrupt to admit of any such thought, but now it becomes more gradual. As I expected, the

proposition is heralded with unanimous shouts of approval, and I
take particular care to stipulate that after this they are to follow me
no farther ; any condition is acceptable to them as long as it includes seeing how the thing is ridden. It is not without certain
misgivings that I mount and start cautiously down the declivity be-
. tween two rows of turbaned and fez-bedecked heads, for I have not
yet forgotten the disagreeable actions of the mob at Adrianople in
running up behind and giving the bicycle vigorous forward pushes,
a proceeding that would be not altogether devoid of danger here,
for besides the gradient, one side of the road is a yawning chasm.
These people, however, confine themselves solely to howling with
delight, proving themselves to be well-meaning and comparatively
well-behaved after all. Having performed my part of the compact, a few of the leading men shake hands, and express their
gratitude and well-wishes ; and after calling back several youngsters
who seem unwilling to abide by the agreement forbidding them
to follow any farther, the whole noisy company proceed along foot-
paths leading down the cliffs to town, which is in plain view almost
immediately below.

The entire distance between Torbali and Keshtobek, where to-
morrow forenoon I cross over into the vilayet of Angora, is through
a rough country for bicycling. Forest-clad mountains, rocky
gorges, and rolling hills characterize the landscape ; rocky passes
lead over mountains where the caravans, engaged in the exportation
of mohair ever since that valuable commodity first began to be ex-
ported, have worn ditch-like trails through ridges of solid rock
three feet in depth ; over the less rocky and precipitous hills beyond a comprehensive view is obtained of the country ahead, and
these time-honored trails are seen leading in many directions,
ramifying the country like veins of one common system, which are
necessarily drawn together wherever there is but one pass. Parts
of these commercial by-ways are frequently found to be roughly
hedged with wild pear and other hardy shrubs indigenous to the
country—the relics of by-gone days, planted when these now
barren hills were cultivated, to protect the growing crops from
depredation.   Old mill-stones with depressions in the centre,
formerly used for pounding corn in, and pieces of hewn masonry
are occasionally seen as one traverses these ancient trails, marking
the site of a village in days long past, when cultivation and centres
of industry were more conspicuous features of Asia Minor than

they are to-day; lone graves and graves in clusters, marked by rude unchiselled headstones or oblong mounds of bowlders, are frequently observed, completing the scene of general decay.

While riding along these tortuous ways, the smooth-worn camel-paths sometimes affording excellent wheeling, the view ahead is often obstructed by the untrimmed hedges on either side, and one sometimes almost comes into collision, in turning a bend, with

The Byways of Asia Minor.

horsemen, wild-looking, armed formidably in the manner peculiar to the country, as though they were assassins stealing forth under cover. Occasionally a female bestriding a donkey suddenly appears but twenty or thirty yards ahead, the narrowness and the crookedness of the hedged-in trail favoring these abrupt meetings; shrouded perhaps in a white *abbas*, and not infrequently riding a white donkey, they seldom fail to inspire thoughts of ghostly equestriennes gliding silently along these now half-deserted pathways. Many a hasty but sincere appeal is made to Allah by these fright-

ened ladies as they fancy themselves brought suddenly face to face
with the evil one ; more than once this afternoon I overhear that
agonizing appeal for providential aid and protection of which I am
the innocent cause. The second thought of the lady—as if it
occurred to her that with any portion of her features visible she
would be adjudged unworthy of divine interference in her behalf
—is to make sure that her *yashmak* is not disarranged, and then
comes a mute appeal to her attendant, if she have one, for some
explanation of the strange apparition so suddenly and unexpectedly
confronting them.

In view of the nature of the country and the distance to Keshtobek, I have no idea of being able to reach that place to-night,
and when I arrive at the ruins of an old mud-built *khan*, at dusk, I
conclude to sup off the memories of my excellent dinner and a
piece of bread I have in my pocket, and avail myself of its shelter
for the night. While eating my frugal repast, up ride three mule-
teers, who, after consulting among themselves some minutes,
finally picket their animals and prepare to join my company ;
whether for all night or only to give their animals a feed of grass,
I am unable to say. Anyhow, not liking the idea of spending the
whole night, or any part of it, in these unfrequented hills with
three ruffianly-looking natives, I again take up my line of march
along mountain mule-paths for some three miles farther, when I
descend into a small valley, and it being too dark to undertake the
task of pitching my tent, I roll myself up in it instead. Soothed
by the music of a babbling brook, I am almost asleep, when a
glorious meteor shoots athwart the sky, lighting up the valley with
startling vividness for one brief moment, and then the dusky pall
of night descends, and I am gathered into the arms of Morpheus.

Toward morning it grows chilly, and I am but fitfully dozing
in the early gray, when I am awakened by the bleating and the
pattering feet of a small sea of Angora goats. Starting up, I discover that I am at that moment the mysterious and interesting
subject of conversation between four goatherds, who have apparently been quietly surveying my sleeping form for some minutes.
Like our covetous friends beyond the Kara Su Pass, these early
morning acquaintances are unlovely representatives of their profession ; their sword-blades are half naked, the scabbards being
rudely fashioned out of two sections of wood, roughly shaped to the
blade, and bound together at top and bottom with twine ; in addi-

tion to which are bell-mouthed pistols, half the size of a Queen Bess blunderbuss. This villainous-looking quartette does not make a very reassuring picture in the foreground of one's waking moments, but they are probably the most harmless mortals imaginable ; anyhow, after seeing me astir, they pass on with their flocks and herds without even submitting me to the customary catechizing.

The morning light reveals in my surroundings a most charming little valley, about half a mile wide, walled in on the south by towering mountains covered with a forest of pine and cedar, and on the north by low, brush-covered hills ; a small brook dances along

Early Morning Callers.

the middle, and thin pasturage and scattered clumps of willow fringe the stream.   Three miles down the valley I arrive at a roadside *khan*, where I obtain some hard bread that requires soaking in water to make it eatable, and some wormy raisins ; and from this choice assortment I attempt to fill the aching void of a ravenous appetite ; with what success I leave to the reader's imagination. Here the *khan-jee* and another man deliver themselves of one of those strange requests peculiar to the Asiatic Turk.   They pool the contents of their respective treasuries, making in all perhaps three medjedis, and, with the simplicity of children whose minds have not yet dawned upon the crooked ways of a wicked world, they offer me the money in exchange for my Whitehouse

leather case with its contents. They have not the remotest idea
of what the case contains ; but their inquisitiveness apparently
overcomes all other considerations. Perhaps, however, their seemingly innocent way of offering me the money may be their own peculiar deep scheme of inducing me to reveal the nature of its contents.

For a short distance down the valley I find road that is generally ridable, when it contracts to a mere ravine, and the only
road is the bowlder - strewn bed of the stream, which is now
nearly dry, but in the spring is evidently a raging torrent. An
hour of this delectable exercise, and I emerge into a region of undulating hills, among which are scattered wheat-fields and clusters
of mud-hovels which it would be a stretch of courtesy to term villages. Here the poverty of the soil, or of the water-supply, is heralded to every observant eye by the poverty-stricken appearance of
the villagers. As I wheel along, I observe that these poor half-
naked wretches are gathering their scant harvest by the laborious
process of pulling it up by the roots, and carrying it to their common threshing-floor on donkeys' backs. Here, also, I come to a
camp of Turkish gypsies ; they are dark-skinned, with an abundance of long black hair dangling about their shoulders, like our
Indians ; the women and larger girls are radiant in scarlet calico
and other high-colored fabrics, and they wear a profusion of bead
necklaces, armlets, anklets, and other ornaments dear to the semi-
savage mind ; the younger children are as wild and as innocent of
clothing as their boon companions, the dogs. The men affect the
fez and general Turkish style of dress, with many unorthodox
trappings and embellishments, however; and with their own wild
appearance, their high-colored females, naked youngsters, wolfish-
looking dogs, picketed horses, and smoke-browned tents, they
make a scene that, for picturesqueness, can give odds even to the
wigwam-villages of Uncle Sam's Crow scouts, on the Little Big
Horn River, Montana Territory, which is saying a good deal.

Twelve miles from my last night's rendezvous, I pass through
Keshtobek, a village that has evidently seen better days. The ruins
of a large stone *khan* take up all the central portion of the place ;
massive gateways of hewn stone, ornamented by the sculptor's
chisel, are still standing, eloquent monuments of a more prosperous
era. The unenterprising descendants of the men who erected this
substantial and commodious retreat for passing caravans and trav-

ellers are now content to house themselves and their families in tumble-down hovels, and to drift aimlessly and unambitiously along on wretched fare and worse clothes, from the cradle to the grave. The Keshtobek people seem principally interested to know why I am travelling without any *zaptieh* escort ; a stranger travelling through these wooded mountains, without guard or guide, and not being able to converse with the natives, seems almost beyond their belief. When they ask me why I have no *zaptieh*, I tell them I *have* one, and show them the Smith & Wesson. They seem to regard this as a very witty remark, and say to each other : " He is right ; an English *effendi* and an American revolver don't require any *zaptiehs* to take care of them, they are quite able to look out for themselves."

From Keshtobek my road leads down another small valley, and before long I find myself in the Angora *vilayet*, bowling briskly eastward over a most excellent road ; not the mule-paths of an hour ago, but a broad, well-graded highway, as good, clear into Nalikhan, as the roads of any New England State. This sudden transition is not unnaturally productive of some astonishment on my part, and inquiries at Nalikhan result in the information that my supposed graded wagon-road is nothing less than the bed of a proposed railway, the preliminary grading for which has been finished between Keshtobek and Angora for some time.

This valley seems to be the gateway into a country entirely different from what I have hitherto traversed. Unlike the forest-crowned mountains and shrubbery hills of this morning, the mountains towering aloft on every hand are now entirely destitute of vegetation ; but they are in nowise objectionable to look upon on that account, for they have their own peculiar features of loveliness. Various colored rocks and clays enter into their composition ; their giant sides are fantastically streaked and seamed with blue, yellow, green, and red ; these variegated masses encompassing one round about on every side are a glorious sight—they are more interesting, more imposing, more grand and impressive even than the piny heights of Kodjaili. Many of these mountains bear evidence of mineral formation, and anywhere in the Occident would be the scene of busy operations. In Constantinople I heard an English mineralist, who has lived many years in the country, express the belief that there is more mineral buried in these Asia Minor hills than in a corresponding area in any other part of the world ; that he knew people who for years have had their eye on certain localities of unusual promise waiting patiently for the advantages of mineral development to dawn upon the sluggish mind of Osmanli statesmen. At present it is useless to attempt prospecting, for there is no guarantee of security; no sooner is anything of value discovered than the finder is embarrassed by imperial taxes, local taxes, backsheesh, and all manner of demands on his resources, often ending in having everything coolly confiscated by the government; which, like the dog in the manger, will do nothing with it, and is perfectly contented and apathetic so long as no one else is reaping any benefit from it.

The general ridableness of this *chemin de fer*, as the natives have been taught to call it, proves not to be without certain disadvantages, for during the afternoon I unwittingly manage to do considerable mischief. Suddenly meeting two horsemen, when bowling at a moderate pace around a bend, the horse of one takes violent exception to my intrusion, and, in spite of the excellent horsemanship of his rider, backs down into a small ravine, both horse and rider coming to grief in some water at the bottom. Fortunately, neither man nor horse sustained any more serious injury than a few scratches and bruises, though it might easily have resulted in broken bones. Soon after this affair, another donkey-rider takes to his heels, or rather to his donkey's heels across country, and his long-eared and generally sure-footed charger ingloriously comes to earth; but I feel quite certain that no damage is sustained in this case, for both steed and rider are instantly on their feet; the bold steeple-chaser looks wildly and apprehensively toward me, but observing that I am giving chase, it dawns upon his mind that I am perhaps after all a human being, whereupon he refrains from further flight.

Wheeling down the gentle declivity of a broad, smooth road that almost deserves the title of boulevard, leading through the vineyards and gardens of Nalikhan's environments, at quite a rattling pace, I startle a quarry of four dears (deers) robed in white mantles, who, the moment they observe the strange apparition approaching them at so vengeful a speed, bolt across a neighboring vineyard like the all-possessed. The rapidity of their movements, notwithstanding the impedimenta of their flowing shrouds, readily suggests the idea of a quarry of dears (deer), but whether they are pretty dears or not, of course, their *yashmaks* fail to reveal; but in return for the beaming smile that lights up our usually solemn-

looking countenance at their ridiculously hasty flight, as a recipro-
cation pure and simple, I suppose we ought to give them the benefit of the doubt.

The evening at Nalikhan is a comparatively happy occasion ; it is
Friday, the Mussulman Sabbath ; everybody seems fairly well-dressed
for a Turkish interior town ; and, more important than all, there is
a good, smooth road on which to satisfy the popular curiosity ; on
this latter fact depends all the difference between an agreeable and
a disagreeable time, and at Nalikhan everything passes off pleasantly
for all concerned. Apart from the novelty of my conveyance, few
Europeans have ever visited these interior places under the same

A Quarry of Startled Dears.

conditions as myself. They have usually provided themselves beforehand with letters of introduction to the pashas and *mudirs* of
the villages, who have entertained them as their guests during their
stay. On the contrary, I have seen fit to provide myself with none
of these way-smoothing missives, and, in consequence of my linguistic shortcomings, immediately upon reaching a town I have to sur-
render myself, as it were, to the intelligence and good-will of the
common people ; to their credit be it recorded, I can invariably
count on their not lacking at least the latter qualification.

The little *khan* I stop at is, of course, besieged by the usual crowd,
but they are a happy-hearted, contented people, bent on lionizing me

the best they know how ; for have they not witnessed my marvellous
performance of riding an *araba*, a beautiful web-like *araba*, more
beautiful than any *makina* they ever saw before, and in a manner
that upsets all their previous ideas of equilibrium ? Have I not
proved how much I esteem them by riding over and over again for
fresh batches of new arrivals, until the whole population has seen
the performance ? And am I not hobnobbing and making myself
accessible to the people, instead of being exclusive and going
straightway to the pasha's, shutting myself up and permitting none
but a few privileged persons to intrude upon my privacy ? All these
things appeal strongly to the better nature of the imaginative Turks,
and not a moment during the whole evening am I suffered to be unconscious of their great appreciation of it all. A bountiful supper
of scrambled eggs fried in butter, and then the *mulazim* of *zaptiehs*
takes me under his special protection and shows me around the
town. He shows me where but a few days ago the Nalikhan bazaar, with all its multifarious merchandise, was destroyed by fire,
and points out the temporary stalls, among the black ruins, that
have been erected by the pasha for the poor merchants who, with
heavy hearts and doleful countenance, are trying to recuperate
their shattered fortunes. He calls my attention to two-story
wooden houses and other modest structures, which, in the simplicity of his Asiatic soul, he imagines are objects of interest ; and
then he takes me to the headquarters of his men, and sends out
for coffee in order to make me literally his guest. Here, in his
office, he calls my attention to a chromo hanging on the wall, which
he says came from Stamboul—Stamboul, where the Asiatic Turk
fondly imagines all wonderful things originate. This chromo is
certainly a wonderful thing in its way. It represents an English
trooper in the late Soudan expedition kneeling behind the shelter
of a dead camel, and with a revolver in each hand keeping at bay
a crowd of Arab spearmen. The soldier is badly wounded, but
with smoking revolvers and an evident determination to die hard,
he has checked, and is still checking, the advance of somewhere
about ten thousand Arab troops. No wonder the people of Keshtobek thought an Englishman and a revolver quite safe in travelling without *zaptiehs ;* some of them had probably been to Nalikhan
and seen this same chromo.

When it grows dark the *mulazim* takes me to the public coffee-
garden, near the burned bazaar, a place which is really no garden at

THROUGH THE ANGORA GOAT COUNTRY.

all, only some broad, rude benches encircling a round water-tank or fountain, and which is fenced in with a low, wabbly picket-fence. Seated crossed-legged on the benches are a score of sober-sided Turks, smoking *nargilehs* and cigarettes, and sipping coffee ; the feeble light dispensed by a lantern on top of a pole in the centre of the tank makes the darkness of the "garden" barely visible ; a continuous splashing of water, the result of the overflow from a pipe projecting three feet above the surface, furnishes the only music ; the sole auricular indication of the presence of patrons is when some customer orders "*kahvay*" or "*nargileh*" in a scarcely audible tone of voice ; and this is the Turk's idea of an evening's enjoyment.

Returning to the *khan*, I find it full of happy people looking at the bicycle ; commenting on the wonderful *marifet* (skill) apparent in its mechanism, and the no less marvellous *marifet* required in riding it. They ask me if I made it myself and *katch-lira ?* (how many liras?) and then requesting the privilege of looking at my *teskeri* they find rare amusement in comparing my personal charms with the description of my form and features as interpreted by the passport officer in Galata. Two men among them have in some manner picked up a sand from the sea-shore of the English language. One of them is a very small sand indeed, the solitary negative phrase, "no ;" nevertheless, during the evening he inspires the attentive auditors with respect for his linguistic accomplishments by asking me numerous questions, and then, anticipating a negative reply, forestalls it himself by querying, "No ?" The other "linguist" has in some unaccountable manner added the ability to say " Good morning " to his other accomplishments ; and when about time to retire, and the crowd reluctantly bestirs itself to depart from the magnetic presence of the bicycle, I notice an extraordinary degree of mysterious whispering and suppressed amusement going on among them, and then they commence filing slowly out of the door with the "linguistic person" at their head ; as that learned individual reaches the threshold he turns toward me, makes a salaam and says, "Good-morning," and everyone of the company, even down to the irrepressible youngster who was cuffed a minute ago for venturing to twirl a pedal, and who now forms the rear-guard of the column, likewise makes a salaam and says, " Good-morning."

Quilts are provided for me, and I spend the night on the divan
20

of the *khan ;* a few roving mosquitoes wander in at the open window and sing their siren songs around my couch, a few entomological specimens sally forth from their permanent abode in the lining of the quilts to attack me and disturb my slumbers ; but later experience teaches me to regard my slumbers to-night as comparatively peaceful and undisturbed. In the early morning I am awakened by the murmuring voices of visitors gathering to see me off ; coffee is handed to me ere my eyes are fairly open, and the savory odor of eggs already sizzling in the pan assail my olfactory nerves. The *khan-jee* is an Osmanli and a good Mussulman, and when ready to depart I carelessly toss him my purse and motion for him to help himself—a thing I would not care to do with the keeper of a small tavern in any other country or of any other nation. Were he entertaining me in a private capacity he would feel injured at any hint of payment ; but being a *khan-jee*, he opèns the purse and extracts a cherik—twenty cents.

CHAPTER XIII.

BEY BAZAAR, ANGORA, AND EASTWARD.

A TRUNDLE of half an hour up the steep slopes leading out of another of those narrow valleys in which all these towns are situated, and then comes a gentle declivity extending with but little interruption for several miles, winding in and out among the inequalities of an elevated table-land. The mountain-breezes blow cool and exhilarating, and just before descending into the little Charkhan Valley I pass some interesting cliffs of castellated rocks, the sight of which immediately wafts my memory back across the thousands of miles of land and water to what they are almost a counterpart of—the famous castellated rocks of Green River, Wyo. Ter.

Another scary youth takes to his heels as I descend into the valley and halt at the village of Charkhan, a mere shapeless cluster of mud-hovels. Before one of these a ragged agriculturist solemnly presides over a small heap of what I unfortunately mistake at the time for pumpkins. I say "unfortunately," because after-knowledge makes it highly probable that they were the celebrated Charhkan musk-melons, famous far and wide for their exquisite flavor; the variety can be grown elsewhere, but, strange to say, the peculiar, delicate flavor which makes them so celebrated is absent when they vegetate anywhere outside this particular locality. It is supposed to be owing to some peculiar mineral properties of the soil. The Charkhan Valley is a wild, weird-looking region, looking as if it were habitually subjected to destructive downpourings of rain, that have washed the grand old mountains out of all resemblance to neighboring ranges round about. They are of a soft, shaly composition, and are worn by the elements into all manner of queer, fantastic shapes; this, together with the same variegated colors observed yesterday afternoon, gives them a distinctive appearance not easily forgotten. They are "grand, gloomy, and peculiar;" especially are they peculiar. The soil of the valley itself seems to be drift-mud from the surrounding hills; a stream furnishes water sufficient to irrigate a number of rice-fields, whose brilliant emerald hue loses none of its brightness from being surrounded by a framework of barren hills.

Ascending from this interesting locality my road now traverses a dreary, monotonous district of whitish, sun-blistered hills, waterless and verdureless for fourteen miles. The cool, refreshing breezes of early morning have been dissipated by the growing heat of the sun ; the road continues fairly good, and while riding I am unconscious of oppressive heat ; but the fierce rays of the sun blisters my neck and the backs of my hands, turning them red and causing the skin to peel off a few days afterward, besides ruining a section of my gossamer coat exposed on top of the Lamson carrier. The air is dry and thirst-creating, there is considerable hill-climbing to be done, and long ere the fourteen miles are covered I become sufficiently warm and thirsty to have little thought of anything else but reaching the means of quenching thirst. Away off in the distance ahead is observed a dark object, whose character is indistinct through the shimmering radiation from the heated hills, but which, upon a nearer approach, proves to be a jujube-tree, a welcome sentinel in those arid regions, beckoning the thirsty traveller to a never-failing supply of water. At the jujube-tree I find a most magnificent fountain, pouring forth at least twenty gallons of delicious cold water to the minute. The spring has been walled up and a marble spout inserted, which gushes forth a round, crystal column, as though endeavoring to compensate for the prevailing aridness and to apologize to the thirsty wayfarer for the inhospitableness of its surroundings.

Miles away to the northward, perched high up among the ravines of a sun-baked mountain-spur, one can see a circumscribed area of luxuriant foliage. This conspicuous oasis in the desert marks the source of the beautiful road-side fountain, which traverses a natural subterranean passage-way between these two distant points. These little isolated clumps of waving trees, rearing their green heads conspicuously above the surrounding barrenness, are an unerring indication of both water and human habitations. Often one sees them suddenly when least expected, nestling in a little depression high up some mountain-slope far away, the little dark-green area looking almost black in contrast with the whitish color of the .hills. These are literally "oases in the desert," on a small scale, and although from a distance no sign of human habitations appear;

since they are but mud-hovels corresponding in color to the hills themselves, a closer examination invariably reveals well-worn donkey-trails leading from different directions to the spot, and perchance a white-turbaned donkey-rider slowly wending his way along a trail.

The heat becomes almost unbearable ; the region of treeless, shelterless hills continues to characterize my way, and when, at two o'clock P.M., I reach the town of Bey Bazaar, I conclude that the thirty-nine miles already covered is the limit of discretion to-day, considering the oppressive heat, and seek the friendly accommodation of a *khan*. There I find that while shelter from the fierce heat of the sun is obtainable, peace and quiet are altogether out of the question. Bey Bazaar is a place of eight thousand inhabitants, and the *khan* at once becomes the objective point of, it seems to me, half the population. I put the machine up on a barricaded *yattack*-divan, and climb up after it ; here I am out of the meddlesome reach of the "madding crowd," but there is no escaping from the bedlam-like clamor of their voices, and not a few, yielding to their uncontrollable curiosity, undertake to invade my retreat ; these invariably "skedaddle" respectfully at my request, but new-comers are continually intruding. The tumult is quite deafening, and I should certainly not be surprised to have the *khan-jee* request me to leave the place, on the reasonable ground that my presence is, under the circumstances, detrimental to his interests, since the crush is so great that transacting business is out of the question. The *khan-jee*, however, proves to be a speculative individual, and quite contrary thoughts are occupying his mind. His subordinate, the *kahvay-jee*, presents himself with mournful countenance and humble attitude, points with a perplexed air to the surging mass of fezzes, turbans, and upturned Turkish faces, and explains—what needs no explanation other than the evidence of one's own eyes—that he cannot transact his business of making coffee.

"This is your *khan*," I reply ; "why not turn them out?"*

"*Mashallah, effendi!* I would, but for everyone I turned out; two others would come in—the sons of burnt fathers!" he says, casting a reproachful look down at the struggling crowd of his fellow-countrymen.

"What do you propose doing, then?" I inquire.

"*Katch para, effendi,*" he answers, smiling approvingly at his own suggestion.

The enterprising *kahvay-jee* advocates charging them an admission fee of five paras (half a cent) each as a measure of protection, both for himself and me, proposing to make a "divvy" of the proceeds. Naturally enough the idea of making a farthing show of either myself or the bicycle is anything but an agreeable proposition, but it is plainly the only way of protecting the *kahvay-jee* and his *khan* from being mobbed all the afternoon and far into the night by a surging mass of inquisitive people ; so I reluctantly give him permission to do whatever he pleases to protect himself. I have no idea of the financial outcome of the speculative *khan-jee's* expedient, but the arrangement secures me to some extent from the rabble, though not to any appreciable extent from being worried. The people nearly drive me out of my seven senses with their peculiar ideas of making themselves agreeable, and honoring me ; they offer me cigarettes, coffee, mastic, cognac, fruit, raw cucumbers, melons, everything, in fact, but the one thing I should really appreciate—a few minutes quiet, undisturbed, enjoyment of my own company ; this is not to be secured by locking one's self in a room, nor by any other expedient I have yet tried in Asia. After examining the bicycle, they want to see my " *Alla Franga* " watch and my revolver ; then they want to know how much each thing costs, and scores of other things that appeal strongly to their excessively inquisitive natures.

One old fellow, yearning for a closer acquaintance, asks me if I ever saw the wonderful " chu, chu, chu ! *chemin de fer* at Stamboul," adding that he has seen it and intends some day to ride on it ; another hands me a Crimean medal, and says he fought against the Muscovs with the "Ingilis," while a third one solemnly introduces himself as a "*makinis*" (machinist), fancying, I suppose, that there is some fraternal connection between himself and me, on account of the bicycle being a *makina*.

I begin to feel uncomfortably like a curiosity in a dime museum —a position not exactly congenial to my nature ; so, after enduring this sort of thing for an hour, I appoint the *kahvay-jee* custodian of the bicycle and sally forth to meander about the bazaar a while, where I can at least have the advantage of being able to move about. Upon returning to the *khan*, an hour later, I find there a man whom I remember passing on the road ; he was riding a donkey, the road was all that could be desired, and I swept past him at racing speed, purely on the impulse of the moment, in order to treat him to the abstract sensation of blank amazement. This impromptu action of mine is now bearing its legitimate fruit, for, surrounded by a most attentive audience, the wonder-struck donkey-rider is endeavoring, by word and gesture, to impress upon them some idea of the speed at which I swept past him and vanished round a bend.

The *kahvay-jee* now approaches me, puffing his cheeks out like a penny balloon and jerking his thumb in the direction of the street door. Seeing that I don't quite comprehend the meaning of this mysterious facial contortion, he whispers confidentially aside, "pasha," and again goes through the highly interesting performance of puffing out his cheeks and winking in a knowing manner ; he then says—also confidentially and aside—"lira," winking even more significantly than before. By all this theatrical by-play, the *kahvay-jee* means that the pasha—a man of extraordinary social, political, and, above all, financial importance—has expressed a wish to see the bicycle, and is now outside ; and the *kahvay-jee*, with many significant winks and mysterious hints of "lira," advises me to take the machine outside and ride it for the pasha's special benefit. A portion of the street near by is " ridable under difficulties ; " so I conclude to act on the *kahvay-jee's* suggestion, simply to see what comes of it. Nothing particular comes of it, whereupon the *kahvay-jee* and his patrons all express themselves as disgusted beyond measure because the Pasha failed—to give me a present.

Shortly after this I find myself hobnobbing with a small company of ex-Mecca pilgrims, holy personages with huge green turbans and flowing gowns ; one of them is evidently very holy indeed, almost too holy for human associations one would imagine, for in addition to his green turban he wears a broad green *kammerbund* and a green undergarment ; he is in fact very green indeed. Then a crazy person pushes his way forward and wants me to cure him of his mental infirmity ; at all events I cannot imagine what else he wants ; the man is crazy as a loon, he cannot even give utterance to his own mother-tongue, but tries to express himself in a series of disjointed grunts beside which the soul-harrowing efforts of a broken-winded donkey are quite melodious. Someone has probably told him that I am a *hakim*, or a wonderful person on general principles, and the fellow is sufficiently conscious of his own condition to come forward and endeavor to grunt himself into my favorable consideration.

Later in the evening a couple of young Turkish dandies come

round to the *khan* and favor me with a serenade ; one of them
twangs a doleful melody on a small stringed instrument, something like the Slavonian tamborica, and the other one sings a doleful, melancholy song (nearly all songs and tunes in Mohammedan
countries seem doleful and melancholy) ; afterwards an Arab camel-
driver joins in with a dance, and furnishes some genuine amusement with his hip play and bodily contortions ; this would scarcely
be considered dancing from our point of view, but it *is* according to
the ideas of the East. The dandies are distinguishable from the
common run of Turkish bipeds, like the same species in other
countries, by the fearful and wonderful cut of their garments.
The Turkish dandy wears a tassel to his fez about three times
larger than the regulation size, and he binds it carefully down to
the fez with a red and yellow silk handkerchief; he wears a jaunty-
looking short jacket of bright blue cloth, cut behind so that it
reaches but little below his shoulder-blades ; the object of this is
apparently to display the whole of the multifold *kammerbund,* a
wonderful, colored waist-scarf that is wound round and round the
waist many times, and which is held at one end by an assistant,
while the wearer spins round like a dancing dervish, the assistant
advancing gradually as the human bobbin takes up the length.
The dandy wears knee-breeches corresponding in color to his
jacket, woollen stockings of mingled red and black, and low, slipper-
like shoes ; he allows his hair to fall about his eyes *à la négligée,*
and affects a reckless, love-lorn air.

The last party of sight-seers for the day call around near midnight, some time after I have retired to sleep ; they awaken me
with their garrulous observations concerning the bicycle, which
they are critically examining close to my head with a classic
lamp ; but I readily forgive them their nocturnal intrusion, since
they awaken me to the first opportunity of hearing women wailing
for the dead. A dozen or so of women are wailing forth their
lamentations in the silent night but a short distance from the
khan ; I can look out of a small opening in the wall near my shake-
down, and see them moving about the house and premises by the
flickering glare of torches. I could never have believed the female
form divine capable of producing such doleful, unearthly music ;
but there is no telling what these shrouded forms are really capable of doing, since the opportunity of passing one's judgment
upon their accomplishments is confined solely to an occasional

Serenaded by Turkish Dandies.

glimpse of a languishing eye. The *kahvay-jee*, who is acting the part of explanatory lecturer to these nocturnal visitors, explains the meaning of the wailing by pantomimically describing a corpse, and then goes on to explain that the smallest imaginable proportion of the lamentations that are making night hideous is genuine grief for the departed, most of the uproar being made by a body of professional mourners hired for the occasion. When I awake in the morning the unearthly.wailing is still going vigorously forward, from which I infer they have been keeping it up all night. Though gradually becoming inured to all sorts of strange scenes and customs, the united wailing and lamentations of a houseful of women, awakening the echoes of the silent night, savor too much of things supernatural and unearthly not to jar unpleasantly on the senses ; the custom is, however, on the eve of being relegated to the musty past by the Ottoman Government.

In the larger cities where there are corpses to be wailed over every night, it has been found so objectionable to the expanding intellects of the more enlightened Turks that it has been prohibited as a public nuisance, and these days it is only in such conservative interior towns as Bey Bazaar that the custom still obtains.

When about starting early on the following morning the *khan-jee* begs me to be seated, and then several men who have been waiting around since before daybreak vanish hastily through the door-way ; in a few minutes I am favored with a small company of leading citizens who, having for various reasons failed to swell yesterday's throng, have taken the precaution to post these messengers to watch my movements and report when I am ready to depart. Our grunting patient, the crazy man, likewise reappears upon the scene of my departure from the *khan*, and, in company with a small but eminently respectable following, accompanies me to the brow of a bluffy hill leading out of the depression in which Bey Bazaar snugly nestles. On the way up he constantly gives utterance to his feelings in guttural gruntings that make last night's lamentations seem quite earthly after all in comparison ; and when the summit is reached, and I mount and glide noiselessly away down a gentle declivity, he uses his vocal organs in a manner that simply defies chirographical description or any known comparison ; it is the despairing howl of a semi-lunatic at witnessing my departure without having exercised my supposed extraordinary powers in some miraculous manner in his behalf.

The road continues as an artificial highway, but is not continuously ridable, owing to the rocky nature of the material used in its construction and the absence of vehicular traffic to wear it smooth ; but it is highly acceptable in the main. From Bey Bazaar eastward it leads for several miles along a stony valley, and then through a region that differs little from yesterday's barren hills in general appearance, but which has the redeeming feature of being traversed here and there by deep cañons or gorges, along which meander tiny streams, and whose wider spaces are areas of remark- ably fertile soil. While wheeling merrily along the valley road I am favored with a "peace-offering" of a splendid bunch of grapes from a bold vintager en route to Bey Bazaar with a grape-laden donkey. When within a few hundred yards the man evinces unmistakable signs of uneasiness concerning my character, and would probably follow the bent of his inclinations and ingloriously flee the field, but his donkey is too heavily laden to accompany him ; he looks apprehensively at my rapidly approaching figure, and then, as if a happy thought suddenly occurs to him, he quickly takes the finest bunch of grapes ready to hand and holds them out toward me while I am yet a good fifty yards away. The grapes are luscious, and the bunch weighs fully an oke, but I should feel uncomfortably like a highwayman, guilty of intimidating the man out of his property, were I to accept them in the spirit in which they are offered ; as it is, the honest fellow will hardly fail to trembling in his tracks should he at any future time again descry the centaur-like form of a mounted wheelman approaching him in the distance.

Later in the forenoon I descend into a cañon-like valley where, among a few scattering vineyards and jujube-trees, nestles Ayash, a place which disputes with the neighboring village of Istanos the honor of being the theatre of Alexander the Great's celebrated exploit of cutting the Gordian knot that disentangled the harness of the Phrygian king. Ayash is to be congratulated upon having its historical reminiscence to recommend it to the notice of the outer world, since it has little to attract attention nowadays ; it is merely the shapeless jumble of inferior dwellings that characterize the average Turkish village. As I trundle through the crooked, ill-paved alley-way that, out of respect to the historical association referred to, may be called its business thoroughfare, with forethought of the near approach of noon I obtain some pears, and hand an *ekmek-jee* a coin for some bread ; he passes over a tough flat cake, abundantly sufficient for my purpose, together with the change. A *zaptieh,* looking on, observes that the man has retained a whole half-penny for the bread, and orders him to fork over another cake ; I refuse to take it up, whereupon the *zaptieh* fulfils his ideas of justice by ordering the *ekmek-jee* to give it to a ragged youth among the spectators.

Continuing on my way I am next halted by a young man of the better class, who, together with the *zaptieh,* endeavors to prevail upon me to stop, going through the pantomime of writing and reading, to express some idea that our mutual ignorance of each other's language prevents being expressed in words. The result is a rather curious *intermezzo.* Thinking they want to examine my *teskeri* merely to gratify their idle curiosity, I refuse to be thus bothered, and, dismissing them quite brusquely, hurry along over the rough cobble-stones in hopes of reaching ridable ground and escaping from the place ere the inevitable "madding crowd" become generally aware of my arrival. The young man disappears, while the *zaptieh* trots smilingly but determinedly by my side, several times endeavoring to coax me into making a halt ; which is, however, promptly interpreted by myself into a paternal plea on behalf of the villagers—a desire to have me stop until they could be generally notified and collected—the very thing I am hurrying along to avoid. I am already clear of the village and trundling up the inevitable acclivity, the *zaptieh* and a small gathering still doggedly hanging on, when the young man reappears, hurriedly approaching from the rear, followed by half the village. The *zaptieh* pats me on the shoulder and points back with a triumphant smile ; thinking he is referring to the rabble, I am rather inclined to be angry with him and chide him for dogging my footsteps, when I observe the young man waving aloft a letter, and at once understand that I have been guilty of an ungenerous misinterpretation of their determined attentions. The letter is from Mr. Binns, an English gentleman at Angora, engaged in the exportation of mohair, and contains an invitation to become his guest while at Angora. A well-deserved backsheesh to the good-natured *zaptieh* and a penitential shake of the young man's hand silence the self-accusations of a guilty conscience, and, after riding a short distance down the hill for the satisfaction of the people, I continue on my way, trundling up the varying gradations of a general acclivity for two miles.

Away up the road ahead I now observe a number of queer, shapeless objects, moving about on the roadway, apparently descending the hill, and resembling nothing so much as animated clumps of brushwood. Upon a closer approach they turn out to be not so very far removed from this conception ; they are a company of poor Ayash peasant-women, each carrying a bundle of camel-thorn shrubs several times larger than herself, which they have been scouring the neighboring hills all morning to obtain for fuel. This camel-thorn is a light, spriggy shrub, so that the size of their burthens is large in proportion to its weight. Instead of being borne on the head, they are carried in a way that forms a complete bushy background, against which the shrouded form of the woman is undistinguishable a few hundred yards away. Instead of keeping a straightforward course, the women seem to be doing an unnecessary amount of erratic wandering about over the road, which, until quite near, gives them the queer appearance of animated clumps of brush dodging about among each other. I ask them whether there is water ahead ; they look frightened and hurry along faster, but one brave soul turns partly round and points mutely in the direction I am going. Two miles of good, ridable road now brings me to the spring, which is situated near a two-acre swamp of rank sword-grass and bulrushes six feet high and of almost inpenetrable thickness, which looks decidedly refreshing in its setting of barren, gray hills ; and I eat my noontide meal of bread and pears to the cheery music of a thousand swamp-frog bands which commence croaking at my approach, and never cease for a moment to twang their tuneful lyre until I depart.

The tortuous windings of the *chemin de fer* finally bring me to a *cul-de-sac* in the hills, terminating on the summit of a ridge overlooking a broad plain ; and a horseman I meet informs me that I am now midway between Bey Bazaar and Angora. While ascending this ridge I become thoroughly convinced of what has frequently occurred to me between here and Nalikhan—that if the road I am traversing is, as the people keep calling it, a *chemin de fer*, then the engineer who graded it must have been a youth of tender age, and inexperienced in railway matters, to imagine that trains can ever round his curve or climb his grades. There is something about this broad, artificial highway, and the tremendous amount of labor that has been expended upon it, when com-

pared with the glaring poverty of the country it traverses, together
with the wellnigh total absence of wheeled vehicles, that seem to
preclude the possibility of its having been made for a wagon-road ;
and yet, notwithstanding the belief of the natives, it is evident
that it can never be the road-bed of a railway.   We must inquire
about it at Angora.

Descending into the Angora Plain, I enjoy the luxury of a continuous coast for nearly a mile, over a road that is simply perfect
for the occasion, after which comes the less desirable performance
of ploughing through a stretch of loose sand and gravel.   While
engaged in this latter occupation I overtake a *zaptieh*, also *en route*
to Angora, who is letting his horse crawl leisurely along while he
concentrates his energies upon a water-melon, evidently the 'spoils
of a recent visitation to a melon-garden somewhere not far off ; he
hands me a portion of the booty, and then requests me to *bin*, and
keeps on requesting me to *bin* at regular three-minute intervals for
the next half-hour.   At the end of that time the loose gravel terminates, and I find myself on a level and reasonably smooth dirt-
road, making a shorter cut across the plain to Angora than the
chemin de fer.   The *zaptieh* is, of course, delighted at seeing me
thus mount, and not doubting but that I will appreciate his company, gives me to understand that he will ride alongside to Angora.
For nearly two miles that sanguine but unsuspecting minion of the
Turkish Government spurs his noble steed alongside the bicycle
in spite of my determined pedalling to shake him off ; but the road
improves ; faster spins the whirling wheels ; the *zaptieh* begins to
lag behind a little, though still spurring his panting horse into
keeping reasonably close behind ; a bend now occurs in the road,
and an intervening knoll hides us from each other ; I put on more
steam, and at the same time the *zaptieh* evidently gives it up and
relapses into his normal crawling pace, for when three miles or
thereabout are covered I look back and perceive him leisurely
heaving in sight from behind the knoll.

Part way across the plain I arrive at a fountain and make a short
halt, for the day is unpleasantly warm, and the dirt-road is covered
with dust ; the government *postaya araba* is also halting here to rest
and refresh the horses.   I have not failed to notice the proneness
of Asiatics to base their conclusions entirely on a person's apparel
and general outward appearance, for the seeming incongruity of my
" Ingilis " helmet and the Circassian moccasins has puzzled them not

a little on more than one occasion. And now one wiseacre among
this party at the road-side fountain stubbornly asserts that I cannot possibly be an Englishman because of my wearing a mustache

Racing with the Zaptieh.

without side whiskers—a feature that seems to have impressed
upon his enlightened mind the unalterable conviction that I am an
" Austrian ; " why an Austrian any more than a Frenchman or an
inhabitant of the moon, I wonder ? and wondering, wonder in vain.

Five P.M., August 16, 1885, finds me seated on a rude stone slab, one of those ancient tombstones whose serried ranks constitute the suburban scenery of Angora, ruefully disburdening my nether garments of mud and water, the results of a slight miscalculation of my abilities at leaping irrigating ditches with the bicycle for a vaulting-pole. While engaged in this absorbing occupation several inquisitives mysteriously collect from somewhere, as they invariably do whenever I happen to halt for a minute, and following the instructions of the Ayash letter I inquire the way to the "Ingilisin Adam" (Englishman's man). They pilot me through a number of narrow, ill-paved streets leading up the sloping hill which Angora occupies—a situation that gives the supposed ancient capital of Galatia a striking appearance from a distance—and into the premises of an Armenian whom I find able to make himself intelligible in English, if allowed several minutes undisturbed possession of his own faculties of recollection between each word—the gentleman is slow but not quite sure. From him I learn that Mr. Binns and family reside during the summer months at a vineyard five miles out, and that Mr. Binns will not be in town before to-morrow morning; also that, "You are welcome to the humble hospitality of our poor family."

This latter way of expressing it is a revelation to me, and the leaden-heeled and labored utterance, together with the general bearing of my volunteer host, is not less striking; if meekness, lowliness, and humbleness, permeating a person's every look, word, and action, constitute worthiness, then is our Armenian friend beyond a doubt the worthiest of men. Laboring under the impression that he is Mr. Binns' "Ingilisin Adam," I have no hesitation about accepting his proffered hospitality for the night; and storing the bicycle away, I proceed to make myself quite at home, in that easy manner peculiar to one accustomed to constant change. Later in the evening imagine my astonishment at learning that I have thus nonchalantly quartered myself, so to speak, not on Mr. Binns' man, but on an Armenian pastor who has acquired his slight acquaintance with my own language from being connected with the American Mission having headquarters at Kaisarieh!

All the evening long, noisy crowds have been besieging the pastorate, worrying the poor man nearly out of his senses on my account; and what makes matters more annoying and lamentable, I learn afterward that his wife has departed this life but

a short time ago, and the bereaved pastor is still bowed down
with sorrow at the affliction—I feel like kicking myself unceremoniously out of his house. Following the Asiatic custom of welcoming a stranger, and influenced, we may reasonably suppose, as
much by their eagerness to satisfy their consuming curiosity as anything else, the people come flocking in swarms to the pastorate
again next morning, filling the house and grounds to overflowing,
and endeavoring to find out all about me and my unheard-of mode
of travelling, by questioning the poor pastor nearly to distraction. That excellent man's thoughts seem to run entirely on missionaries and mission enterprises ; so much so, in fact, that several negative assertions from me fail to entirely disabuse his mind
of an idea that I am in some way connected with the work of
spreading the Gospel in Asia Minor ; and coming into the room
where I am engaged in the interesting occupation of returning the
salaams and inquisitive gaze of fifty ceremonious visitors, in slow,
measured words he asks, "Have you any words for these people?"
as if quite expecting to see me rise up and solemnly call upon the
assembled Mussulmans, Greeks, and Armenians to forsake the religion of the False Prophet in the one case, and mend the error
of their ways in the other. I know well enough what they all
want, though, and dismiss them in a highly satisfactory manner by
promising them that they shall all have an opportunity of seeing
the bicycle ridden before I leave Angora.

About ten o'clock Mr. Binns arrives, and is highly amused at the
ludicrous mistake that brought me to the Armenian pastor's instead
of to his man, with whom he had left instructions concerning me,
should I arrive after his departure in the evening for the vineyard ;
in return he has an amusing story to tell of the people waylaying
him on his way to his office, telling him that an Englishman had
arrived with a wonderful *araba*, which he had immediately locked
up in a dark room and would allow nobody to look at it, and begging him to ask me if they might come and see it. We spend the
remainder of the forenoon looking over the town and the bazaar,
Mr. Binns kindly announcing himself as at my service for the day,
and seemingly bent on pointing out everything of interest.

One of the most curious sights, and one that is peculiar to Angora, owing to its situation on a hill where little or no water is
obtainable, is the bewildering swarms of *su-katirs* (water donkeys)
engaged in the transportation of that important necessary up into

21

the city from a stream that flows near the base of the hill. These unhappy animals do nothing from one end of their working lives to the other but toil, with almost machine-like regularity and un-eventfulness, up the crooked, stony streets with a dozen large earthen-ware jars of water, and down again with the empty jars. The donkey is sandwiched between two long wooden troughs sus-pended to a rude pack-saddle, and each trough accommodates six jars, each holding about two gallons of water; one can readily im-agine the swarms of these novel and primitive conveyances required

Angora Water-works.

to supply a population of thirty-five thousand people. Upon in-quiring what they do in case of a fire, I learn that they don't even think of fighting the devouring element with its natural enemy, but, collecting on the adjoining roofs, they smother the flames by pelting the burning building with the soft, crumbly bricks of which Angora is chiefly built; a house on fire, with a swarm of half-naked natives on the neighboring housetops bombarding the leaping flames with bricks, would certainly be an interesting sight.

Other pity-exciting scenes besides the patient little water-carry-ing donkeys are not likely to be wanting on the streets of an Asiatic city ; one case I notice merits particular mention. A youth with both arms amputated at the shoulder, having not so much as the stump of an arm, is riding a donkey, and persuading the unwilling animal along quite briskly—with a stick. All Christendom could never guess how a person thus afflicted could possibly wield a stick so as to make any impression upon a donkey ; but this ingenious person holds it quite handily between his chin and right shoulder, and from constant practice has acquired the ability to visit his long-eared steed with quite vigorous thwacks.

Near noon we repair to the government house to pay a visit to Sirra Pasha, the Vali or governor of the *vilayet*, who, having heard of my arrival, has expressed a wish to have us call on him. We happen to arrive while he is busily engaged with an important legal decision, but upon our being announced he begs us to wait a few minutes, promising to hurry through with the business. We are then requested to enter an adjoining apartment, where we find the Mayor, the Cadi, the Secretary of State, the Chief of the Angora *zaptiehs*, and several other functionaries, signing documents, affix-ing seals, and otherwise variously occupied. At our entrance, doc-uments, pens, seals, and everything are relegated to temporary oblivion, coffee and cigarettes are produced, and the journey *dunia-nin-athrafana* (around the world) I am making with the wonderful *araba* becomes the all-absorbing subject. These wise men of state entertain queer, Asiatic notions concerning the probable object of my journey ; they cannot bring themselves to believe it possible that I am performing so great a journey "merely as the *Outing* correspondent ;" they think it more probable, they say, that my real incentive is to "spite an enemy"—that, having quarrelled with another wheelman about our comparative skill as riders, I am wheeling entirely around the globe in order to prove my superior-ity, and at the same time leave no opportunity for my hated rival to perform a greater feat—Asiatic reasoning, sure enough ! Rea-soning thus, and commenting in this wise among themselves, their curiosity becomes worked up to the highest possible pitch, and they commence plying Mr. Binns with questions concerning the mechanism and general appearance of the bicycle. To facilitate Mr. Binns in his task of elucidation, I produce from my inner coat-pocket a set of the earlier sketches illustrating the tour across

America, and for the next few minutes the set of sketches are of more importance than all the State documents in the room. Curiously enough, the sketch entitled "A Fair Young Mormon " attracts more attention than any of the others.

The Mayor is Suleiman Effendi, the same gentleman mentioned at some length by Colonel Burnaby in his " On Horseback Through Asia Minor," and one of his first questions is whether I am acquainted with "my friend Burnaby, whose tragic death in the Soudan will never cease to make me feel unhappy." Suleiman Effendi appears to be remarkably intelligent, compared with many Asiatics, and, moreover, of quite a practical turn of mind ; he inquires what I should do in case of a serious break-down somewhere in the far interior, and his curiosity to see the bicycle is not a little increased by hearing that, notwithstanding the extreme airiness of my strange vehicle, I have had no serious mishap on the whole journey across two continents. Alluding to the bicycle as the latest product of that Western ingenuity that appears so marvellous to the Asiatic mind, he then remarks, with some animation, "The next thing we shall see will be Englishmen crossing over to India in balloons, and dropping down at Angora for refreshments."

A uniformed servant now announces that the Vali is at liberty, and waiting to receive us in private audience. Following the attendant into another room, we find Sirra Pasha seated on a richly cushioned divan, and upon our entrance he rises smilingly to receive us, shaking us both cordially by the hand. As the distinguished visitor of the occasion, I am appointed to the place of honor next to the governor, while Mr. Binns, with whom, of course, as a resident of Angora, His Excellency is already quite well acquainted, graciously fills the office of interpreter, and enlightener of the Vali's understanding concerning bicycles in general, and my own wheel and wheel journey in particular. Sirra Pasha is a full-faced man of medium height, black-eyed, black-haired, and, like nearly all Turkish pashas, is rather inclined to corpulency. Like many prominent Turkish officials, he has discarded the Turkish costume, retaining only the national fez ; a head-dress which, by the by, is without one single merit to recommend it save its picturesqueness. In sunny weather it affords no protection to the eyes, and in rainy weather its contour conducts the water in a trickling stream down one's spinal column. It is too thin to protect the scalp from the fierce sun-rays, and too close-fitting and close in texture to afford

any ventilation, yet with all this formidable array of disadvantages
it is universally worn.

I have learned during the morning that I have to thank Sirra
Pasha's energetic administration for the artificial highway from
Keshtobek, and that he has constructed in the *vilayet* no less than
two hundred and fifty miles of this highway, broad and reasonably
well made, and actually macadamized in localities where the necessary material is to be obtained. The amount of work done in con-
structing this road through so mountainous a country is, as before
mentioned, plainly out of all proportion to the wealth and population of a second-grade *vilayet* like Angora, and its accomplishment
has been possible only by the employment of forced labor. Every
man in the whole *vilayet* is ordered out to work at the road-making
a certain number of days every year, or provide a substitute ; thus,
during the present summer there have been as many as twenty thousand men, besides donkeys, working on the roads at one time. Unaccustomed to public improvements of this nature, and, no doubt,
failing to see their advantages in a country practically without vehicles, the people have sometimes ventured to grumble at the rather
arbitrary proceeding of making them work for nothing, and board
themselves ; and it has been found expedient to make them believe
that they were doing the preliminary grading for a railway that
was shortly coming to make them all prosperous and happy ; beyond being credulous enough to swallow the latter part of the bait,
few of them have the least idea of what sort of a looking thing a
railroad would be.

When the Vali hears that the people all along the road have
been telling me it was a *chemin de fer*, he fairly shakes in his boots
with laughter. Of course I point out that no one can possibly appreciate the road improvements any more than a wheelman, and
explain the great difference I have found between the mule-paths
of Kodjaili and the broad highways he has made through Angora,
and I promise him the universal good opinion of the whole world
of 'cyclers. In reply, His Excellency hopes this favorable opinion
will not be jeopardized by the journey to Yuzgat, but expresses
the fear that I shall find heavier wheeling in that direction, as the
road is newly made, and there has been no vehicular traffic to pack
it down.

The Governor invites me to remain over until Thursday and
witness the ceremony of laying the corner-stone of a new school, of

the founding of which he has good reason to feel proud, and which
ought to secure him the esteem of right-thinking people everywhere. He has determined it to be a common school in which no
question of Mohammedan, Jew, or Christian, will be allowed to enter, but where the young ideas of Turkish, Christian, and Jewish
youths shall be taught to shoot peacefully and harmoniously together. Begging to be excused from this, he then invites me to
take dinner with him to-morrow evening; but this I also decline,
excusing myself for having determined to remain over no longer
than a day on account of the approaching rainy season and my
anxiety to reach Teheran before it sets in. Yet a third time the
pasha rallies to the charge, as though determined not to let me off
without honoring me in some way ; and this time he offers to furnish me a *zaptieh* escort, but I tell him of the *zaptieh's* inability to
keep up yesterday, at which he is immensely amused. His Excellency then promises to be present at the starting-point to-morrow
morning, asking me to name the time and place, after which we
finish the cigarettes and coffee and take our leave.

We next take a survey of the mohair caravansary, where buyers
and sellers and exporters congregate to transact business, and I
watch with some interest the corps of half-naked sorters seated
before large heaps of mohair, assorting it into the several classes
ready for exportation. Here Mr. Binns' office is situated, and
we are waited upon by several of his business acquaintances ; among
them a member of the celebrated—celebrated in Asia Minor—Tifticjeeoghlou family, whose ancestors have been prominently engaged
in the mohair business for so long that their very name is significa-
tory of their profession—Tifticjee-oghlou, literally, "Mohair-dealer's
son." The Smiths, Bakers, and Hunters of Occidental society are
not a whit more significative than are many prominent names of
the Orient. Prominent among the Angorians is a certain Mr. Al-
tentopoghlou, the literal interpretation of which is, " Son of the
golden ball," and the origin of whose family name Eastern tradition
has surrounded by the following little interesting anecdote :

Ages ago it pleased one of the Sultans to issue a proclamation
throughout the empire, promising to present a golden ball to
whichever among all his subjects should prove himself the biggest
liar, giving it to be understood beforehand that no " merely improbable story " would stand the ghost of a chance of winning,
since he himself was to be the judge, and nothing short of a story

that was simply impossible would secure the prize. The proclamation naturally made quite a stir among the great prevaricators
of the realm, and hundreds of stories came pouring in from competitors everywhere, some even surreptitiously borrowing "whop-
pers" from the Persians, who are well known as the greatest
economizers of the truth in all Asia; but they were one and all adjudged by the astute monarch—who was himself a most experienced prevaricator—probably the noblest Roman of them all—as
containing incidents that might under extraordinary circumstances
have been true. The coveted golden ball still remained unawarded,
when one day there appeared before the gate of the Sultan's
palace, requesting an audience, an old man with travel-worn
appearance, as though from a long pilgrimage, and bearing on his
stooping shoulders an immense earthen-ware jar. The Sultan received the aged pilgrim kindly, and asked him what he could do
for him.

"Oh, Sultan, may you live forever!" exclaimed the old man,
"for your Imperial Highness is loved and celebrated throughout
all the empire for your many virtues, but most of all for your well-
known love of justice."

"Inshallah!" replied the monarch, reverently.

"May it please Your Imperial Majesty," continued the old man,
calling the monarch's attention to the jar, "Your Highness' most
excellent father—may his bones rest in peace!—borrowed from
my father this jar full of gold coins, the conditions being that
Your Majesty was to pay the same amount back to me."

"Absurd, impossible!" exclaimed the astonished Sultan, eying the huge vessel in question.

"If the story be true," gravely continued the pilgrim, "pay
your father's debt; if it is as you say, impossible, I have fairly won
the golden ball." And the Sultan immediately awarded him the
prize.

In the cool of the evening we ride out on horseback through
vineyards and yellow-berry gardens to Mr. Binns' country residence, a place that formerly belonged to an old pasha, a veritable
Bluebeard, who built the house and placed the windows of his
harem, even closely latticed as they always are, in a position that
would not command so much as a glimpse of passers-by on the
road, hundreds of yards away. He planted trees and gardens, and
erected marble fountains at great cost. Surrounding the whole

with a wall, and purchasing three beautiful young wives, the old
Turk fondly fancied he had created for himself an earthly paradise ;
but as love laughs at locksmiths, so did these three frisky dames
laugh at latticed windows, and lay their heads together against
being prevented from watching passers-by through the windows of
the harem. With nothing else to do, they would scheme and plot
all day long against their misguided husband's tranquillity and
peace of mind. One day, while sunning himself in the garden, he
discovered that they had managed to detach a section of the
lattice-work from a window, and were in the habit of sticking out
their heads—awful discovery ! Flying into a righteous rage at
this act of flagrant disobedience, he seized a thick stick and sought
their apartments, only to find the lattice-work skilfully replaced,
and to be confronted with a general denial of what he had witnessed with his own eyes. This did not prevent them from all
three getting a severe chastisement ; but as time wore on he
found the life these three caged-up young women managed to lead
him anything but the earthly paradise he thought he was creating;
and, financial troubles overtaking him at the same time, the old
fellow fairly died of a broken heart in less than twelve months
after he had so hopefully installed himself in his self-created
heaven.

There is a moral in the story somewhere, I think, for anybody
caring to analyze it. Mr. Binns says the old Mussulman was also
an inveterate hater of unbelievers, and that the old fellow's bones
would fairly rattle in his coffin were he conscious that a family of
Christians are now actually occupying the house he built with such
careful regard for the Mussulman's ideas of a material heaven, with
trees and fountains and black-eyed houris.

Near ten o'clock on Tuesday morning finds Angora the scene
of more excitement than it has seen for some time. I am trundling through the narrow streets toward the appointed starting-
place, which is at the commencement of a half-mile stretch of excellent level macadam, just beyond the tombstone-planted suburbs
of the city. Mr. Binns is with me, and a squad of *zaptiehs* are engaged in the lively occupation of protecting us from the crush of
people following us out ; they are armed especially for the occasion with long switches, with which they unsparingly lay about
them, seemingly only too delighted at the chance of making the
dust fly from the shoulders of such unfortunate wights as the

pressure of the throng forces anywhere near the magic cause of the commotion. The time and place of starting have been proclaimed by the Vali and have become generally noised abroad, and near three thousand people are already assembled when we arrive; among them is seen the genial face of Suleiman Effendi, who, in his capacity of mayor, is early on the ground with a force of *zaptiehs* to maintain order; and with a little knot of friends, behold, is also our humble friend the Armenian pastor, the irresistible attractions of the wicked bicycle having temporarily overcome his contempt of the pomps and vanities of secular displays.

"Englishmen are always punctual!" says Suleiman Effendi, looking at his watch; and, upon consulting our own, sure enough we have happened to arrive precisely to the minute. An individual named Mustapha, a blacksmith who has acquired an enviable reputation for skill on account of the beautiful horseshoes he turns out, now presents himself and begs leave to examine the mechanism of the bicycle, and the question arises among the officers standing by as to whether Mustapha would be able to make one; Mustapha himself thinks he could, providing he had mine always at hand to copy from.

"Yes," suggests the practical-minded Suleiman Effendi, "yes, Mustapha, you may have *marifet* enough to make one; but when you have finished it, who among all of us will have *marifet* enough to ride it?"

"True, effendi," solemnly assents another, "we would have to send for an Englishman to ride it for us, after Mustapha had turned it out!"

The Mayor now requests me to ride along the road once or twice to appease the clamor of the multitude until the Vali arrives. The crowd along the road is tremendous, and on a neighboring knoll, commanding a view of the proceedings, are several carriage-loads of ladies, the wives and female relatives of the officials. The Mayor is indulgent to his people, allowing them to throng the roadway, simply ordering the *zaptiehs* to keep my road through the surging mass open. While on the home-stretch from the second spin, up dashes the Vali in the state equipage with quite an imposing bodyguard of mounted *zaptiehs*, their chief being a fine military-looking Circassian in the picturesque military costume of the Caucasus. These horsemen the Governor at once orders to clear the people entirely off the road-way—an order no sooner given than executed;

and after the customary interchange of salutations, I mount and wheel briskly up the broad, smooth macadam between two compact masses of delighted natives ; excitement runs high, and the people clap their hands and howl approvingly at the performance, while the horsemen gallop briskly to and fro to keep them from intruding on the road after I have wheeled past, and obstructing the Governor's view.   After riding back and forth a couple of times, I dismount at the Vali's carriage ; a mutual interchange of adieus and well-wishes all around, and I take my departure, wheeling along at a ten-mile pace amid the vociferous plaudits of at least four thousand people, who watch my retreating figure until I disappear over the brow of a hill.   At the upper end of the main crowd are stationed the "irregular cavalry" on horses, mules, and donkeys ; and among the latter I notice our ingenious friend, the armless youth of yesterday, whom I now make happy by a nod of recognition, having scraped up a backsheesh acquaintance with him yesterday.

For some miles the way continues fairly smooth and hard, leading through a region of low vineyard-covered hills, but ere long I arrive at the newly made road mentioned by the Vali.

After which, like the course of true love, my forward career seldom runs smooth for any length of time, though ridable donkey-trails occasionally run parallel with the bogus *chemin de fer*.   For mile after mile I now alternately ride and trundle along donkey-paths, by the side of an artificial highway that would be an enterprise worthy of a European State.   The surface of the road is either gravelled or of broken rock, and well rounded for self-drainage ; it is graded over the mountains, and wooden bridges, with substantial rock supports, are built across the streams ; nothing is lacking except the vehicles to utilize it.   In the absence of these it would almost seem to have been an unnecessary and superfluous expenditure of the people's labor to make such a road through a country most of which is fit for little else but grazing goats and buffaloes.   Aside from some half-dozen carriages at Angora, and a few light government *postaya arabas*—an innovation from horses for carrying the mail, recently introduced as a result of the improved roads, and which make weekly trips between such points as Angora, Yuzgat, and Tokat—the only vehicles in the country are the buffalo-carts of the larger farmers, rude home made *arabas* with solid wooden wheels, whose infernal creaking can be heard for a mile, and which they seldom take any distance from home, preferring their pack-donkeys and cross-country trails when going to town with produce. Perhaps in time vehicular traffic may appear as a result of suitable roads ; but the natives are slow to adopt new improvements.

About two hours from Angora I pass through a swampy upland basin, containing several small lakes, and then emerge into a much less mountainous country, passing several mud villages, the inhabitants of which are a dark-skinned people—Turkoman refugees, I think—who look several degrees less particular about their personal cleanliness than the villagers west of Angora. Their wretched mud hovels would seem to indicate the last degree of poverty, but numerous flocks of goats and herds of buffalo grazing near apparently tell a somewhat different story. The women and children seem mostly engaged in manufacturing cakes of *tezek* (large flat cakes of buffalo manure mixed with chopped straw, which are "dobbed" on the outer walls to dry ; it makes very good fuel, like the "buffalo chips" of the far West), and stacking it up on the house-tops, with provident forethought, for the approaching winter.

Just as darkness is beginning to settle down over the landscape I arrive at one of these unpromising-looking clusters, which, it seems, are now peculiar to the country, and not characteristic of any particular race, for the one I arrive at is a purely Turkish village. After the usual preliminaries of pantomime and *binning*, I am conducted to a capacious flat roof, the common covering of several dwellings and stables bunched up together. This roof is as smooth and hard as a native threshing-floor, and well knowing, from recent experiences, the *modus operandi* of capturing the hearts of these bland and childlike villagers, I mount and straightway secure their universal admiration and applause by riding a few times round the roof. I obtain a supper of fried eggs and *yaort* (milk soured with rennet), eating it on the house-top, surrounded by the whole population of the village, on this and adjoining roofs, who watch my every movement with the most intense curiosity. It is the raggedest audience I have yet been favored with. There are not over half a dozen decently clad people among them all, and two of these are horsemen, simply remaining over night, like myself. Everybody has a fearfully flea-bitten appearance, which augurs ill for a refreshing night's repose.

Here, likewise I am first introduced to a peculiar kind of bread, that I straightway condemn as the most execrable of the many varieties my everchanging experiences bring me in contact with, and which I find myself mentally, and half unconsciously, naming—"blotting-paper *ekmek*"—a not inappropriate title to convey its ap-

Genuine Ekmek.

pearance to the civilized mind; but the sheets of blotting-paper must be of a wheaten color and in circular sheets about two feet in diameter. This peculiar kind of bread is, we may suppose, the natural result of a great scarcity of fuel, a handful of *tezek*, beneath the large, thin sheet-iron griddle, being sufficient to bake many

cakes of this bread. At first I start eating it something like a
Shantytown goat would set about consuming a political poster, if
it—not the political poster, but the Shantytown goat—had a pair
of hands. This outlandish performance creates no small merriment among the watchful on-lookers, who forthwith initiate me
into the mode of eating it à la Turque, which is, to roll it up like
a scroll of paper and bite mouthfuls off the end. I afterwards find
this particular variety of ekmek quite handy when seated around a
communal bowl of yaort with a dozen natives ; instead of taking
my turn with the one wooden spoon in common use, I would form
pieces of the thin bread into small handleless scoops, and, dipping up the yaort, eat scoop and all. Besides sparing me from
using the same greasy spoon in common with a dozen natives,
none of them overly squeamish as regards personal cleanliness, this
gave me the appreciable advantage of dipping into the dish as often
as I choose, instead of waiting for my regular turn at the wooden
spoon.

Though they are Osmanli Turks, the women of these small villages appear to make little pretence of covering their faces. Among
themselves they constitute, as it were, one large family gathering,
and a stranger is but seldom seen. They are apparently simple-
minded females, just a trifle shame-faced in their demeanor before
a stranger, sitting apart by themselves while listening to the conversation between myself and the men. This, of course, is very
edifying, even apart from its pantomimic and monosyllabic character, for I am now among a queer people, a people through the
unoccupied chambers of whose unsophisticated minds wander
strange, fantastic thoughts. One of the transient horsemen, a contemplative young man, the promising appearance of whose upper
lip proclaims him something over twenty, announces that he likewise is on the way to Yuzgat ; and after listening attentively to my
explanations of how a wheelman climbs mountains and overcomes
stretches of bad road, he solemnly inquires whether a 'cycler could
scurry up a mountain slope all right if some one were to follow behind and touch him up occasionally with a whip, in the persuasive
manner required in driving a horse. He then produces a rawhide
"persuader," and ventures the opinion that if he followed close
behind me to Yuzgat, and touched me up smartly with it whenever
we came to a mountain, or a sandy road, there would be no necessity of trundling any of the way. He then asks, with the innocent

simplicity of a child, whether in case he made the experiment, I would get angry and shoot him.

The Unspeakable Oriental.

The other transient appears of a more speculative turn of mind, and draws largely upon his own pantomimic powers and my limited

knowledge of Turkish, to ascertain the difference between the
katch lira of a bicycle at retail, and the *katch lira* of its manufacture. From the amount of mental labor he voluntarily inflicts
upon himself to acquire this particular item of information, I apprehend that nothing less than wild visions of acquiring a rapid
fortune by starting a bicycle factory at Angora, are flitting through
his imaginative mind. The villagers themselves seem to consider
me chiefly from the standpoint of their own peculiar ideas concerning the nature of an Englishman's feelings toward a Russian.
My performance on the roof has put them in the best of humor,
and has evidently whetted their appetites for further amusement.
Pointing to a stolid-looking individual, of an apparently taciturn
disposition, and who is one of the respectably-dressed few, they
accuse him of being a Russian ; and then all eyes are turned towards me, as though they quite expect to see me rise up wrathfully
and make some warlike demonstration against him. My undemon-
strative disposition forbids so theatrical a proceeding, however,
and I confine myself to making a pretence of falling into the trap,
casting furtive glances of suspicion towards the supposed hated
subject of the Czar, and making whispered inquiries of my immediate neighbors concerning the nature of his mission in Turkish territory. During this interesting comedy the "audience" are fairly
shaking in their rags with suppressed merriment ; and when the
taciturn individual himself—who has thus far retained his habitual
self-composure—growing restive under the hateful imputation of
being a Muscov and my supposed bellicose sentiments toward him
in consequence, finally repudiates the part thus summarily assigned
him, the whole company bursts out into a boisterous roar of
laughter. At this happy turn of sentiment I assume an air of intense relief, shake the taciturn man's hand, and, borrowing the
speculative transient's fez, proclaim myself a Turk, an act that fairly
"brings down the house."

Thus the evening passes merrily away until about ten o'clock,
when the people begin to slowly disperse to the roofs of their respective habitations, the whole population sleeping on the house-
tops, with no roof over them save the star-spangled vault—the
arched dome of the great mosque of the universe, so often adorned
with the pale yellow, crescent-shaped emblem of their religion.
Several families occupy the roof which has been the theatre of the
evening's social gathering, and the men now consign me to a com-

fortable couch made up of several quilts, one of the transients
thoughtfully cautioning me to put my moccasins under my pillow,
as these articles were the object of almost universal covetousness
during the evening. No sooner am I comfortably settled down,
than a wordy warfare breaks out in my immediate vicinity, and
an ancient female makes a determined dash at my coverlet, with
the object of taking forcible possession; but she is seized and
unceremoniously hustled away by the men who assigned me my
quarters. It appears that, with an eye singly and disinterestedly
to my own comfort, and regardless of anybody else's, they have,
without taking the trouble to obtain her consent, appropriated to
my use the old lady's bed, leaving her to shift for herself any way
she can, a high-handed proceeding that naturally enough arouses
her virtuous indignation to the pitch of resentment.

Upon this fact occurring to me, I of course immediately vacate
the property in dispute, and, with true Western gallantry, arraign
myself on the rightful owner's side by carrying my wheel and other
effects to another position; whereupon a satisfactory compromise
is soon arranged between the disputants, by which another bed is
prepared for me, and the ancient dame takes triumphant possession
of her own. Peace and tranquillity being thus established on a
firm basis, the several families tenanting our roof settle themselves
snugly down. The night is still and calm, and naught is heard
save my nearer neighbors' scratching, scratching, scratching. This
—not the scratching, but the quietness—doesn't last long, however,
for it is customary to collect all the four-footed possessions of the
village together every night and permit them to occupy the interspaces between the houses, while the humans are occupying the
roofs, the horde of watch-dogs being depended upon to keep
watch and ward over everything. The hovels are more underground than above the surface, and often, when the village occupies sloping ground, the upper edge of the roof is practically but
a continuation of the solid ground, or at the most there is but a
single step-up between them. The goats are of course permitted
to wander whithersoever they will, and equally, of course, they abuse
their privileges by preferring the roofs to the ground and wandering incessantly about among the sleepers. Where the roof comes
too near the ground some temporary obstruction is erected, to
guard against the intrusion of venturesome buffaloes.

No sooner have the humans quieted down, than several goats

promptly invade the roof, and commence their usual nocturnal promenade among the prostrate forms of their owners, and further indulge their well-known goatish propensities by nibbling away the edges of the roof. (They would, of course, prefer a square meal off a patchwork quilt, but from their earliest infancy they are taught that meddling with the bedclothes will bring severe punishment.) A buffalo occasionally gives utterance to a solemn, prolonged " m-o-o-o ; " now and then a baby wails its infantile disapproval of the fleas, and frequent noisy squabbles occur among the dogs. Under these conditions, it is not surprising that one should woo in vain the drowsy goddess ; and near midnight some person within a few yards of my couch begins groaning fearfully, as if in great pain —probably a case of the stomach-ache, I mentally conclude, though this hasty conclusion may not unnaturally result from an inner consciousness of being better equipped for curing that particular affliction than any other. From the position of the sufferer, I am inclined to think it is the same ancient party that ousted me out of her possessions two hours ago, and I lay here as far removed from the realms of unconsciousness as the moment I retired, expecting every minute to see her appear before me in a penitential mood, asking me to cure her, for the inevitable *hakim* question had been raised during the evening. She doesn't present herself, however ; perhaps the self-accusations of her conscience, for having in the moment of her wrath attempted to appropriate my coverlet in so rude a manner, prevent her appealing to me now in the hour of distress.

These people are early risers ; the women are up milking the goats and buffaloes before daybreak, and the men hieing them away to the harvest fields and threshing-floors. I, likewise, bestir myself at daylight, intending to reach the next village before breakfast.

22

CHAPTER XIV.

ACROSS THE KIZIL IRMAK RIVER TO YUZGAT.

The country continues much the same as yesterday, with the road indifferent for wheeling. Reaching the expected village about eight o'clock, I breakfast off *ekmek* and new buffalo milk, and at once continue on my way, meeting nothing particularly interesting, save a lively bout occasionally with goat-herds' dogs—the reminiscences of which are doubtless more vividly interesting to myself than they would be to the reader—until high noon, when I arrive at another village, larger, but equally wretched-looking, on the Kizil Irmak River, called Jas-chi-khan. On the west bank of the stream are some ancient ruins of quite massive architecture, and standing on the opposite side of the road, evidently having some time been removed from the ruins with a view to being transported elsewhere, is a couchant lion of heroic proportions, carved out of a solid block of white marble ; the head is gone, as though its would-be possessors, having found it beyond their power to transport the whole animal, have made off with what they could. An old and curiously arched bridge of massive rock spans the river near its entrance to a wild, rocky gorge in the mountains ; a primitive grist mill occupies a position to the left, near the entrance to the gorge, and a herd of camels are slaking their thirst or grazing near the water's edge to the right—a genuine Eastern picture, surely, and one not to be seen every day, even in the land where to see it occasionally is quite possible.

Riding into Jas-chi-khan, I dismount at a building which, from the presence of several "do-nothings," I take to be a *khan* for the accommodation of travellers. In a partially open shed-like apartment are a number of demure looking maidens, industriously employed in weaving carpets by hand on a rude, upright frame, while two others, equally demure-looking, are seated on the ground cracking wheat for *pillau*, wheat being substituted for rice where the latter is not easily obtainable, or is too expensive. Waiving all

considerations of whether I am welcome or not, I at once enter
this abode of female industry, and after watching the interesting
process of carpet-weaving for some minutes, turn my attention to
the preparers of cracked wheat.   The process is the same primitive
one that has been employed among these people from time immemorial, and the same that is referred to in the passage of Scripture
which says: "Two women were grinding corn in the field;" it consists of a small upper and nether millstone, the upper one being

A Sketch on the Kizil Irmak.

turned round by two women sitting facing each other; they both
take hold of a perpendicular wooden handle with one hand, employing the other to feed the mill and rake away the cracked grain.
These two young women have evidently been very industrious this
morning; they have half-buried themselves in the product of their
labors, and are still grinding away as though for their very lives,
while the constant "click-clack" of the carpet weavers prove them
likewise the embodiment of industry.

They seem rather disconcerted by the abrupt intrusion and scrutinizing attentions of a Frank and a stranger; however, the fascinating search for bits of interesting experience forbids my retirement on that account, but rather urges me to make the most of fleeting opportunities. Picking up a handful of the cracked wheat, I inquire of one of the maidens if it is for *pillau;* the maiden blushes at being thus directly addressed, and with downcast eyes vouchsafes an affirmative nod in reply ; at the same time an observant eye happens to discover a little brown big-toe peeping out of the heap of wheat, and belonging to the same demure maiden with the downcast eyes. I know full well that I am stretching a point of Mohammedan etiquette, even by coming among these industrious damsels in the manner I am doing, but the attention of the men is fully concentrated on the bicycle outside, and the temptation of trying the experiment of a little jocularity, just to see what comes of it, is under the circumstances irresistible.   Conscious of venturing where angels fear to tread, I stoop down, and take hold of the peeping little brown big-toe, and addressing the demure maiden with the downcast eyes, inquire, "Is this also for *pillau ?* "   This proves entirely too much for the risibilities of the industrious *pillau* grinders, and letting go the handle of the mill, they both give themselves up to uncontrollable laughter ; the carpet-weavers have been watching me out of the corners of their bright, black eyes, and catching the infection, the click-clack of the carpet-weaving machines instantly ceases, and several of the weavers hurriedly retreat into an adjoining room to avoid the awful and well-nigh unheard-of indiscretion of laughing in the presence of a stranger.   Having thus yielded to the temptation and witnessed the results, I discreetly retire, meeting at the entrance a gray-bearded Turk coming to see what the merriment and the unaccountable stopping of the carpet-weaving frames is all about.

A sheep has been slaughtered in Jas-chi-khan this morning, and I obtain a nice piece of mutton, which I hand to a bystander, asking him to go somewhere and cook it ; in five minutes he returns with the meat burnt black outside and perfectly raw within. Seeing my evident disapproval of its condition, the same ancient person who recently appeared upon the scene of my jocular experiment and who has now squatted himself down close beside me, probably to make sure against any further indiscretions, takes the

meat, slashes it across in several directions with his dagger, orders
the afore-mentioned bystander to try it over again, and then coolly
wipes his blackened and greasy fingers on my sheet of *ekmek* as
though it were a table napkin. I obtain a few mouthfuls of eatable
meat from the bystander's second culinary effort, and then buy a
water-melon from a man happening along with a laden donkey ;
cutting into the melon I find it perfectly green all through, and
toss it away ; the men look surprised, and some youngsters
straightway pick it up, eat the inside out until they can scoop out
no more, and then, breaking the rind in pieces, they scrape it out
with their teeth until it is of egg-shell thinness. They seem to do
these things with impunity in Asia.

The grade and the wind are united against me on leaving Jas-
chi-khan, but it is ridable, and having made such a dismal failure
about getting dinner, I push on toward a green area at the base of
a rocky mountain spur, which I observed an hour ago from a point
some distance west of the Kizil Irmak, and concluded to be a
cluster of vineyards. This conjecture turns out quite correct, and,
what is more, my experience upon arriving there would seem to indicate that the good *genii* detailed to arrange the daily programme
of my journey had determined to recompense me to-day for having seen nothing of the feminine world of late but *yashmaks* and
shrouds, and momentary monocular evidence ; for here again am
I thrown into the society of a bevy of maidens, more interesting,
if anything, than the nymphs of industry at Jas-chi-khan.

There is apparently some festive occasion at the little vineyard-
environed village, which stands back a hundred yards or so from
the road, and which is approached by a narrow foot-way between
thrifty-looking vineyards. Three blooming damsels, in all the bravery of holiday attire, with necklaces and pendants of jingling coins
to distinguish them from the matrons, come hurrying down the pathway toward the road at my approach. Seeing me dismount, upon
arriving opposite the village, the handsomest and gayest dressed
of the three goes into one of the vineyards, and with charming
grace of manner, presents herself before me with both hands overflowing with bunches of luscious black grapes. Their abundant
black tresses are gathered in one long plait behind ; they wear
bracelets, necklaces, pendants, brow-bands, head ornaments, and
all sorts of wonderful articles of jewelry, made out of the common
silver and metallic coins of the country ; they are small of stature

and possess oval faces, large black eyes, and warm, dark complexions. Their. manner and dress prove rather a puzzle in determining their nationality; they are not Turkish, nor Greek, nor Armenian, nor Circassian; they may possibly be sedentary Turkomans; but they possess rather a Jewish cast of countenance, and my first impression of them is, that they are "Bible people," the original inhabitants of the country, who have somehow managed to cling to their little possessions here, in spite of Greeks, Turks, and Persians, and other conquering races who have at times over-run the country; perhaps they have softened the hearts of everybody undertaking to oust them by their graceful manners.

Other villagers soon collect, making a picturesque and interesting group around the bicycle; but the maiden with the grapes makes too pretty and complete a picture for any of the others to attract more than passing notice. One of her two companions whisperingly calls her attention to the plainly evident fact that she is being regarded with admiration by the stranger. She blushes perceptibly through her nut-brown cheeks at hearing this, but she is also quite conscious of her claims to admiration, and likes to be admired; so she neither changes her attitude of respectful grace, nor raises her long drooping eyelashes, while I eat and eat grapes, taking them bunch after bunch from her overflowing hands, until ashamed to eat any more. I confess to almost falling in love with that maiden, her manners were so easy and graceful; and when, with ever-downcast eyes and a bewitching manner that leaves not the slightest room for considering the doing so a bold or forward action, she puts the remainder of the grapes in my coat pockets, a peculiar fluttering sensation—but I draw a veil over my feelings, they are too sacred for the garish pages of a book. I do not inquire about their nationality, I would rather it remain a mystery, and a matter for future conjecture; but before leaving I add something to her already conspicuous array of coins that have been increasing since her birth, and which will form her modest dowry at marriage.

The road continues of excellent surface, but rather hilly for a few miles, when it descends into the Valley of the Delijeh Irmak, where the artificial highway again deteriorates into the unpacked condition of yesterday; the donkey trails are shallow trenches of dust, and are no longer to be depended upon as keeping my general course, but are rather cross-country trails leading from one

mountain village to another.   The well-defined caravan trail leading from Ismidt to Angora comes no farther eastward than the latter city, which is the central point where the one exportable commodity of the *vilayet* is collected for barter and transportation to
the seaboard.   The Delijeh Irmak Valley is under partial cultiva-

Grapes and Grace.

tion, and occasionally one passes through small areas of melon
gardens far away from any permanent habitations ; temporary huts
or dug-outs are, however, an invariable adjunct to these isolated
possession of the villagers, in which some one resides day and
night during the melon season, guarding their property with gun

and dog from unscrupulous wayfarers, who otherwise would not hesitate to make their visit to town profitable as well as pleasurable, by surreptitiously confiscating a donkey-load of salable melons from their neighbor's roadside garden. Sometimes I essay to purchase a musk-melon from these lone sentinels, but it is impossible to obtain one fit to eat; these wretched preyers on Nature's bounty evidently pluck and devour them the moment they develop from the bitterness of their earliest growth. No villages are passed on the road after leaving the vintagers' cluster at noon, but bunches of mud hovels are at intervals descried a few miles to the right, perched among the hills that form the southern boundary of the valley; being of the same color as the general surface about them, they are not easily distinguishable at a distance. There seems to be a decided propensity among the natives for choosing the hills as an habitation, even when their arable lands are miles away in the valley; the salubrity of the more elevated location may be the chief consideration, but a swiftly flowing mountain rivulet near his habitation is to the Mohammedan a source of perpetual satisfaction.

I travel along for some time after nightfall, in hopes of reaching a village, but none appearing, I finally decide to camp out. Choosing a position behind a convenient knoll, I pitch the tent where it will be invisible from the road, using stones in lieu of tent-pegs; and inhabiting for the first time this unique contrivance, I sup off the grapes remaining over from the bountiful feast at noon- and, being without any covering, stretch myself without undressing beside the upturned bicycle; notwithstanding the gentle reminders of unsatisfied hunger, I am enjoying the legitimate reward of constant exercise in the open air ten minutes after pitching the tent. Soon after midnight I am awakened by the chilly influence of the " wee sma' hours," and recognizing the likelihood of the tent proving more beneficial as a coverlet than a roof, in the absence of rain, I take it down and roll myself up in it; the thin, oiled cambric is far from being a blanket, however, and at daybreak the bicycle and everything is drenched with one of the heavy dews of the country.

Ten miles over an indifferent road is traversed next morning; the comfortless reflection that anything like a " square meal" seems out of the question anywhere between the larger towns scarcely tends to exert a soothing influence on the ravenous attacks of a most awful appetite; and I am beginning to think seriously of making a detour of several miles to reach a mountain village, when

I meet a party of three horsemen, a Turkish Bey, with an escort of two *zaptiehs*. I am trundling at the time, and without a moment's hesitancy I make a dead set at the Bey, with the single object of satisfying to some extent my gastronomic requirements.

" Bey Effendi, have you any *ekmek* ? " I ask, pointing inquiringly to his saddle-bags on a *zaptieh's* horse, and at the same time giving him to understand by impressive pantomime the uncontrollable condition of my appetite. With what seems to me, under the circumstances, simply cold-blooded indifference to human suffering, the Bey ignores my inquiry altogether, and concentrating his whole attention on the bicycle, asks, " What is that ? "

" An Americanish *araba*, Effendi ; have you any *ekmek?* " toying suggestively with the tell-tale slack of my revolver belt.

" Where have you come from ? "

" Stamboul ; have you *ekmek* in the saddle-bags, Effendi? " this time boldly beckoning the *zaptieh* with the Bey's effects to approach nearer.

" Where are you going ? "

" Yuzgat ! *ekmek ! ekmek !* " tapping the saddle-bags in quite an imperative manner. This does not make any outward impression upon the Bey's aggravating imperturbability, however ; he is not so indifferent to my side of the question as he pretends ; aware of his inability to supply my want, and afraid that a negative answer would hasten my departure before he has fully satisfied his curiosity concerning me, he is playing a little game of diplomacy in his own interests.

" What is it for ? " he now asks, with soul-harrowing indifference to all my counter inquiries.

" To *bin*," I reply, desperately, curt and indifferent, beginning to see through his game.

" *Bin, bin! bacalem !* " he says ; supplementing the request with a coaxing smile. At the same moment my long-suffering digestive apparatus favors me with an unusually savage reminder, and nettled beyond the point where forbearance ceases to be any longer a virtue, I return an answer not exactly complimentary to the Bey's ancestors, and continue my hungry way down the valley. A couple of miles after leaving the Bey, I intercept a party of peasants traversing a cross-country trail, with a number of pack-donkeys loaded with rock-salt, from whom I am fortunately able to obtain several thin sheets of *ekmek*, which I sit down and devour immediately, without even water to moisten the repast ; it seems one of the most tasteful and soul-satisfying breakfasts I ever ate.

Like misfortunes, blessings never seem to come singly, for, an hour after thus breaking my fast I happen upon a party of villagers working on an unfinished portion of the new road ; some of them are eating their morning meal of *ekmek* and *yaort*, and no sooner do I appear upon the scene than I am straightway invited to partake, a seat in the ragged circle congregated around the large bowl of clabbered milk being especially prepared with a bunch of pulled grass for my benefit. The eager hospitality of these poor villagers is really touching ; they are working without so much as "thank you" for payment, there is not a garment amongst the gang fit for a human covering ; their unvarying daily fare is the "blotting-paper *ekmek*" and *yaort*, with a melon or a cucumber occasionally as a luxury ; yet, the moment I approach, they assign me a place at their "table," and two of them immediately bestir themselves to make me a comfortable seat. Neither is there so much as a mercenary thought among them in connection with the invitation ; these poor fellows, whose scant rags it would be a farce to call clothing, actually betray embarrassment at the barest mention of compensation ; they fill my pockets with bread, apologize for the absence of coffee, and compare the quality of their respective pouches of native tobacco in order to make me a decent cigarette.

Never, surely, was the reputation of Dame Fortune for fickleness so completely proved as in her treatment of me this morning —ten o'clock finds me seated on a pile of rugs in a capacious black tent, "wrassling" with a huge bowl of savory mutton *pillau*, flavored with green herbs, as the guest of a Koordish sheikh ; shortly afterwards I meet a man taking a donkey-load of muskmelons to the Koordish camp, who insists on presenting me with the finest melon I have tasted since leaving Constantinople ; and high noon finds me the guest of another Koordish sheikh ; thus does a morning, which commenced with a fair prospect of no breakfast, following after yesterday's scant supply of unsuitable food, end in more hospitality than I know what to do with.

These nomad tribes of the famous "black-tents" wander up toward Angora every summer with their flocks, in order to be near a market at shearing time ; they are famed far and wide for their hospitality. Upon approaching the great open-faced tent of the Sheikh, there is a hurrying movement among the attendants to prepare a suitable raised seat, for they know at a glance that I am an Englishman, and likewise are aware that an Englishman cannot sit cross-legged like an Asiatic ; at first, I am rather surprised at their evident ready recognition of my nationality, but I soon afterwards discover the reason. A hugh bowl of *pillau*, and another of excellent *yaort* is placed before me without asking any questions, while the dignified old Sheikh fulfils one's idea of a gray-bearded nomad patriarch to perfection, as he sits cross-legged on a rug, solemnly smoking a nargileh, and watching to see that no letter of his generous code of hospitality toward strangers is overlooked by the attendants. These latter seem to be the picked young men of the tribe ; fine, strapping fellows, well-dresed, six-footers, and of athletic proportions ; perfect specimens of semi-civilized manhood, that would seem better employed in a grenadier regiment than in hovering about the old Sheikh's tent, attending to the filling and lighting of his nargileh, the arranging of his cushions by day and his bed at night, the serving of his food, and the proper reception of his guests ; and yet it is an interesting sight to see these splendid young fellows waiting upon their beloved old chieftain, fairly bounding, like great affectionate mastiffs, at his merest look or suggestion.

Most of the boys and young men are out with the flocks, but the older men, the women and children, gather in a curious crowd before the open tent ; they maintain a respectful silence so long as I am their Sheikh's guest, but they gather about me without reserve when I leave the hospitable shelter of that respected person's quarters. After examining my helmet and sizing up my general appearance, they pronounce me an " English *zaptieh*," a distinction for which I am indebted to the circumstance of Col. N——, an English officer, having recently been engaged in Koordistan organizing a force of native *zaptiehs*. The women of this particular camp seem, on the whole, rather unprepossessing specimens ; some of them are hooked-nosed old hags, with piercing black eyes, and hair dyed to a flaming " carrotty " hue with *henna ;* this latter is supposed to render them beautiful, and enhance their personal appearance in the eyes of the men ; they need something to enhance their personal appearance, certainly, but to the untutored and inartistic eye of the writer it produces a horrid, unnatural effect. According to our ideas, flaming red hair looks uncanny and of vulgar, uneducated taste, when associated with coal-black eyes and a complexion like gathering darkness. These vain mortals seem in-

clined to think that in me they have discovered something to be
petted and made much of, treating me pretty much as a troop of
affectionate little girls would treat a wandering kitten that might
unexpectedly appear in their midst. Giddy young things of about ·
fifty summers cluster around me in a compact body, examining my
clothes from helmet to moccasins, and critically feeling the texture of my coat and shirt, they take off my helmet, reach over each
other's shoulders to stroke my hair, and pat my cheeks in the most
affectionate manner; meanwhile expressing themselves in soft,
purring comments, that require no linguistic abilities to interpret
into such endearing remarks as, "Ain't he a darling, though?"
"What nice soft hair and pretty blue eyes?" "Don't you wish the
dear old Sheikh would let us keep him?" Considering the source
whence it comes, it requires very little of this to satisfy one, and as
soon as I can prevail upon them to let me escape, I mount and
wheel away, several huge dogs escorting me, for some minutes,
in the peculiar manner Koordish dogs have of escorting stray
'cyclers.

CHAPTER XV.

FROM the Koordish encampment my route leads over a low mountain spur by easy gradients, and by a winding, unridable trail down into the valley of the eastern fork of the Delijah Irmak. The road improves as this valley is reached, and noon finds me the wonder and admiration of another Koordish camp, where I remain a couple of hours in deference to the powers of the midday sun. One has no scruples about partaking of the hospitality of the nomad Koords, for they are the wealthiest people in the country, their flocks covering the hills in many localities ; they are, as a general thing, fairly well dressed, are cleaner in their cooking than the villagers, and hospitable to the last degree. Like the rest of us, however, they have their faults as well as their virtues ; they are born freebooters, and in unsettled times, when the Turkish Government, being handicapped by weightier considerations, is compelled to relax its control over them, they seldom fail to promptly respond to their plundering instincts and make no end of trouble. They still retain their hospitableness, but after making a traveller their guest for the night, and allowing him to depart with everything he has, they will intercept him on the road and rob him. They have some objectionable habits, even in these peaceful times, which will better appear when we reach their own Koordistan, where we shall, doubtless, have better opportunities for criticising them. Whatever their faults or virtues, I leave this camp, hoping that the termination of the day may find me the guest of another sheikh for the night. An hour after leaving this camp I pass through an area of vineyards, out of which people come running with as many grapes among them as would feed a dozen people ; the road is ridable, and I hurry along to avoid their bother. Verily it would seem that I am being hounded down by retributive justice for sundry evil thoughts and impatient remarks, associated with my hungry experiences of early morning ; then I was wondering where the next mouthful of food was going to overtake me, this afternoon finds me pedalling determinedly to prevent being overtaken by *it*.

The afternoon is hot and with scarcely a breath of air moving; the little valley terminates in a region of barren, red hills, on which the sun glares fiercely; some toughish climbing has to be accomplished in scaling a ridge, and then I emerge into an upland lava plateau, where the only vegetation is sun-dried weeds and thistles. Here a herd of camels are contentedly browsing, munching the dry, thorny herbage with a satisfaction that is evident a mile away. From casual observations along the route, I am inclined to think a camel not far behind a goat in the depravity of its appetite; a camel will wander uneasily about over a greensward of moist, succulent grass, scanning his surroundings in search of giant thistles, frost-bitten tumble-weeds, tough, spriggy camel thorns, and odds and ends of unpalatable vegetation generally. Of course, the "ship of the desert" never sinks to such total depravity as to hanker after old gum overshoes and circus posters, but if permitted to forage around human habitations for a few generations, I think they would eventually degenerate to the goat's disreputable level. The expression of utter astonishment that overspreads the angular countenance of the camels browsing near the roadside, at my appearance, is one of the most ludicrous sights imaginable; they seem quite intelligent enough to recognize in a wheelman and his steed something inexplicable and foreign to their country, and their look of timid inquiry seems ridiculously unsuited to their size and the general ungainliness of their appearance, producing a comical effect that is worth going miles to see.

It is approaching sun-down, when, ascending a ridge overlooking another valley, I am gratified at seeing it occupied by several Koordish camps, their clusters of black tents being a conspicuous feature of the landscape. With a fair prospect of hospitable quarters for the night before me, and there being no distinguishable signs of a road, I make my way across country toward one of the camps that seems to be nearest my proper course. I have arrived within a mile of my objective point, when I observe, at the base of a mountain about half the distance to my right, a large, white two-storied building, the most pretentious structure, by long odds, that has been seen since leaving Angora. My curiosity is, of course, aroused concerning its probable character; it looks like a bit of civilization that has in some unaccountable manner found its way to a region where no other human habitations are visible, save the tents of wild tribesmen, and I at once shape my course toward it. It turns out to be a rock-salt mine or quarry, that supplies the whole region for scores of miles around with salt, rock-salt being the only kind obtainable in the country ; it was from this mine that the donkey party from whom I first obtained bread this morning fetched their loads. Here I am invited to remain over night, am provided with a substantial supper, the *menu* including boiled mutton, with cucumbers for desert. The managers and employees of the quarry make their cucumbers tasteful by rubbing the end with a piece of rock-salt each time it is cut off or bitten, each person keeping a select little square for the purpose. The salt is sold at the mine, and owners of transportation facilities in the shape of pack animals make money by purchasing it here at six paras an *oke*, and selling it at a profit in distant towns.

Two young men seem to have charge of transacting the business; one of them is inordinately inquisitive, he even wants to try and unstick the envelope containing a letter of introduction to Mr. Tifticjeeoghlou's father in Yuzgat, and read it out of pure curiosity to see what it says; and he offers me a lira for my Waterbury watch, notwithstanding its Alla Franga face is beyond his Turkish comprehension. The loud, confident tone in which the Waterbury ticks impresses the natives very favorably toward it, and the fact of its not opening at the back like other time-pieces, creates the impression that it is a watch that never gets cranky and out of order ; quite different from the ones they carry, since their curiosity leads them to be always fooling with the works. American clocks are found all through Asia Minor, fitted with Oriental faces and there is little doubt but the Waterbury, with its resonant tick, if similiarly prepared, would find here a ready market.

The other branch of the managerial staff is a specimen of humanity peculiarly Asiatic Turkish, a melancholy-faced, contemplative person, who spends nearly the whole evening in gazing in silent wonder at me and the bicycle ; now and then giving expression to his utter inability to understand how such things can possibly be by shaking his head and giving utterance to a peculiar clucking of astonishment. He has heard me mention having come from Stamboul, which satisfies him to a certain extent; for, like a true Turk, he believes that at Stamboul all wonderful things originate ; whether the bicycle was made there, or whether it originally came from somewhere else, doesn't seem to enter into his speculations ; the simple knowledge that I have come from Stamboul is all-sufficient for him ; so far as he is concerned, the bicycle is simply another wonder from Stamboul, another proof that the earthly paradise of the Mussulman world on the Bosphorus is all that he has been taught to believe it. When the contemplative young man ventures away from the

The Contemplative Young Man.

dreamy realms of his own imaginations, and from the society of his inmost thoughts, far enough to make a remark, it is to ask me something about Stamboul ; but being naturally. taciturn and retiring, and moreover, anything but an adept at pantomimic language, he prefers mainly to draw his own conclusions in silence. He manages to make me understand, however, that he intends before long making a journey to see Stamboul for himself ; like many another Turk from the barren hills of the interior, he will visit the Ottoman capital ; he will recite from the Koran under the glorious mosaic dome of St. Sophia ; wander about that wonder of the Orient, the Stamboul bazaar ; gaze for hours on the matchless beauties of the Bosphorus ; ride on one of the steamboats; see the railway, the tramway, the Sultan's palaces, and the shipping, and return to his native hills thoroughly convinced that in all the world there is no place fit to be compared with Stamboul; no place so full of wonders ; no place so beautiful ; and wondering how even the land of the *kara ghuz kiz*, the material paradise of the Mohammedans, can possibly be more lovely. The contemplative young man is tall and slender, has large, dreamy, black eyes, a downy upper lip, a melancholy cast of countenance, and wears a long print wrapper of neat dotted pattern, gathered at the waist with a girdle *à la* dressing-gown.

The inquisitive partner makes me up a comfortable bed of quilts on the divan of a large room, which is also occupied by several salt traders remaining over night, and into which their own small private apartments open. A few minutes after they have retired to their respective rooms, the contemplative young man reappears with silent tread, and with a scornful glance at my surroundings, both human and inanimate, gathers up my loose effects, and bids me bring bicycle and everything into his room ; here, I find, he has already prepared for my reception quite a downy couch, having contributed, among other comfortable things, his wolf-skin overcoat ; after seeing me comfortably established on a couch more appropriate to my importance as a person recently from Stamboul than the other, he takes a lingering look at the bicycle, shakes his head and clucks, and then extinguishes the light.

Sunrise on the following morning finds me wheeling eastward from the salt quarry, over a trail well worn by salt caravans, to Yuzgat; the road leads for some distance down a grassy valley, covered with the flocks of the several Koordish camps round about ; the wild herdsmen come galloping from all directions across the valley toward me, their uncivilized garb and long swords giving them more the appearance of a ferocious gang of cut-throats advancing to the attack than shepherds. Hitherto, nobody has seemed any way inclined to attack me ; I have almost wished somebody would undertake a little devilment of some kind, for the sake of livening things up a little, and making my narrative more stirring; after venturing everything, I have so far nothing to tell but a story of being everywhere treated with the greatest consideration, and much of the time even petted. I have met armed men far away from any habitations, whose appearance was equal to our most ferocious conception of *bashi bazouks,* and merely from a disinclination to be bothered, perhaps being in a hurry at the time, have met their curious inquiries with imperious gestures to be gone ; and have been guilty of really inconsiderate conduct on more than one occasion, but under no considerations have I yet found them guilty of anything worse than casting covetous glances at my effects. But there is an apparent churlishness of manner, and an overbearing demeanor, as of men chafing under the restraining influences that prevent them gratifying their natural freebooting instincts, about these Koordish herdsmen whom I encounter this morning, that forms quite a striking contrast to the almost childlike harmlessness and universal respect toward me observed in the disposition of the villagers.

It requires no penetrating scrutiny of these fellows' countenances to ascertain that nothing could be more uncongenial to them than the state of affairs that prevents them stopping me and looting me of everything I possess ; a couple of them order me quite imperatively to make a detour from my road to avoid approaching too near their flock of sheep, and their general behavior is pretty much as though seeking to draw me into a quarrel, that would afford them an opportunity of plundering me. Continuing on the even tenor of my way, affecting a lofty unconsciousness of their existence, and wondering whether, in case of being molested, it would be advisable to use my Smith & Wesson in defending my effects, or taking the advice received in Constantinople, offer no resistance whatever, and trust to being able to recover them through the authorities, I finally emerge from their vicinity. Their behavior simply confirms what I have previously understood of their character ; that while they will invariably extend hospitable treatment to a stranger visiting their camps, like unreliable explosives, they require to be handled quite "gingerly" when encountered on the road, to prevent disagreeable consequences.

Passing through a low, marshy district, peopled with solemn-looking storks and croaking frogs, I meet a young sheikh and his personal attendants returning from a morning's outing at their favorite sport of hawking ; they carry their falcons about on small perches, fastened by the leg with a tiny chain. I try to induce them to make a flight, but for some reason or other they refuse ; an Osmanli Turk would have accommodated me in a minute. Soon I arrive at another Koordish camp, fording a stream in order to reach their tents, for I have not yet breakfasted, and know full well that no better opportunity of obtaining one will be likely to turn up. Entering the nearest tent, I make no ceremony of calling for refreshments, knowing well enough that a heaping dish of *pillau* will be forthcoming, and that the hospitable Koords will regard the ordering of it as the most natural thing in the world. The *pillau* is of rice, mutton, and green herbs, and is brought in a large pewter dish ; and, together with sheet bread and a bowl of excellent *yaort*, is brought on a massive pewter tray, which has possibly belonged to the tribe for centuries. These tents are divided into several compartments ; one end is a compartment where the men congregate in the daytime, and the younger men sleep at night, and where guests are received and entertained ; the central space is the commissary and female industrial department ; the others are female and family sleeping places. Each compartment is partitioned off with a hanging carpet partition ; light portable railing of small, upright willow sticks bound closely together protects the central compartment from a horde of dogs hungrily nosing about the camp, and small " coops " of the same material are usually built inside as a further protection for bowls of milk, *yaort*, butter, cheese, and cooked food ; they also obtain fowls from the villagers, which they keep cooped up in a similar manner, until the hapless prisoners are required to fulfil their destiny in chicken *pillau ;* the capacious covering over all is strongly woven goats'-hair material of a black or smoky brown color. In a wealthy tribe, the tent of their sheikh is often a capacious affair, twenty-five by one hundred feet, containing, among other compartments, stabling and hay-room for the sheikh's horses in winter.

My breakfast is brought in from the culinary department by a young woman of most striking appearance, certainly not less than six feet in height ; she is of slender, willowy build, and straight as an arrow ; a wealth of auburn hair is surmounted by a small, gay-colored turban ; her complexion is fairer than common among Koordish woman, and her features are the queenly features of a Juno ; the eyes are brown and lustrous, and, were the expression but of ordinary gentleness, the picture would be perfect ; but they are the round, wild-looking orbs of a newly-caged panther—grimal-kin-like eyes, that would, most assuredly, turn green and luminous in the dark. Other women come to take a look at the stranger, gathering around and staring at me, while I eat, with all their eyes —and such eyes! I never before saw such an array of " wild-ani-mal eyes ;" no, not even in the Zoo! Many of them are magnifi-cent types of womanhood in every other respect, tall, queenly, and symmetrically perfect ; but the eyes—oh, those wild, tigress eyes! Travellers have told queer, queer stories about bands of these wild-eyed Koordish women waylaying and capturing them on the roads through Koordistan, and subjecting them to barbarous treatment. I have smiled, and thought them merely " travellers' tales ; " but I can see plain enough, this morning, that there is no improbability in the stories, for, from a dozen pairs of female eyes, behold, there gleams not one single ray of tenderness : these women are capable of anything that tigresses are capable of, beyond a doubt.

Almost the first question asked by the men of these camps is whether the English and Muscovs are fighting ; they have either heard of the present (summer of 1885) crisis over the Afghan boundary question, or they imagine that the English and Russians maintain a sort of desultory warfare all the time. When I tell them that the Muscov is *fenna* (bad) they invariably express their ap-proval of the sentiment by eagerly calling each other's attention to my expression. It is singular with what perfect faith and confi-dence these rude tribesmen accept any statement I choose to make, and how eagerly they seem to dwell on simple statements of facts that are known to every school-boy in Christendom. I entertain them with my map, showing them the position of Stamboul, Mecca, Erzeroum, and towns in their own Koordistan, which they recog-nize joyfully as I call them by name. They are profoundly im-pressed at the " extent of my knowledge," and some of the more deeply impressed stoop down and reverently kiss Stamboul and Mecca, as I point them out.

While thus pleasantly engaged, an aged sheikh comes to the tent and straightway begins "kicking up a blooming row" about me. It seems that the others have been guilty of trespassing on the sheikh's prerogative, in entertaining me themselves, instead of conducting me to his own tent. After upbraiding them in un-measured terms, he angrily orders several of the younger men to make themselves beautifully scarce forthwith. The culprits—some of them abundantly able to throw the old fellow over their shoulders—instinctively obey ; but they move off at a snail's pace, with lowering brows, and muttering angry growls that betray fully their untamed, intractable dispositions.

A two-hours' road experience among the constantly varying slopes of rolling hills, and then comes a fertile valley, abounding in villages, wheat-fields, orchards, and melon-gardens. These days I find it incumbent on me to turn washer-woman occasionally, and, halting at the first little stream in this valley, I take upon myself the onerous duties of Wah Lung in Sacramento City, having for an interested and interesting audience two evil-looking kleptomaniacs, buffalo-herders dressed in next to nothing, who eye my garments drying on the bushes with lingering covetousness. It is scarcely necessary to add that I watch them quite as interestingly myself ; for, while I pity the scantiness of their wardrobe, I have nothing that I could possibly spare among mine. A network of irrigating ditches, many of them overflowed, render this valley difficult to traverse with a bicycle, and I reach a large village about noon, myself and wheel plastered with mud, after traversing a section where the normal condition is three inches of dust.

Bread and grapes are obtained here, a light, airy dinner, that is seasoned and made interesting by the unanimous worrying of the entire population. Once I make a desperate effort to silence their clamorous importunities, and obtain a little quiet, by attempting to ride over impossible ground, and reap the well-merited reward of permitting my equanimity to be thus disturbed in the shape of a header and a slightly-bent handle-bar. While I am eating, the gazing-stock of a wondering, commenting crowd, a respectably dressed man elbows his way through the compact mass of humans around me, and announces himself as having fought under Osman Pasha at Plevna. What this has to do with me is a puzzler ; but the man himself, and every Turk of patriotic age in the crowd, is evidently expecting to see me make some demonstration of approval ; so, not knowing what else to do, I shake the man cordially by the hand, and modestly inform my attentively listening audience that Osman Pasha and myself are brothers, that Osman yielded only when the overwhelming numbers of the Muscovs proved that it was his *kismet* to do so ; and that the Russians would never be permitted to occupy Constantinople ; a statement, that probably makes my simple auditors feel as though they were inheriting a new lease of national life; anyhow, they seem not a little gratified at what I am saying.

After this the people seem to find material for no end of amusement among themselves, by contrasting the *marifet* of the bicycle with the *marifet* of their creaking *arabas*, of which there seems to be quite a number in this valley. They are used chiefly in harvesting, are roughly made, used, and worn out in these mountain-environed valleys without ever going beyond the hills that encompass them in on every side. From these villages the people begin to evince an alarming disposition to follow me out some distance on donkeys. This undesirable trait of their character is, of course, easily counteracted by a short spurt, where spurting is possible, but it is a soul-harrowing thing to trundle along a mile of unridable road, in company with twenty importuning *katir-jees*, their diminutive donkeys filling the air with suffocating clouds of dust. There is nothing on all this mundane sphere that will so effectually subdue the proud, haughty spirit of a wheelman, or that will so promptly and completely snuff out his last flickering ray of dignity; it is one of the pleasantries of 'cycling through a country where the people have been riding donkeys and camels since the flood.

A few miles from the village I meet another candidate for medical treatment; this time it is a woman, among a merry company of donkey-riders, bound from Yuzgat to the salt-mines; they are laughing, singing, and otherwise enjoying themselves, after the manner of a New England berrying party. The woman's affliction, she says, is " *fenna ghuz*," which, it appears, is the term used to denote ophthalmia, as well as the "evil-eye;" but of course, not being a *ghuz hakim*, I can do nothing more than express my sympathy. The fertile valley gradually contracts to a narrow, rocky defile, leading up into a hilly region, and at five o'clock I reach Yuzgat, a city claiming a population of thirty thousand, that is situated in a depression among the mountains that can scarcely be called a valley. I have been three and a half days making the one hundred and thirty miles from Angora.

Everybody in Yuzgat knows Youvanaki Effendi Tifticjeeoghlou, to whom I have brought a letter of introduction; and, shortly after reaching town, I find myself comfortably installed on the cushioned divan of honor in that worthy old gentleman's large reception room, while half a dozen serving-men are almost knocking each other over in their anxiety to furnish me coffee, vishner-

su, cigarettes, etc. They seem determined upon interpreting the slightest motion of my hand or head into some want which I am unable to explain, and, fancying thus, they are constantly bobbing up before me with all sorts of surprising things. Tevfik Bey, general superintendent of the Regie (a company having the monopoly of the tobacco trade in Turkey, for which they pay the government a fixed sum per annum), is also a guest of Tifticjeeoghlou Effendi's hospitable mansion, and he at once despatches a messenger to his Yuzgat agent, Mr. G. O. Tchetchian, a vivacious Greek, who speaks English quite fluently. After that gentleman's arrival, we soon come to a more perfect understanding of each other all round, and a very pleasant evening is spent in receiving crowds of visitors in a ceremonious manner, in which I really seem to be holding a sort of a levee, except that it is evening instead of morning. Open door is kept for everybody, and mine host's retinue of pages and serving men are kept pretty busy supplying coffee right and left ; beggars in their rags are even allowed to penetrate into the reception-room, to sip a cup of coffee and take a curious peep at the Ingilisin and his wonderful *araba*, the fame of which has spread like wildfire through the city. Mine host himself is kept pretty well occupied in returning the salaams of the more distinguished visitors, besides keeping his eye on the servants, by way of keeping them well up to their task of dispensing coffee in a manner satisfactory to his own liberal ideas of hospitality ; but he presides over all with a bearing of easy dignity that it is a pleasure to witness.

The street in front of the Tifticjeeoghlou residence is swarmed with people next morning ; keeping open house is, under the circumstances, no longer practicable ; the entrance gate has to be guarded, and none permitted to enter but privileged persons. During the forenoon the *Caimacan* and several officials call round and ask me to favor them by riding along a smooth piece of road opposite the municipal *konak ;* as I intend remaining over here to-day, I enter no objections, and accompany them forthwith. The rabble becomes wildly excited at seeing me emerge with the bicycle, in company with the *Caimacan* and his staff, for they know that their curiosity is probably on the eve of being gratified. It proves no easy task to traverse the streets, for, like in all Oriental cities, they are narrow, and are now jammed with people. Time and again the *Caimacan* is compelled to supplement the exertions of an inadequate force of *zaptiehs* with his authoritative voice, to keep down the excitement and the wild shouts of "*Bin bacalem! bin bacalem!*" (Ride, so that we can see—an innovation on *bin, bin,* that has made itself manifest since crossing the Kizil Irmak River) that are raised, gradually swelling into the tumultuous howl of a multitude. The uproar is deafening, and, long before reaching the place, the *Caimacan* repents having brought me out. As for myself, I certainly repent having come out, and have still better reasons for doing so before reaching the safe retreat of Tifticjeeoghlou Effendi's house, an hour afterward.

The most that the inadequate squad of *zaptiehs* present can do, when we arrive opposite the muncipal *konak,* is to keep the crowd from pressing forward and overwhelming me and the bicycle. They attempt to keep open a narrow passage through the surging sea of humans blocking the street, for me to ride down ; but ten yards ahead the lane terminates in a mass of fez-crowned heads. Under the impression that one can mount a bicycle on the stand, like mounting a horse, the *Caimacan* asks me to mount, saying that when the people see me mounted and ready to start, they will themselves yield a passage-way. Seeing the utter futility of attempting explanations under existing conditions, amid the defeaning clamor of "*Bin bacalem! bin bacalem!*" I mount and slowly pedal along a crooked "fissure" in the compact mass of people, which the *zaptiehs* manage to create by frantically flogging right and left before me. Gaining, at length, more open ground, and the smooth road continuing on, I speed away from the multitude, and the *Caimacan* sends one fleet-footed *zaptieh* after me, with instructions to pilot me back to Tifticjeeoghlou's by a roundabout way, so as to avoid returning through the crowds.

The rabble are not to be so easily deceived and shook off as the *Caimacan* thinks, however ; by taking various short cuts, they manage to intercept us, and, as though considering the having detected and overtaken us in attempting to elude them, justifies them in taking liberties, their "*Bin bacalem!*" now develops into the imperious cry of a domineering majority, determined upon doing pretty much as they please. It is the worst mob I have seen on the journey, so far ; excitement runs high, and their shouts of "*Bin bacalem!*" can, most assuredly, be heard for miles. We are enveloped by clouds of dust, raised by the feet of the multitude ; the hot sun glares down savagely upon us ; the poor *zaptieh,* in heavy top-boots and a brand-new uniform, heavy enough for winter, works like a beaver to protect the bicycle, until, with perspiration and dust, his face is streaked and tattooed like a South Sea Islander's. Unable to proceed, we come to a stand-still, and simply occupy ourselves in protecting the bicycle from the crush, and reasoning with the mob ; but the only satisfaction we obtain in reply to anything we say is " *Bin bacalem.*"

One or two pig-headed, obstreperous young men near us, emboldened by our apparent helplessness, persist in handling the bicycle. After being pushed away several times, one of them even assumes a menacing attitude toward me the last time I thrust his meddlesome hand away. Under such circumstances retributive justice, prompt and impressive, is the only politic course to pursue ; so, leaving the bicycle to the *zaptieh* a moment, in the absence of a stick, I feel justified in favoring the culprit with a brief, pointed lesson in the noble art of self-defence, the first boxing lesson ever given in Yuzgat. In a Western mob this would have been anything but an act of discretion, probably, but with these people it has a salutary effect ; the idea of attempting retaliation is the farthest of anything from their thoughts, and in all the obstreperous crowd there is, perhaps, not one but what is quite delighted at either seeing or hearing of me having thus chastised one of their number, and involuntarily thanks Allah that it didn't happen to be himself.

It would be useless to attempt a description of how we finally managed, by the assistance of two more *zaptiehs*, to get back to Tifticjeeoghlou Effendi's, both myself and the *zaptieh* simply unrecognizable from dust and perspiration. The *zaptieh*, having first washed the streaks and tattooing off his face, now presents himself, with the broad, honest smile of one who knows he well deserves what he is asking for, and says, " Effendi, backsheesh ! "

There is nothing more certain than that the honest fellow merits backsheesh from somebody ; it is also equally certain that I am the only person from whom he stands the ghost of a chance of getting any ; nevertheless, the idea of being appealed to for backsheesh, after what I have just undergone, merely as an act of accommodation, strikes me as just a trifle ridiculous, and the opportunity of engaging the grinning, good-humored *zaptieh* in a little banter concerning the abstract preposterousness of his expectations is too good to be lost. So, assuming an air of astonishment, I reply :

"Backsheesh! where is *my* backsheesh? I should think it's *me* that deserves backsheesh if anybody does!" This argument is entirely beyond the *zaptieh's* child-like comprehension, however; he only understands by my manner that there is a "hitch" somewhere; and never was there a more broadly good-humored countenance, or a smile more expressive of meritoriousness, nor an utterance more coaxing in its modulations than his "E-f-fendi, backsheesh!" as he repeats the appeal ; the smile and the modulation is well worth the backsheesh.

In the afternoon, an officer appears with a note saying that the *Mutaserif* and a number of gentlemen would like to see me ride inside the municipal *konak* grounds. This I very naturally promise to do, only, under conditions that an adequate force of *zaptiehs* be provided. This the *Mutaserif* readily agrees to, and once more I venture into the streets, trundling along under a strong escort of *zaptiehs* who form a hollow square around me. The people accumulate rapidly, as we progress, and, by the time we arrive at the *konak* gate there is a regular crush. In spite of the frantic exertions of my escort, the mob press determinedly forward, in an attempt to rush inside when the gate is opened ; instantly I find myself and bicycle wedged in among a struggling mass of natives ; a cry of "*Sakin araba! sakin araba!*" (Take care! the bicycle!) is raised ; the *zaptiehs* make a supreme effort, the gate is opened, I am fairly carried in, and the gate is closed. A couple of dozen happy mortals have gained admittance in the rush. Hundreds of the better class natives are in the inclosure, and the walls and neighboring house-tops are swarming with an interested audience.

There is a small plat of decently smooth ground, upon which I circle around for a few minutes, to as delighted an audience as ever collected in Barnum's circus. After the exhibition, the *Mutaserif* eyes the swarming multitude on the roofs and wall, and looks perplexed ; some one suggests that the bicycle be locked up for the present, and, when the crowds have dispersed, it can be removed without further excitement. The *Mutaserif* then places the municipal chamber at my disposal, ordering an officer to lock it up and give me the key. Later in the afternoon I am visited by the Armenian pastor of Yuzgat, and another young Armenian, who can speak a little English, and together we take a strolling peep at the city. The American missionaries at Kaizarieh have a small book store here, and the pastor kindly offers me a New Testament

My Yuzgat Audience.

to carry along. We drop in on several Armenian shopkeepers, who are introduced as converts of the mission. Coffee is supplied wherever we call. While sitting down a minute in a tailor's stall, a young Armenian peeps in, smiles, and indulges in the pantomime of rubbing his chin. Asking the meaning of this, I am informed by the interpreter that the fellow belongs to the barber shop next door, and is taking this method of reminding me that I stand in need of his professional attentions, not having shaved of late.

There appears to be a large proportion of Circassians in town ; a group of several wild-looking bipeds, armed à la Anatolia, ragged and unkempt-haired for Circassians, who are generally respectable in their personal appearance, approach us, and want me to show them the bicycle, on the strength of their having fought against the Russians in the late war. " I think they are liars," says the young Armenian, who speaks English ; " they only say they fought against the Russians because you are an Englishman, and they think you will show them the bicycle." Some one comes to me with old coins for sale, another brings a stone with hieroglyphics on it, and the inevitable genius likewise appears ; this time it is an Armenian ; the tremendous ovation I have received has filled his mind with exaggerated ideas of making a fortune, by purchasing the bicycle and making a two-piastre show out of it. He wants to know how much I will take for it.

Early daylight finds me astir on the following morning, for I have found it a desirable thing to escape from town ere the populace is out to crowd about me. Tifticjeeoghlou Effendi's better half has kindly risen at an unusually early hour, to see me off, and provides me with a dozen circular rolls of hard bread—rings the size of rope quoits aboard an Atlantic steamer, which I string on Igali's cerulean waist-scarf, and sling over one shoulder. The good lady lets me out of the gate, and says, " Bin bacalem, Effendi." She hasn't seen me ride yet. She is a motherly old creature, of Greek extraction, and I naturally feel like an ingrate of the meanest type, at my inability to grant her modest request. Stealing along the side streets, I manage to reach ridable ground, gathering by the way only a small following of worthy early risers, and two katir-jees, who essay to follow me on their long-eared chargers ; but, the road being smooth and level from the beginning, I at once discourage them by a short spurt. A half-hour's trundling up a steep hill, and then comes a coastable descent into lower territory. A conscription party collected from the neighboring Mussulman villages, *en route* to Samsoon, the nearest Black Sea port, is met while riding down this declivity. In anticipation of the Sultan's new uniforms awaiting them at Constantinople, they have provided themselves for the journey with barely enough rags to cover their nakedness. They are in high glee at their departure for Stamboul, and favor me with considerable good-natured chaff as I wheel past. "Human nature is everywhere pretty much alike the world over," I think to myself. There is little difference between this regiment of ragamuffins chaffing me this morning and the well-dressed troopers of Kaiser William, bantering me the day I wheeled out of Strassburg.

CHAPTER XVI.

THROUGH THE SIVAS VILAYET INTO ARMENIA.

It is six hours distant from Yuzgat to the large village of Koehne, as distance is measured here, or about twenty-three English miles ; but the road is mostly ridable, and I roll into the village in about three hours and a half. Just beyond Koehne, the roads fork, and the *mudir* kindly sends a mounted *zaptieh* to guide me aright, for fear I shouldn't quite understand by his pantomimic explanations. I understand well enough, though, and the road just here happening to be excellent wheeling, to the delight of the whole village, I spurt ahead, outdistancing the *zaptieh's* not over sprightly animal, and bowling briskly along the right road within their range of vision, for over a mile. Soon after leaving Koehne my attention is attracted by a small cluster of civilized-looking tents, pitched on the bank of a running stream near the road, and from whence issues the joyous sounds of mirth and music. The road continues ridable, and I am wheeling leisurely along, hesitating about whether to go and investigate or not, when a number of persons, in holiday attire, present themselves outside the tents, and by shouting and gesturing, invite me to pay them a visit. It turns out to be a reunion of the Yuzgat branch of the Pampasian-Pamparsan family—an Armenian name whose representatives in Armenia and Anatolia, it appears, correspond in comparative numerical importance to the great and illustrious family of Smiths in the United States. Following—or doubtless, more properly, setting—a worthy example, they likewise have their periodical reunions, where they eat, drink, spin yarns, sing, and twang the tuneful lyre in frolicsome consciousness of always having a howling majority over their less prolific neighbors.

Refreshments in abundance are tendered, and the usual pantomimic explanations exchanged between us ; some of the men have been honoring the joyful occasion by a liberal patronage of the flowing bowl, and are already mildly hilarious ; stringed instruments are twanged by the musical members of the great family, while several others, misinterpreting the inspiration of *raki* punch for terpsichorean talent are prancing wildly about the tent. Middle-aged matrons are here in plenty, housewifely persons, finding their chief enjoyment in catering to the gastronomic pleasures of the others ; while a score or two of blooming maidens stand coyly aloof, watching the festive merry-makings of the men ; their heads and necks are resplendent with bands and necklaces of gold coins, it still being a custom of the East to let the female mem-

An Armenian Family Reunion.

bers of a family wear the surplus wealth about them in the shape of gold ornaments and jewels, a custom resulting from the absence of safe investments and the unstability of national affairs.   Yuzgat enjoys among neighboring cities a reputation for beautiful women, and this auspicious occasion gives me an excellent opportunity for drawing my own conclusions.   It is not fair perhaps to pass judgment on Yuzgat's pretensions, by the damsels of one family connection, not even the great and numerous Pampasian-Pamparsan family, but still they ought to be at least a fair average.   They have beautiful large black eyes, and usually a luxuriant head of

24

hair ; but their faces are, on the whole, babyish and expressionless. The Yuzgat maiden of "sweet sixteen" is a coy, babyish
creature, possessed of a certain doll-like prettiness, but at twenty-
three is a rapidly fading flower, and at thirty is already beginning
to get wrinkled and old.

Happening to fall in with this festive gathering this morning
is quite a gratifying and enlivening surprise ; besides the music
and dancing and a substantial breakfast of chicken, boiled mutton,
and rice *pillau*, it gives me an opportunity of witnessing an Armenian family reunion
under primitive conditions. Watching
over this peaceful and
gambolling flock of Armenian lambkins is a
lone Circassian watch-
dog ; he is of a stalwart, warlike appearance ; and although
wearing no arms—except a cavalry sword,
a shorter broad-sword,
a dragoon revolver, a
two-foot horse-pistol,
and a double-barrelled
shot-gun slung at his
back—the Armenians
seem to feel perfectly
safe under his protection. They probably

Slightly Armed.

don't require any such protection really ; they are nevertheless wise
in employing a Circassian to guard them, if for nothing else for
the sake of freeing their own unwarlike minds of all disquieting apprehensions, and enjoying their family reunion in the calm atmosphere of perfect security ; some lawless party passing along the
road might peradventure drop in and abuse their hospitality, or
partaking too freely of *raki*, make themselves obnoxious, were
they unprotected ; but with one Circassian patrolling the camp,
they are doubly sure against anything of the kind.

These people invite me to remain with them until to-morrow ;

but of course I excuse myself from this, and, after spending a very
agreeable hour in their company, take my departure. The country develops into an undulating plateau, which is under general
cultivation, as cultivation goes in Asiatic Turkey. A number of
Circassian villages are scattered over this upland plain ; most of
them are distant from my road, but many horsemen are encountered ; they ride the finest animals in the country, and one naturally falls to wondering how they manage to keep so well-dressed
and well-mounted, while rags and poverty and diminutive donkeys
seem to be the well-nigh universal rule among their neighbors.
The Circassians betray more interest in my purely personal affairs—
whether I am Russian or English, whither I am bound, etc.—and
less interest in the bicycle, than either Turks or Armenians, and
seem altogether of a more reserved disposition ; I generally have
as little conversation with them as possible, confining myself to
letting them know I am English and not Russian, and replying
" *Turkchi binmus* " (I don't understand) to other questions ; they
have a look about them that makes one apprehensive as to the dis-
interestedness of their wanting to know whither I am bound—apprehensive that their object is to find out where three or four of them
could " see me later." I see but few Circassian women ; what few
I approach sufficiently near to observe are all more or less pleasant-
faced, prepossessing females ; many have blue eyes, which is very
rare among their neighbors; the men average quite as handsome
as the women, and they have a peculiar dare-devil expression of
countenance that makes them distinguishable immediately from
either Turk or Armenian ; they look like men who wouldn't hesitate about undertaking any devilment they felt themselves equal
to for the sake of plunder. They are very like their neighbors,
however, in one respect ; such among them as take any great interest in my extraordinary outfit find it entirely beyond their comprehension ; the bicycle is a Gordian knot too intricate for their
semi-civilized minds to unravel, and there are no Alexanders
among them to think of cutting it. Before they recover from their
first astonishment I have disappeared.

The road continues for the most part ridable until about 2 P.M.,
when I arrive at a mountainous region of rocky ridges, covered
chiefly with a growth of scrub-oak. Upon reaching the summit
of one of these ridges, I observe some distance ahead what appears
to be a tremendous field of large cabbages, stretching away in

a northeasterly direction almost to the horizon of one's vision;
the view presents the striking appearance of large compact cab-
bage-heads, thickly dotting a well-cultivated area of clean black
loam, surrounded on all sides by rocky, uncultivatable wilds. Fifteen minutes later I am picking my way through this "cultivated
field," which, upon closer acquaintance, proves to be a smooth
lava-bed, and the "cabbages" are nothing more or less than boul-
ders of singular uniformity; and what is equally curious, they are
all covered with a growth of moss, while the volcanic bed they
repose on is perfectly naked.

Beyond this singular area, the country continues wild and mountainous, with no habitations near the road; and thus it continues until some time after night-fall, when I emerge upon a few
scattering wheat-fields. The baying of dogs in the distance indicates the presence of a village somewhere around; but having
plenty of bread on which to sup I once again determine upon
studying astronomy behind a wheat-shock. It is a glorious moonlight night, but the altitude of the country hereabouts is not less
than six thousand feet, and the chilliness of the atmosphere, already apparent, bodes ill for anything like a comfortable night;
but I scarcely anticipate being disturbed by anything save atmos-
pheric conditions. I am rolled up in my tent instead of under it,
slumbering as lightly as men are wont to slumber under these unfavorable conditions, when, about eleven o'clock, the unearthly
creaking of native *arabas* approaching arouses me from my lethar-
gical condition. Judging from the sounds, they appear to be making a bee-line for my position; but not caring to voluntarily reveal
my presence, I simply remain quiet and listen. It soon becomes
evident that they are a party of villagers, coming to load up their
buffalo *arabas* by moonlight with these very shocks of wheat. One
of the *arabas* now approaches the shock which conceals my recum-
bent form, and where the pale moonbeams are coquettishly ogling
the nickel-plated portions of my wheel, making it conspicuously
scintillant by their attentions.

Hoping the *araba* may be going to pass by, and that my presence may escape the driver's notice, I hesitate even yet to reveal myself; but the *araba* stops, and I can observe the driver's frightened
expression as he suddenly becomes aware of the presence of strange,
supernatural objects. At the same moment I rise up in my wind-
ing-sheet-like covering; the man utters a wild yell, and abandoning

the *araba*, vanishes like a deer in the direction of his companions. It is an unenviable situation to find one's self in; if I boldly approach them, these people, not being able to ascertain my character in the moonlight, would be quite likely to discharge their fire-arms at me in their fright; if, on the contrary, I remain under cover, they might also try the experiment of a shot before venturing to approach the deserted buffaloes, who are complacently chewing the cud on the spot where their chicken-hearted driver took to his heels.

Under the circumstances I think it best to strike off toward the road, leaving them to draw their own conclusions as to whether I am Sheitan himself, or merely a plain, inoffensive hobgoblin. But while gathering up my effects, one heroic individual ventures to approach part way and open up a shouting inquiry ; my answers, though unintelligible to him in the main, satisfy him that I am at all events a human being ; there are six of them, and in a few minutes after the ignominious flight of the driver, they are all gathered around me, as much interested and nonplussed at the appearance of myself and bicycle as a party of Nebraska homesteaders might be had they, under similar circumstances, discovered a turbaned old Turk complacently enjoying a nargileh.

No sooner do their apprehensions concerning my probable warlike character and capacity become allayed, than they get altogether too familiar and inquisitive about my packages ; and I detect one venturesome kleptomaniac surreptitiously unfastening a strap when he fancies I am not noticing. Moreover, laboring under the impression that I don't understand a word they are saying, I observe they are commenting in language smacking unmistakably of covetousness, as to the probable contents of my Whitehouse leather case ; some think it is sure to contain *chokh para* (much money), while others suggest that I am a *postaya* (courier), and that it contains letters. Under these alarming circumstances there is only one way to manage these overgrown children ; that is, to make them afraid of you forthwith ; so, shoving the strap-unfastener roughly away, I imperatively order the whole covetous crew to "haidi ! " Without a moment's hesitation they betake themselves off to their work, it being an inborn trait of their character to mechanically obey an authoritative command. Following them to their other *arabas*, I find that they have brought quilts along, intending, after loading up to sleep in the field until daylight. Se-

lecting a good heavy quilt with as little ceremony as though it were
my own property, I take it and the bicycle to another shock, and
curl myself up warm and comfortable ; once or twice the owner of
the coverlet approaches quietly, just near enough to ascertain that
I am not intending making off with his property, but there is not
the slightest danger of being disturbed or molested in any way till
morning ; thus, in this curious round-about manner, does fortune
provide me with the wherewithal to pass a comparatively comfortable night.   "Rather arbitrary proceedings to take a quilt without
asking permission," some might think ; but the owner thinks nothing of the kind ; it is quite customary for travellers of their own
nation to help themselves in this way, and the villagers have come
to regard it as quite a natural occurrence.

At daylight I am again on the move, and sunrise finds me busy
making an outline sketch of the ruins of an ancient castle, that occupies, I should imagine, one of the most impregnable positions in
all Asia Minor ; a regular Gibraltar.   It occupies the summit of a
precipitous detached mountain peak, which is accessible only from
one point, all the other sides presenting a sheer precipice of rock ;
it forms a conspicuous feature of the landscape for many miles
around, and situated as it is amid a wilderness of rugged brush-
covered heights, admirably suited for ambuscades, it was doubtless
a very important position at one time.   It probably belongs to the
Byzantine period, and if the number of old graves scattered among
the hills indicate anything, it has in its day been the theatre of stirring tragedy.   An hour after leaving the frowning battlements of
the grim old relic behind, I arrive at a cluster of four rock houses,
which are apparently occupied by a sort of a patriarchal family consisting of a turbaned old Turk and his two generations of descendants.   The old fellow is seated on a rock, smoking a cigarette and
endeavoring to coax a little comfort from the slanting rays of the
morning sun, and I straightway approach him and broach the all-
important subject of refreshments.

He turns out to be a fanatical old gentleman, one of those old-
school Mussulmans who have neither eye nor ear for anything but
the Mohammedan religion ; I have irreverently interrupted him in
his morning meditations, it seems, and he administers a rebuke in
the form of a sidewise glance, such as a Pharisee might be expected
to bestow on a Cannibal Islander venturing to approach him, and
delivers himself of two deep-fetched sighs of "Allah, Allah !"

Anybody would think from his actions that the sanctimonious old man—ikin (five feet three) had made the pilgrimage to Mecca a dozen times, whereas he has evidently not even earned the privilege of wearing a green turban ; he has neither been to Mecca himself during his whole unprofitable life nor sent a substitute, and he now thinks of gaining a nice numerous harem, and a walled-in garden, with trees and fountains, cucumbers and *carpooses*, in the land of the *kara ghuz kiz*, by cultivating the spirit of fanaticism at the eleventh hour. I feel too independent this morning to sacrifice any of the wellnigh invisible remnant of dignity remaining from the respectable quantity with which I started into Asia, for I still have a couple of the wheaten "quoits" I brought from Yuzgat ; so, leaving the ancient Mussulman to his meditations, I push on over the hills, when, coming to a spring, I eat my frugal breakfast, soaking the unbiteable "quoits" in the water.

After getting beyond this hilly region, I emerge upon a level plateau of considerable extent, across which very fair wheeling is found ; but before noon the inevitable mountains present themselves again, and some of the acclivities are trundleable only by repeating the stair-climbing process of the Kara Su Pass. Necessity forces me to seek dinner at a village where abject poverty, beyond anything hitherto encountered, seems to exist. A decently large fig-leaf, without anything else, would be eminently preferable to the tattered remnants hanging about these people, and among the smaller children *puris naturalis* is the rule. It is also quite evident that few of them ever take a bath ; as there is plenty of water about them, this doubtless comes of the pure contrariness of human nature in the absence of social obligations. Their religion teaches these people that they ought to bathe every day ; consequently, they never bathe at all. There is a small threshing-floor handy, and, taking pity on their wretched condition, I hesitate not to "drive dull care away" from them for a few minutes, by giving them an exhibition ; not that there is any "dull care" among them, though, after all ; for, in spite of desperate poverty, they know more contentment than the well-fed, respectably-dressed mechanic of the Western World. It is, however, the contentment born of not realizing their own condition, the bliss that comes of ignorance.

They search the entire village for eatables, but nothing is readily obtainable but bread. A few gaunt, angular fowls are scratching about, but they have a beruffled, disreputable appearance, as

though their lives had been a continuous struggle against being
caught and devoured ; moreover, I don't care to wait around three
hours on purpose to pass judgment on these people's cooking.
Eggs there are none ; they are devoured, I fancy, almost before
they are laid. Finally, while making the best of bread and water,
which is hardly made more palatable by the appearance of the people watching me feed—a woman in an airy, fairy costume, that is
little better than no costume at all, comes forward, and contributes a
small bowl of *yaort ;* but, unfortuntaely, this is *old yaort, yaort* that
is in the sere and yellow stage of its usefulness as human food ;
and although these people doubtless consume it thus, I prefer to
wait until something more acceptable and less odoriferous turns
up. I miss the genial hospitality of the gentle Koords to-day ! Instead of heaping plates of *pillau,* and bowls of wholesome new *yaort,*
fickle fortune brings me nothing but an exclusive diet of bread and
water.

My road, this afternoon, is a tortuous donkey-trail, intersecting
ravines with well-nigh perpendicular sides, and rocky ridges, covered with a stunted growth of cedar and scrub-oak. The higher
mountains round about are heavily timbered with pine and cedar.
A large forest on a mountain-slope is on fire, and I pass a camp of
people who have been driven out of their permanent abode by the
flames.   Fortunately, they have saved everything except their
naked houses and their grain. They can easily build new houses,
and their neighbors will give or lend them sufficient grain to tide
them over till another harvest.

Toward sundown the hilly country terminates, and I descend
into a broad cultivated valley, through which is a very good
wagon-road ; and I have the additional satisfaction of learning that
it will so continue clear into Sivas, a wagon-road having been
made from Sivas into this forest to enable the people to haul wood
and building-timber on their *arabas.* Arriving at a good-sized
and comparatively well-to-do Mussulman village, I obtain an ample
supper of eggs and *pillau,* and, after *binning* over and over again
until the most unconscionable Turk among them all can bring himself to importune me no more, I obtain a little peace. Supper for
two, together with the tough hill-climbing to-day, and insufficient
sleep last night, produces its natural effect ; I quietly doze off to
sleep while sitting on the divan of a small *khan,* which might very
appropriately be called an open shed. Soon I am awakened ; they

want me to accommodate them by *binning* once more before they retire for the night. As the moon is shining brightly, I offer no objections, knowing that to grant the request will be the quickest way to get rid of their worry. They then provide me with quilts, and I spend the night in the *khan* alone. I am soon asleep, but one habitually sleeps lightly under these strange and ever-varying conditions, and several times I am awakened by dogs invading the *khan* and sniffing about my couch.

My daily experience among these people is teaching me the commendable habit of rising with the lark ; not that I am an enthusiastic student, or even a willing one—be it observed that few people are—but it is a case of either turning out and sneaking off before the inhabitants are astir, or to be worried from one's waking moments to the departure from the village, and of the two evils one comes finally to prefer the early rising. One can always obtain something to eat before starting by waiting till an hour after sunrise, but I have had quite enough of these people's importunities to make breakfasting with them a secondary consideration, and so pull out at early daylight. The road is exceptionally good, but an east wind rises with the sun and quickly develops into a stiff breeze that renders riding against it anything but child's play ; no rose is to be expected without a thorn, nevertheless it is rather aggravating to have the good road and the howling head-wind happen together, especially in traversing a country where good roads are the exception instead of the rule.

About eight o'clock I reach a village situated at the entrance to a rocky defile, with a babbling brook dancing through the space between its two divisions. Upon inquiring for refreshments, a man immediately orders his wife to bring me *pillau.* For some reason or other—perhaps the poor woman has none prepared ; who knows? —the woman, instead of obeying the command like a "guid wifey," enters upon a wordy demurrer, whereupon her husband borrows a hoe-handle from a bystander and advances to chastise her for daring to thus hesitate about obeying his orders ; the woman retreats precipitately into the house, heaping Turkish epithets on her devoted husband's head. This woman is evidently a regular termagant, or she would never have used such violent language to her husband in the presence of a stranger and the whole village ; some day, if she doesn't be more reasonable, her husband, instead of satisfying his outraged feelings by chastising her with a hoe-handle, will, in a moment of passion, bid her begone from his house, which in Turkish law constitutes a legal separation; if the command be given in the presence of a competent witness it is irrevocable. Seeing me thus placed, as it were, in an embarrassing situation, another woman—dear, thoughtful creature!—fetches me enough wheat *pillau* to feed a mule, and a nice bowl of *yaort*, off which I make a substantial breakfast.

Near by where I am eating are five industrious maidens, preparing cracked or broken wheat by a novel and interesting process, that has hitherto failed to come under my observation; perhaps it is peculiar to' the Sivas vilayet, which I have now entered. A large rock is hollowed out like a shallow druggist's mortar; wheat is put in, and several girls (sometimes as many as eight, I am told by the American missionaries at Sivas) gather in a circle about it, and pound the wheat with light, long-headed mauls or beetles, striking in regular succession, as the reader has probably seen a gang of circus roustabouts driving tent-pins. When I first saw circus tent-pins driven in this manner, a few years ago, I remember hearing on-lookers remarking it as quite novel and wonderful how so many could be striking the same peg without their swinging sledges coming into collision; but that very same performance has been practised by the maidens hereabout, it seems, from time immemorial—another proof that there is nothing new under the sun.

Ten miles of good riding, and I wheel into the considerable town of Yennikhan, a place sufficiently important to maintain a public coffee-*khan* and several small shops. Here I take aboard a pocketful of fine large pears, and after wheeling a couple of miles to a secluded spot, halt for the purpose of shifting the pears from my pocket to where they will be better appreciated. Ere I have finished the second pear, a gentle goatherd, who from an adjacent hill observed me alight, appears upon the scene and waits around, with the laudable intention of further enlightening his mind when I remount. He is carrying a musical instrument something akin to a flute; it is a mere hollow tube with the customary finger-holes, but it is blown at the end; having neither reed nor mouth-piece of any description, it requires a peculiar sidewise application of the lips, and is not to be blown readily by a novice. When properly played, it produces soft, melodious music that, to say nothing else, must exert a gentle soothing in-

fluence on the wild, turbulent souls of a herd of goats. The goatherd offers me a cake of *ekmek* out of his wallet, as a sort of a
peace-offering, but thanks to a generous breakfast, music hath
more charms at present than dry *ekmek*, and handing him a pear,
I strike up a bargain by which he is to entertain me with a solo
until I am ready to start, when of course he will be amply recom-
pensed by seeing me *bin ;* the bargain is agreed to, and the solo
duly played.

East of Yennikhan, the road develops into an excellent macadamized highway, on which I find plenty of genuine amusement
by electrifying the natives whom I chance to meet or overtake.
Creeping noiselessly up behind an unsuspecting donkey-driver,
until quite close, I suddenly reveal my presence. Looking round
and observing a strange, unearthly combination, apparently swoop-
ing down upon him, the affrighted *katir-jee's* first impulse is to
seek refuge in flight, not infrequently bolting clear off the roadway, before venturing upon taking a second look. Sometimes I
simply put on a spurt, and whisk past at a fifteen mile pace.
Looking back, the *katir-jee* generally seems rooted to the spot with
astonishment, and his utter inability to comprehend. These men
will have marvellous tales to tell in their respective villages concerning what they saw ; unless other bicycles are introduced, the
time the "Ingilisin" went through the country with his wonderful *araba* will become a red-letter event in the memory of the people along my route through Asia Minor. Crossing the Yeldez
Irmak River, on a stone bridge, I follow along the valley of the
head-waters of our old acquaintance, the Kizil Irmak, and at three
o'clock in the afternoon, roll into Sivas, having wheeled nearly
fifty miles to-day, the last forty of which will compare favorably in
smoothness, though not in levelness, with any forty-mile stretch I
know of in the United States. From Angora I have brought a
letter of introduction to Mr. Ernest Weakley, a young Englishman,
engaged, together with Mr. Rodigas, a Belgian gentleman, for the
Ottoman Government, in collecting the Sivas vilayet's proportion of
the Russian indemnity ; and I am soon installed in hospitable quarters. Sivas artisans enjoy a certain amount of celebrity among
their compatriots of other Asia Minor cities for unusual skilfulness,
particularly in making filigree silver work. Toward evening myself
and Mr. Weakley take a stroll through the silversmiths' quarters.
The quarters consist of twenty or thirty small wooden shops, sur-

rounding an oblong court; spreading willows and a tiny rivulet running through it give the place a semi-rural appearance. In the little open-front workshops, which might more appropriately be called stalls, Armenian silversmiths are seated cross-legged, some working industriously at their trade, others gossiping and sipping coffee with friends or purchasers.

"Doesn't it call up ideas of what you conceive the quarters of the old alchemists to have been hundreds of years ago?" asks my companion. "Precisely what I was on the eve of suggesting to you," I reply, and then we drop into one of the shops, sip coffee with the old silversmith, and examine his filigree jewelry. There is nothing denoting remarkable skill about any of it; an intricate pattern of their jewelry simply represents a great expenditure of time and Asiatic patience, and the finishing of clasps, rivetting, etc., is conspicuously rough. Sivas was also formerly a seat of learning; the imposing gates, with portions of the fronts of the old Arabic universities are still standing, with sufficient beautiful arabesque designs in glazed tile-work still undestroyed, to proclaim eloquently of departed glories. The squalid mud hovels of refugees from the Caucasus now occupy the interior of these venerable edifices; ragged urchins romp with dogs and baby buffaloes where pashas' sons formerly congregated to learn wisdom from the teachings of their prophet, and now what remains of the intricate arabesque designs, worked out in small, bright-colored tiles, that once formed the glorious ceiling of the dome, seems to look down reproachfully, and yet sorrowfully, upon the wretched heaps of *tezek* placed beneath it for shelter.

I am remaining over one day at Sivas, and in the morning we call on the American missionaries. Mr. Perry is at home, and hopes I am going to stay a week, so that they can "sort of make up for the discomforts of journeying through the country;" Mr. Hubbard and the ladies of the Mission are out of town, but will be back this evening. After dinner we go round to the government *konak* and call on the Vali, Hallil Rifaat Pasha, whom Mr. Weakley describes beforehand as a very practical man, fond of mechanical contrivances; and who would never forgive him if he allowed me to leave Sivas with the bicycle without paying him a visit. The usual rigmarole of salaams, cigarettes, coffee, compliments, and questioning are gone through with; the Vali is a jolly-faced, good-natured man, and is evidently much interested in my companion's description of the bicycle and my journey.

Of course I don't forget to praise the excellence of the road from Yennikhan ; I can conscientiously tell him that it is superior to anything I have wheeled over south of the Balkans ; the Pasha is delighted at hearing this, and beaming joyously over his spectacles, his fat jolly face a rotund picture of satisfaction, he says to Mr. Weakley : "You see, he praises up our roads ; and he ought to know, he has travelled on wagon roads half way round the world." The interview ends by the Vali inviting me to ride the bicycle out to his country residence this evening, giving the order for a squad of *zaptiehs* to escort me out of town at the appointed time. "The Vali is one of the most energetic pashas in Turkey," says Mr. Weakley, as we take our departure. "You would scarcely believe that he has established a small weekly newspaper here, and makes it self-supporting into the bargain, would you?"

"I confess I don't see how he manages it among these people," I reply, quite truthfully, for these are anything but newspaper-supporting people ; "how does he manage to make it self-supporting?"

"Why, he makes every employé of the government subscribe for a certain number of copies, and the subscription price is kept back out of their salaries ; for instance, the *mulazim* of *zaptiehs* would have to take half a dozen copies, the *mutaserif* a dozen, etc. ; if from any unforeseen cause the current expenses are found to be more than the income, a few additional copies are saddled on each 'subscriber.'" Before leaving Sivas, I arrive at the conclusion that Hallil Rifaat Pasha knows just about what's what ; while administering the affairs of the Sivas vilayet in a manner that has gained him the good-will of the population at large, he hasn't neglected his opportunities at the Constantinople end of the rope ; more than one beautiful Circassian girl has, I am told, been forwarded to the Sultan's harem by the enterprising and sagacious Sivas Vali ; consequently he holds "trump cards," so to speak, both in the province and the palace.

Promptly at the hour appointed the squad of *zaptiehs* arrive ; Mr. Weakley mounts his servant on a prancing Arab charger, and orders him to manœuvre the horse so as to clear the way in front ; the *zaptiehs* commence their flogging, and in the middle of the

cleared space I trundle the bicycle. While making our way
through the streets, Mr. Hubbard, who, with the ladies, has just
returned to the city, is encountered on the way to invite Mr.
Weakley and myself to supper ; as he pushes his way through the
crowd and reaches my side, he pronounces it the worst rabble he
ever saw in the streets of Sivas, and he has been stationed here
over twelve years. Once clear of the streets, I mount and soon
outdistance the crowd, though still followed by a number of horsemen. Part way out we wait for the Vali's state carriage, in which
he daily rides between the city and his residence. While waiting,

A Harem Beauty.

a terrific squall of wind and dust comes howling from the direction
we are going, and while it is still blowing great guns, the Vali
and his mounted escort arrive. His Excellency alights and examines the Columbia with much interest, and then requests me to
ride on immediately in advance of the carriage. The grade is
slightly against me, and the whistling wind seems to be shrieking
a defiance; but by superhuman efforts, almost, I pedal ahead and
manage to keep in front of his horses all the way. The distance
from Sivas is four and a quarter miles by the cyclometer ; this is
the first time it has ever been measured.

We are ushered into a room quite elegantly furnished, and light refreshments served. Observing my partiality for vishner-*su*, the Governor kindly offers me a flask of the syrup to take along; which I am, however, reluctantly compelled to refuse, owing to my inability to carry it. Here, also, we meet Djaved Bey, the Pasha's son, who has recently returned from Constantinople, and who says he saw me riding at Prinkipo. The Vali gets down on his hands and knees to examine the route of my journey on a map of the world which he spreads out on the carpet; he grows quite enthusiastic, and exclaims, "Wonderful!"   "Very wonderful!" says Djaved Bey; "when you get back to America they will—build you a statue."   Mr. Hubbard has mounted a horse and followed us to the Vali's residence, and at the approach of dusk we take our departure; the wind is favorable for the return, as is also the gradient; ere my two friends have unhitched their horses,

I mount and am scudding before the gale half a mile away.

"Hi hi hi-hi! you'll never overtake him!" the Vali shouts enthusiastically to the two horsemen as they start at full gallop after me, and

The Vali on Floor with Map.

which they laughingly repeat to me shortly afterward. A very pleasant evening is spent at Mr. Hubbard's house; after supper the ladies sing "Sweet Bye and Bye," "Home, Sweet Home," and other melodious reminders of the land of liberty and song that gave them birth. Everything looks comfortable and homelike, and they have English ivy inside the dining-room trained up the walls and partly covering the ceiling, which produces a wonderfully pleasant effect. The usual extraordinary rumors of my wonderful speeding ability have circulated about the city during the day and evening, some of which have happened to come to the ears of the missionaries. One story is that I came from the port of Samsoon, a distance of nearly three hundred miles, in six hours, while an imaginative *katir-jee*, whom I whisked past on the road, has been telling the Sivas people an exaggerated story of how a *genii* had ridden past him with lightning-like speed on a shining wheel; but whether it was a good or an evil *genii* he said he didn't have time to determine, as I went past like a flash and vanished in

the distance. The missionaries have four hundred scholars attending their school here at Sivas, which would seem to indicate
a pretty flourishing state of affairs. Their recruiting ground is, of
course, among the Armenians, who, though professedly Christians,
really stand in more need of regeneration than their Mohammedan
neighbors. The characteristic condition of the average Armenian
villager's mind is deep, dense ignorance and moral gloominess ;
it requires more patience and perseverance to ingraft a new idea
on the unimpressionable trunk of an Armenian villager's intellect
than it does to put up second-hand stove-pipe ; and it is a generally admitted fact—*i.e.*, west of the Missouri River—that anyone
capable of setting up three joints of second-hand stove-pipe without using profane language deserves a seat in Paradise.

"Come in here a minute," says Mr. Hubbard, just before our
departure for the night, leading the way into an adjoining room ;
"here's shirts, under-clothing, socks, handkerchiefs—everything ;
help yourself to anything you require ; I know something about
travelling through this country *myself !* " But not caring to impose too much on good nature, I content myself with merely
pocketing a strong pair of socks, that I know will come in handy.
I leave the bicycle at the mission over night, and in the morning,
at Miss Chamberlain's request, I ride round the school-house yard
a few times for the edification of the scholars. The greatest difficulty, I am informed, with Armenian pupils is to get them to take
sufficient interest in anything to ask questions ; it is mainly because
the bicycle will be certain to awaken interest, and excite the spirit
of inquiry among them, that I am requested to ride for their benefit.
Thus is the bicycle fairly recognized as a valuable aid to missionary
work. Moral : let the American and Episcopal boards provide
their Asia Minor and Persian missionaries with nickel-plated bicycles ; let them wheel their way into the empty wilderness of the
Armenian mind, and light up the impenetrable moral darkness
lurking therein with the glowing and mist-dispelling orbs of cycle
lamps.

Messrs. Perry, Hubbard, and Weakley accompany me out some
distance on horseback, and at parting I am commissioned to carry
salaams to the brethren in China. This is the first opportunity
that has ever presented of sending greetings overland to far-off
China, they say, and such rare occasions are not to be lightly overlooked. They also promise to send word to the Erzeroum mission

to expect me ; the chances are, however, that I shall reach Erzeroum before their letter ; there are no lightning mail-trains in
Asia Minor. The road eastward from Sivas is an artificial highway, and affords reasonably good wheeling, but is somewhat inferior to the road from Yennikhan. Before long I enter a region of
low hills, dales, and small lakes, beyond which the road again descends into the valley of the Kizil Irmak. All day long the roadway averages better wheeling than I ever expected to find in Asiatic
Turkey ; but the prevailing east wind offers strenuous opposition
to my progress every inch of the way along the hundred miles or
so of ridable road from Yennikhan to Zara, a town at which I arrive near sundown. Zara is situated at the entrance to a narrow
passage between two mountain spurs, and although the road is
here a dead level and the surface smooth, the wind comes roaring
from the gorge with such tremendous pressure that it is only by
extraordinary exertions that I am able to keep the saddle.

Tifticjeeoghlou Effendi was a gentleman of Greek descent. At
Zara I have an opportunity of seeing and experiencing something
of what hospitality is like among the better class Armenians, for I
have brought from Sivas a letter of introduction to Kirkor-agha
Vartarian, the most prominent Armenian gentleman in Zara. I have
no difficulty whatever in finding the house, and am at once installed
in the customary position of honor, while five serving-men hover
about, ready to wait on me ; some take a hand in the inevitable
ceremony of preparing and serving coffee and lighting cigarettes,
while others stand watchfully by awaiting word or look from myself or mine host, or from the privileged guests that immediately
begin to arrive. The room is of cedar planking throughout, and is
absolutely without furniture, save the carpeting and the cushioned
divan on which I am seated. Mr. Vartarian sits crossed-legged on
the carpet to my left, smoking a nargileh ; his younger brother occupies a similar position on my right, rolling and smoking cigarettes ; while the guests, as they arrive, squat themselves on the carpet in positions varying in distance from the divan, according to
their respective rank and social importance. No one ventures to
occupy the cushioned divan alongside myself, although the divan is
fifteen feet long, and it makes me feel uncomfortably like the dog
in the manger to occupy its whole length alone.

In a farther corner, and off the slightly raised and carpeted floor
on which are seated the guests, is a small brick fire-place, on which

25

a charcoal fire is brightly burning, and here Mr. Vartarian's private
kahvay-jee is kept busily employed in brewing tiny cups of strong
black coffee ; another servant constantly visits the fire to ferret out
pieces of glowing charcoal with small pipe-lighting tongs, with which
he circulates among the guests, supplying a light to the various
smokers of cigarettes. A third youth is kept pretty tolerably busy
performing the same office for Mr. Vartarian's nargileh, for the gentleman is an inveterate smoker, and in all Turkey there can scarcely
be another nargileh requiring so much tinkering with as his. All
the livelong evening something keeps getting wrong with that
wretched pipe ; mine host himself is continually rearranging the
little pile of live coals on top of the dampened tobacco (the tobacco
smoked in a nargileh is dampened, and live coals are placed on top),
taking off the long coiled tube and blowing down it, or prying
around in the tobacco receptacle with an awl-like instrument in his
efforts to make it draw properly, but without making anything like
a success ; while his nargileh-boy is constantly hovering over it with
a new supply of live coals. "Job himself could scarcely have been
possessed of more patience," I think at first; but before the evening
is over I come to the conclusion that my worthy host wouldn't exchange that particular hubble-bubble with its everlasting contrariness for the most perfectly drawing nargileh in Turkey : like certain devotees of the weed among ourselves, who never seem to be
happier than when running a broom-straw down the stem of a pipe
that chronically refuses to draw, so Kirkor-agha Vartarian finds his
chief amusement in thus tinkering from one week's end to another
with his nargileh.

At the supper table mine host and his brother both lavish attentions upon me ; knives and forks of course there are none, these
things being seldom seen in Asia Minor, and to a cycler who has
spent the day in pedalling against a stiff breeze, their absence is a
matter of small moment. I am ravenously hungry, and they both
win my warmest esteem by transferring choice morsels from their
own plates into mine with their fingers. From what I know of
strict *haut ton* Zaran etiquette, I think they should really pop these
tid-bits in my mouth, and the reason they don't do so is, perhaps,
because I fail to open it in the customary *haut ton* manner ; however, it is a distasteful thing to be always sticking up for one's individual rights. A pile of quilts and mattresses, three feet thick,
and feather pillows galore are prepared for me to sleep on. An

Armenian Hospitality.

attendant presents himself with a wonderful night-shirt, on the
ample proportions of which are displayed bewildering colors and
figures ; and following the custom of the country, shapes himself for
undressing me and assisting me into bed.   This, however, I prefer
to do without assistance, owing to a large stock of native modesty.

I never fell among people more devoted in their attentions; their
only thought during my stay is to make me comfortable ; but they
are very ceremonious and great sticklers for etiquette.   I had intended making my usual early start, but mine host receives with
open disapproval—I fancy even with a showing of displeasure—my
proposition to depart without first par-
taking of refreshments, and it is nearly
eight o'clock before I finally get started.
Immediately after rising comes the inevitable coffee and early morning visi-

tors ; later an attend-
ant arrives with break-
fast for myself on a

At Kirkor-agha Vartarian's.

small wooden tray.   Mr. Vartarian occupies precisely the same
position, and is engaged in precisely the same occupation as yesterday evening, as is also his brother.   No sooner does the hapless
attendant make his appearance with the eatables than these two
persons spring simultaneously to their feet, apparently in a towering rage, and chase him back out of the room, meanwhile pursuing
him with a torrent of angry words ; they then return to their respective positions and respective occupations.   Ten minutes later
the attendant reappears, but this time bringing a larger tray with
an ample spread for three persons ; this, it afterward appears, is
not because mine host and his brother intends partaking of any,

THROUGH THE SIVAS VILAYET INTO ARMENIA. 389

but because it is Armenian etiquette to do so, and Armenian etiquette therefore becomes responsible for the spectacle of a solitary
feeder seated at breakfast with dishes and everything prepared for
three, while of the other two, one is smoking a nargileh, the other
cigarettes, and both of them regarding my evident relish of scrambled eggs and cold fowl with intense satisfaction.

Having by this time determined to merely drift with the current
of mine host's intentions concerning the time of my departure, I
resume my position on the divan after breakfasting, simply hinting
that I would like to depart as soon as possible. To this Mr. Vartarian complacently nods assent, and his brother, with equal complacency rolls me a cigarette, after which a good half-hour is consumed in preparing for me a letter of introduction to their friend
Mùdura Ohana in the village of Kachahurda, which I expect to
reach somewhere near noon ; mine host dictates while his brother
writes. Visitors continue coming in, and I am beginning to get
a trifle impatient about starting ; am beginning in fact to wish all
their nonsensical ceremoniousness at the bottom of the deep blue
sea or some equally unfathomable quarter, when, at a signal from
Mr. Vartarian himself, his brother and the whole roomful of visitors rise simultaneously to their feet, and equally simultaneously
put their hands on their respective stomachs, and, turning toward
me, salaam ; mine host then comes forward, shakes hands, gives
me the letter to Mùdura Ohana, and permits me to depart.

He has provided two *zaptiehs* to escort me outside the town, and
in a few minutes I find myself bowling briskly along a beautiful
little valley ; the pellucid waters of a purling brook dance merrily
alongside an excellent piece of road ; birds are singing merrily in
the willow-trees, and dark rocky crags tower skyward immediately
around. The lovely little valley terminates all too soon, for in fifteen
minutes I am footing it up another mountain ; but it proves to be
the entrance gate of a region containing grander pine-clad mountain
scenery than anything encountered outside the Sierra Nevadas ; in
fact the famous scenery of Cape Horn, California, almost finds its
counterpart at one particular point I traverse this morning ; only
instead of a Central Pacific Railway winding around the gray old
crags and precipices, the enterprising Sivas Vali has built an *araba*
road. One can scarce resist the temptation of wheeling down some
of the less precipitous slopes, but it is sheer indiscretion, for the
roadway makes sharp turns at points where to continue straight

ahead a few feet too far would launch one into eternity ; a broken brake, a wild "coast" of a thousand feet through mid-air into the dark depths of a rocky gorge, and the "tour around the world" would abruptly terminate.

For a dozen miles I traverse a tortuous road winding its way among wild mountain gorges and dark pine forests ; Circassian horsemen are occasionally encountered : it seems the most appropriate place imaginable for robbers, and I have again been cautioned against these freebooting mountaineers at Sivas.   They eye me curiously, and generally halt after they have passed, and watch my progress for some minutes.   Once I am overtaken by a couple of them ; they follow close behind me up a mountain slope ; they are heavily armed and look capable of anything, and I plod along, mentally calculating how to best encompass their destruction with the Smith & Wesson, without coming to grief myself, should their intentions toward me prove criminal.   It is not exactly comfortable or reassuring to have two armed horsemen, of a people who are regarded with universal fear and mistrust by everybody around them, following close upon one's heels, with the disadvantage of not being able to keep an eye on their movements ; however, they have little to say ; and as none of them attempt any interference, it is not for me to make insinuations against them on the barren testimony of their outward appearance and the voluntary opinions of their neighbors.

My route now leads up a rocky ravine, the road being fairly under cover of over-arching rocks at times, thence over a billowy region of mountain summits—an elevated region of pine-clad ridges and rocky peaks—to descend again into a cultivated country of undulating hills and dales, checkered with fields of grain.   These low rolling hills appear to be in a higher state of cultivation than any district I have traversed in Asia Minor ; from points of vantage the whole country immediately around looks like a swelling sea of golden grain ; harvesting is going merrily on ; men and women are reaping side by side in the fields, and the songs of the women come floating through the air from all directions.   They are Armenian peasants, for I am now in Armenia proper ; the inhabitants of this particular locality impress me as a light hearted, industrious people ; they have an abundant harvest, and it is a pleasure to stand and see them reap, and listen to the singing of the women ; moreover they are more respectably clothed than the lower class natives round about them, barring, of course, our unfathomable acquaintances, the Circassians.

Toward the eastern extremity of this peaceful, happy scene is the village of Kachahurda, which I reach soon after noon, and where resides Mûdura Ohana, to whom I bring a letter. Picturesquely speaking, Kachahurda is a disgrace to the neighborhood in which it stands; its mud hovels are combined cow-pens, chicken-coops, and human habitations, and they are bunched up together without any

Apprehensive of Danger.

pretence to order or regularity; yet the light-hearted, decently-clad people, whose songs come floating from the harvest-fields, live contentedly in this and other equally wretched villages round about. Mûdura Ohana provides me with a repast of bread and *yaort*, and endeavors to make my brief halt comfortable. While I am discussing these refreshments, himself and another unwashed, unkempt old party come to high, angry words about me; but whatever it is about I haven't the slightest idea. Mine host seems a regular old savage when angry. He is the happy possessor of a pair of powerful lungs, which are ably seconded by a fog-horn voice, and he howls at the other man like an enraged bull. The other man doesn't seem to mind it, though, and keeps up his end of the controversy—or whatever it is—in a comparatively cool and aggravating manner, that seems to feed Mùdura Ohana's righteous wrath, until I quite expect to see that outraged person reach down one of the swords off the wall and hack his opponent into sausage-meat. Once I venture to inquire, as far as one can inquire by pantomime, what they are quarrelling so violently about me for, being really inquisitive to find out. They both immediately cease hostilities to assure me that it is nothing for which I am in any way personally responsible ; and then they straightway fall to glaring savagely at each other again, and renew their vocal warfare more vigorously, if anything, from having just drawn a peaceful breath. Mine host of Kachahurda can scarcely be called a very civilized or refined individual ; he has neither the gentle kindliness of Kirkoragha Vartarian, nor the dignified, gentlemanly bearing of Tifticjeeoghlou Effendi ; but he grabs a club, and roaring like the hoarse whistle of a Mississippi steamboat, chases a crowd of villagers out of the room who venture to come in on purpose to stare rudely at his guest ; and for this charitable action alone he deserves much credit ; nothing is so annoying as to have these unwashed crowds standing gazing and commenting while one is eating. A man is sent with me to direct me aright where the road forks, a mile or so from the village ; from the forks it is a newly made road, in fact, unfinished ; it resembles a ploughed field for looseness and depth ; and when, in addition to this, one has to climb a gradient of twenty metres to the hundred, a bicycle is anything but a comforting thing to possess.

The country becomes broken and more mountainous than ever, and the road winds about fearfully. Often a part of the road that is but a mile away as the crow flies requires an hour's steady going to reach it ; but the mountain scenery is glorious. Occasionally I round a point, or reach a summit, from whence a magnificent and comprehensive view bursts upon the vision, and it really requires an effort to tear one's self away, realizing that in all probability I shall never see it again. At one point I seem to be overlooking a vast amphitheatre which encompasses within itself the physical geography of a continent. It is traversed by whole mountain-ranges of lesser degree ; it contains tracts of stony desert and fertile valley,

lakes, and a river, not excepting even the completing element of a
fine forest, and encompassing it round about, like an impenetrable
palisade protecting it against invasion, are scores of grand old
mountains—grim sentinels that nothing can overcome. The road,
though still among the mountains, is now descending in a general
way from the elevated divide, down toward Enderes and the valley
of the Gevmeili Chai River ; and toward evening I enter an Armenian village.

The custom from here eastward appears to be to have the
threshing-floors in or near the village ; there are sometimes several
different floors, and when they are winnowing the grain on windy
days the whole village becomes covered with an inch or two of
chaff. I am glad to find these threshing-floors in the villages, because they give me an excellent opportunity to ride and satisfy the
people, thus saving me no end of worry and annoyance.

The air becomes chilly after sundown, and I am shown into a
close room containing one small air-hole, and am provided with a
quilt and pillow. Later in the evening a Turkish Bey arrives with
an escort of *zaptiehs* and occupies the same apartment, which would
seem to be a room especially provided for the accommodation of
travellers. The moment the officer arrives, behold, there is a hurrying to and fro of the villagers to sweep out the room, kindle a
fire to brew his coffee, and to bring him water and a vessel for his
ablutions before saying his evening prayers. Cringing servility
characterizes the demeanor of these Armenian villagers toward the
Turkish officer, and their hurrying hither and thither to supply him
ere they are asked looks to me wonderfully like a "propitiating of
the gods." The Bey himself seems to be a pretty good sort of a
fellow, offering me a portion of his supper, consisting of bread,
olives, and onions ; which, however, I decline, having already ordered
eggs and *pillau* of a villager. The Bey's company is highly acceptable, since it saves me from the annoyance of being surrounded by
the usual ragged, unwashed crowd during the evening, and secures
me a refreshing sleep, undisturbed by visions of purloined straps
or moccasins. He appears to be a very pious Mussulman ; after
washing his head, hands, and feet, he kneels toward Mecca on the
wet towel, and prays for nearly twenty minutes by my timepiece ;
and his sighs of Allah ! are wonderfully deep-fetched, coming apparently from clear down in his stomach. While he is thus devotionally engaged, his two *zaptiehs* stand respectfully by, and divide their

time between eying myself and the bicycle with wonder and the
Bey with mingled reverence and awe.

At early dawn I steal noiselessly away, to avoid disturbing the
peaceful slumbers of the Bey. For several miles my road winds
around among the foot-hills of the range I crossed yesterday, but
following a gradually widening depression, which finally terminates
in the Gevmeili Chai Valley ; and directly ahead and below me lies
the considerable town of Enderes, surrounded by a broad fringe of
apple-orchards, and walnut and jujube groves. Here I obtain a
substantial breakfast of Turkish kabobs (tid-bits of mutton, spitted
on a skewer, and broiled over a charcoal fire) at a public eating *khan*,
after which the *mudir* kindly undertakes to explain to me the best
route to Erzingan, giving me the names of several villages to inquire
for as a guidance. While talking to the *mudir*, Mr. Pronatti, an
Italian engineer in the employ of the Sivas Vali, makes his appearance, shakes hands, reminds me that Italy has recently volunteered
assistance to England in the Soudan campaign, and then conducts
me to his quarters in another part of the town. Mr. Pronatti can
speak almost any language but English ; I speak next to nothing *but*
English ; nevertheless, we manage to converse quite readily, for, besides proficiency in pantomimic language acquired by daily practice,
I have necessarily picked up a few scattering words of the vernacular of the several countries traversed on the tour. While discussing
a nice ripe water-melon with this gentleman, several respectable-
looking people enter and introduce themselves through Mr. Pronatti
as Osmanli Turks, not Armenians, expecting me to regard them
more favorably on that account. Soon afterward a party of Armenians arrive, and take labored pains to impress upon me that they
are not Turks, but Christian Armenians. Both parties seem desirous of winning my favorable opinion. One party thinks the
surest plan is to let me know that they are Turks ; the others, to let
me know that they are *not* Turks. "I have told both parties to go
to Gehenna," says my Italian friend. "These people will worry
you to death with their foolishness if you make the mistake of
treating them with consideration."

Donning an Indian pith-helmet that is three sizes too large, and
wellnigh conceals his features, Mr. Pronatti orders his horse, and
accompanies me some distance out, to put me on the proper course
to Erzingan. My route from Enderes leads along a lovely fertile
valley, between lofty mountain ranges ; an intricate net-work of irri-

gating ditches, fed by mountain streams, affords an abundance of water for wheat-fields, vineyards, and orchards; it is the best, and yet the worst watered valley I ever saw—the best, because the irrigating ditches are so numerous; the worst, because most of them are overflowing and converting my road into mud-holes and shallow pools. In the afternoon I reach somewhat higher ground, where the road becomes firmer, and I bowl merrily along eastward, interrupted by nothing save the necessity of dismounting and shedding my nether garments every few minutes to ford a broad, swift feeder to the lesser ditches lower down the valley. In this fructiferous vale my road sometimes leads through areas of vineyards surrounded by low mud walls, where grapes can be had for the reaching, and where the proprietor of an orchard will shake down a shower of delicious yellow pears for whatever you like to give him, or for nothing if one wants him to. I suppose these villagers have established prices for their commodities when dealing with each other, but they almost invariably refuse to charge me anything; some will absolutely refuse any payment, and my only plan of recompensing them is to give money to the children; others accept, with as great a show of gratitude as if I were simply giving it to them without having received an equivalent, whatever I choose to give.

The numerous irrigating ditches have retarded my progress to an appreciable extent to-day, so that, notwithstanding the early start and the absence of mountain-climbing, my cyclometer registers but a gain of thirty-seven miles, when, having continued my eastward course for some time after nightfall, and failing to reach a village, I commence looking around for somewhere to spend the night. The valley of the Gevmeili Chai has been left behind, and I am again traversing a narrow, rocky pass between the hills. Among the rocks I discover a small open cave, in which I determine to spend the night. The region is elevated, and the night air chilly; so I gather together some dry weeds and rubbish and kindle a fire. With something to cook and eat, and a pair of blankets, I could have spent a reasonably comfortable night; but a pocketful of pears has to suffice for supper, and when the unsubstantial fuel is burned away, my airy chamber on the bleak mountain-side and the thin cambric tent affords little protection from the insinuating chilliness of the night air. Variety is said to be the spice of life; no doubt it is, under certain conditions, but I think it all depends on the conditions whether it is spicy or not spicy. For instance, the vicissitudes of fortune that favor me with bread and sour milk for dinner, a few pears for supper, and a wakeful night of shivering discomfort in a cave, as the reward of wading fifty irrigating ditches and traversing thirty miles of ditch-bedevilled donkey-trails during the day, may look spicy, and even romantic, from a distance ; but when one wakes up in a cold shiver about 1.30 A.M. and realizes that several hours of wretchedness are before him, his waking thoughts are apt to be anything but thoughts complimentary of the spiciness of the situation. Inshallah! fortune will favor me with better dues to-morrow ; and if not to-morrow, then the next day, or the next.

CHAPTER XVII.

For mile after mile, on the following morning, my route leads through broad areas strewn with bowlders and masses of rock that appear to have been brought down from the adjacent mountains by the annual spring floods, caused by the melting winter's snows ; scattering wheat-fields are observed here and there on the higher patches of ground, which look like small yellow oases amid the desert-like area of loose rocks surrounding them. Squads of diminutive donkeys are seen picking their weary way through the bowlders, toiling from the isolated fields to the village threshing-floors beneath small mountains of wheat-sheaves. Sometimes the donkeys themselves are invisible below the general level of the bowlders, and nothing is to be seen but the head and shoulders of a man, persuading before him several animated heaps of straw. Small lakes of accumulated surface-water are passed in depressions having no outlet ; thickets and bulrushes are growing around the edges, and the surfaces of some are fairly black with multitudes of wild-ducks. Soon I reach an Armenian village ; after satisfying the popular curiosity by riding around their threshing-floor, they bring me some excellent wheat-bread, thick, oval cakes that are quite acceptable, compared with the wafer-like sheets of the past several days, and five boiled eggs. The people providing these will not accept any direct payment, no doubt thinking my having provided them with the only real entertainment most of them ever saw, a fair equivalent for their breakfast ; but it seems too much like robbing paupers to accept anything from these people without returning something, so I give money to the children. These villagers seem utterly destitute of manners, standing around and watching my efforts to eat soft-boiled eggs with a pocket-knife with undisguised merriment. I inquire for a spoon, but they evidently prefer to extract amusement from watching my interesting attempts with the pocket-knife. One of them finally fetches a clumsy wooden ladle, three times broader than an egg, which, of course, is worse than nothing.

I now traverse a mountainous country with a remarkably clear atmosphere. The mountains are of a light cream-colored shaly composition ; wherever a living stream of water is found, there also is a village, with clusters of trees. From points where a comprehensive view is obtainable the effect of these dark-green spots, scattered here and there among the whitish hills, seen through the clear, rarefied atmosphere, is most beautiful. It seems a peculiar feature of everything in the East—not only the cities themselves,

The Armenian Egg-spoon.

but even of the landscape—to look beautiful and enchanting at a distance ; but upon a closer approach all its beauty vanishes like an illusory dream. Spots that from a distance look, amid their barren, sun-blistered surroundings, like lovely bits of fairyland, upon closer investigation degenerate into wretched habitations of a ragged, poverty-stricken people, having about them a few neglected orchards and vineyards, and a couple of dozen straggling willows and jujubes.

For many hours again to-day I am traversing mountains, mountains, nothing but mountains ; following tortuous camel-paths far up their giant slopes. Sometimes these camel-paths are splendidly smooth, and make most excellent riding. At one place, particularly, where they wind horizontally around the mountain-side, hundreds of feet above a village immediately below, it is as though the villagers were in the pit of a vast amphitheatre, and myself were wheeling around a semicircular platform, five hundred feet above them, but in plain view of them all. I can hear the wonder-struck villagers calling each other's attention to the strange apparition, and can observe them swarming upon the house-tops. What won-derful stories the inhabitants of this particular village will have to recount to their neighbors, of this marvellous sight, concerning which their own unaided minds can give no explanation! Noontide comes and goes without bringing me any dinner, when I emerge upon a small, cultivated plateau, and descry a co-terie of industrious females reaping together in a field near by, and straightway turn my footsteps thitherward with a view of ascer-taining whether they happen to have any eatables. No sooner do they observe me trundling toward them than they ingloriously flee the field, thoughtlessly leaving bag and baggage to the tender mercies of a ruthless invader. Among their effects I find some bread and a cucumber, which I forthwith confiscate, leaving a two and a half piastre *métallique* piece in its stead ; the affrighted women are watching me from the safe distance of three hundred yards ; when they return and discover the coin they will wish some 'cycler would happen along and frighten them away on similar conditions every day. Later in the afternoon I find myself wandering along the wrong trail ; not a very unnatural occurrence hereabout, for since leaving the valley of the Gevmeili Chai, it has been difficult to distinguish the Erzingan trail from the numerous other trails intersecting the country in every direction. On such a journey as this one seems to acquire a certain amount of instinct concerning roads ; certain it is, that I never traverse a wrong trail any dis-tance these days ere, without any tangible evidence whatever, I feel instinctively that I am going astray. A party of camel-drivers direct me toward the lost Erzingan trail, and in an hour I am fol-lowing a tributary of the ancient Lycus River, along a valley where everything looks marvellously green and refreshing ; it is as though I have been suddenly transferred into an entirely different country.

This innovation from barren rocks and sun-baked shale to a valley where the principal crops seem to be alfalfa and clover, and which is flanked on the south by dense forests of pine, encroaching downward from the mountain slopes clear on to the level green-sward, is rather an agreeable surprise ; the secret of the magic change does not remain a secret long ; it reveals itself in the shape of sundry broad snow-patches still lingering on the summits of a higher mountain range beyond. These pine forests, the pleasant greensward, and the lingering snow-banks, tell an oft-repeated tale ; they speak eloquently of forests preserved and the winter snow-fall thereby increased; they speak all the more eloquently because of being surrounded by barren, parched-up hills which, under like conditions, might produce similar happy results, but which now produce nothing. While traversing this smiling valley I meet a man asleep on a buffalo *araba*; an irrigating ditch runs parallel with the road and immediately alongside; the meek-eyed buffaloes swerve into the ditch in deference to their awe of the bicycle, and upset their drowsy driver into the water. The man evidently stands in need of a bath, but somehow he doesn't seem to appreciate it; perhaps it happened a trifle too *impromptu*, as it were, to suit his easy-going Asiatic temperament. He returns my rude, unsympathetic smile with a prolonged stare of bewilderment, but says nothing.

Soon I meet a boy riding on a donkey, and ask him the *postaya* distance to Erzingan; the youth looks frightened half out of his senses, but manages to retain sufficient presence of mind to elevate one finger, by which I understánd him to mean that it is one hour, or about four miles. Accordingly I pedal perseveringly ahead, hoping to reach the city before dusk, at the same time feeling rather surprised at finding it so near, as I haven't been expecting to reach there before to-morrow. Five miles beyond where I met the boy, and just after sundown, I overtake some *katir-jees en route* to Erzingan with donkey-loads of grain, and ask them the same question. From them I learn that instead of one, it is not less than twelve hours distant, also that the trail leads over a fearfully mountainous country. Nestling at the base of the mountains, a short distance to the northward, is the large village of Merriserriff, and not caring to tempt the fates into giving me another supperless night in a cold, cheerless cave, I wend my way thither.

Fortune throws me into the society of an Armenian whose chief anxiety seems to be, first, that I shall thoroughly understand that he is an Armenian, and not a Mussulman; and, secondly, to hasten me into the presence of the *mudir*, who *is* a Mussulman, and a Turkish Bey, in order that he may bring himself into the *mudir's* favorable notice by personally introducing me as a rare novelty on to his (the *mudir's*) threshing-floor. The official and a few friends are sipping coffee in one corner of the threshing floor, and, although I don't much relish my position of the Armenian's puppet-show, I give the *mudir* an exhibition of the bicycle's use, in the expectation that he will invite me to remain his guest over night.

He proves uncourteous, however, not even inviting me to partake of coffee ; evidently, he has become so thoroughly accustomed to the abject servility of the Armenians about him—who would never think of expecting reciprocating courtesies from a social superior —that he has unconsciously come to regard everybody else, save those whom he knows as his official superiors, as tarred, more or less, with the same feather.   In consequence of this belief I am not a little gratified when, upon the point of leaving the threshing-floor, an occasion offers of teaching him different.

Other friends of the *mudir's* appear upon the scene just as I am leaving, and he beckons me to come back and *bin* for the enlighten-ment of the new arrivals.   The Armenian's countenance fairly beams with importance at thus being, as it were, *encored*, and the collected villagers murmur their approval ; but I answer the *mudir's* beck-oned invitation by a negative wave of the hand, signifying that I can't bother with him any further.   The common herd around re-gard this self-assertive reply with open-mouthed astonishment, as though quite too incredible for belief ; it seems to them an act of almost criminal discourtesy, and those immediately about me seem almost inclined to take me back to the threshing-floor like a cul-prit.   But the *mudir* himself is not such a blockhead but that he realizes the mistake he has made.   He is too proud to acknowledge it, though ; consequently his friends miss, perhaps, the only op-portunity in their uneventful lives of seeing a bicycle ridden.

Owing to my ignorance of the vernacular, I am compelled to drift more or less with the tide of circumstances about me, upon entering one of these villages, for accommodation, and make the best of whatever capricious chance provides.   My Armenian "man-ager" now delivers me into the hands of one of his compatriots, from whom I obtain supper and a quilt, sleeping, from a not over extensive choice, on some straw, beneath the broad eaves of a log granary adjoining the house.

I am for once quite mistaken in making an early, breakfastless start, for it proves to be eighteen weary miles over a rocky moun-tain pass before another human habitation is reached, a region of jagged rocks, deep gorges, and scattered pines.   Fortunately, how-ever, I am not destined to travel the whole eighteen miles in a breakfastless condition—not quite a breakfastless condition.   Per-haps half the distance is traversed, when, while trundling up the ascent, I meet a party of horsemen, a turbaned old Turk, with an

2G

escort of three *zaptiehs*, and another traveller, who is keeping pace
with them for company and safety. The old Turk asks me to *bin
bacalem*, supplementing the request by calling my attention to his
turban, a gorgeously spangled affair that would seem to indicate
the wearer to be a personage of some importance ; I observe, also,
that the butt of his revolver is of pearl inlaid with gold, another
indication of either rank or opulence. Having turned about and
granted his request, I in turn call his attention to the fact that
mountain climbing on an empty stomach is anything but satisfactory or agreeable, and give him a broad hint by inquiring how far
it is before *ekmek* is obtainable. For reply, he orders a *zaptieh* to
produce a wheaten cake from his saddle-bags, and the other traveller voluntarily contributes three apples, which he ferrets out from
the ample folds of his kammerbund and off this I make a breakfast.

Toward noon, the highest elevation of the pass is reached, and I
commence the descent toward the Erzingan Valley, following for a
number of miles the course of a tributary of the western fork of
the Euphrates, known among the natives in a general sense as the
"Frat ;" this particular branch is locally termed the Kara Su, or
black water. The stream and my road lead down a rocky defile
between towering hills of rock and slaty formation, whose precipitous slopes vegetable nature seems to shun, and everything looks
black and desolate, as though some blighting curse had fallen upon
the place. Up this same rocky passage-way, eight summers ago,
swarmed thousands of wretched refugees from the seat of war in
Eastern Armenia ; small oblong mounds of loose rocks and bowlders are frequently observed all down the ravine, mournful reminders of one of the most heartrending phases of the Armenian
campaign ; green lizards are scuttling about among the rude
graves, making their habitations in the oblong mounds.

About two o'clock I arrive at a road-side *khan*, where an ancient
Osmanli dispenses feeds of grain for travellers' animals, and brews
coffee for the travellers themselves, besides furnishing them with
whatever he happens to possess in the way of eatables to such as
are unfortunately obliged to patronize his cuisine or go without anything ; among this latter class belongs, unhappily, my hungry self.
Upon inquiring for refreshments the *khan-jee* conducts me to a rear
apartment and exhibits for my inspection the contents of two jars,
one containing the native idea of butter and the other the native
conception of a soft variety of cheese ; what difference is discover-

able between these two kindred products is chiefly a difference in
the degree of rancidity and odoriferousness, in which respect the
cheese plainly carries off the honors ; in fact these venerable and
esteemable qualities of the cheese are so remarkably developed
that after one cautious peep into its receptacle I forbear to investigate their comparative excellencies any further ; but obtaining
some bread and a portion of the comparatively mild and inoffensive
butter, I proceed to make the best of circumstances.   The old
khan-jee proves himself a thoughtful, considerate landlord, for as

I eat he busies himself
picking the most glar-
ingly conspicuous hairs
out of my butter with
the point of his dagger.
One is usually somewhat
squeamish regarding
hirsute butter, but all
such little refinements of

The Native Idea of Butter.

civilized life as hairless butter or strained milk have to be winked
at to a greater or less extent in Asiatic travelling, especially when
depending solely on what happens to turn up from one meal to another.

The narrow, lonely defile continues for some miles eastward
from the *khan*, and ere I emerge from it altogether I encounter a
couple of ill-starred natives, who venture upon an effort to intimi-
date me into yielding up my purse.   A certain Mahmoud Ali
and his band of enterprising freebooters have been terrorizing the
villagers and committing highway robberies of late around the
country ; but from the general appearance of these two, as they

approach, I take them to be merely villagers returning home from
Erzingan afoot.   They are armed with Circassian guardless swords
and flint-lock horse-pistols ; upon meeting they address some question to me in Turkish, to which I make my customary reply of
Turkchi binmus; one of them then demands *para* (money) in a
manner that leaves something of a doubt whether he means it for
begging, or is ordering me to deliver.   In order to the better discover their intentions, I pretend not to understand, whereupon

" Stand and Deliver ! "

the spokesman reveals their meaning plain enough by reiterating
the demand in a tone meant to be intimidating, and half unsheaths
his sword in a significant manner.   Intuitively the precise situation of affairs seems to reveal itself in a moment ; they are but or-
dinarily inoffensive villagers returning from Erzingan, where they
have sold and squandered even the donkeys they rode to town ;
meeting me alone, and, as they think in the absence of outward
evidence that I am unarmed, they have become possessed of the
idea of retrieving their fortunes by intimidating me out of money.

Never were men more astonished and taken aback at finding me
armed, and they both turn pale and fairly shiver with fright
as I produce the Smith & Wesson from its inconspicuous position
at my hip, and hold it on a level with the bold spokesman's head;
they both look as if they expected their last hour had arrived and
both seem incapable either of utterance or of running away; in
fact, their embarrassment is so ridiculous that it provokes a smile
and it is with anything but a threatening or angry voice that I bid
them *haidy!* The bold highwaymen seem only too thankful of a
chance to "*haidy,*" and they look quite confused, and I fancy even
ashamed of themselves, as they betake themselves off up the ravine.
I am quite as thankful as themselves at getting off without the
necessity of using my revolver, for had I killed or badly wounded
one of them it would probably have caused no end of trouble
or vexatious delay, especially in case they prove to be what
I take them for, instead of professional robbers; moreover, I
might not have gotten off unscathed myself, for while their ancient
flint-locks were in all probability not even loaded, being worn
more for appearances by the native than anything else, these fellows sometimes do desperate work with their ugly and ever-handy
swords when cornered up, in proof of which we have the late das-
tardly assault on the British Consul at Erzeroum, of which we
shall doubtless hear the particulars upon reaching that city.

Before long the ravine terminates, and I emerge upon the broad
and smiling Erzingan Valley; at the lower extremity of the ravine
the stream has cut its channel through an immense depth of con-
glomerate formation, a hundred feet of bowlders and pebbles ce-
mented together by integrant particles which appear to have been
washed down from the mountains—probably during the subsidence
of the deluge, for even if that great catastrophe were a comparatively
local occurrence, instead of a universal flood, as some profess to believe, we are now gradually creeping up toward Ararat, so that this
particular region was undoubtedly submerged. What appear to
be petrified chunks of wood are interspersed through the mass.
There is nothing new under the sun, they say; peradventure they may
be sticks of cooking-stove wood indignantly cast out of the kitchen
window of the ark by Mrs. Noah, because the absent-minded patriarch habitually persisted in cutting them three inches too long for
the stove; who knows? I now wheel along a smooth, level road
leading through several orchard-environed villages; general cul-

tivation and an atmosphere of peace and plenty seems to pervade the valley, which, with its scattering villages amid the foliage of their orchards, looks most charming upon emerging from the gloomy environments of the rock-ribbed and verdureless ravine ; a fitting background is presented on the south by a mountain-chain of considerable elevation, upon the highest peaks of which still linger tardy patches of snow.

Since the occupation of Kars by the Russians the military mantle of that important fortress has fallen upon Erzeroum and Erzingan ; the booming of cannon fired in honor of the Sultan's birthday is awakening the echoes of the rock-ribbed mountains as I wheel eastward down the valley, and within about three miles of the city I pass the headquarters of the garrison. Long rows of hundreds of white field-tents are ranged about the position on the level greensward ; the place presents an animated scene, with the soldiers, some in the ordinary blue, trimmed with red, others in cool, white uniforms especially provided for the summer, but which they are not unlikely to be found also wearing in winter, owing to the ruinous state of the Ottoman exchequer, and one and all wearing the picturesque but uncomfortable fez ; cannons are booming, drums beating, and bugles playing. From the military headquarters to the city is a splendid broad macadam, converted into a magnificent avenue by rows of trees ; it is a general holiday with the military, and the avenue is alive with officers and soldiers going and returning between Erzingan and the camp. The astonishment of the valiant warriors of Islam as I wheel briskly down the thronged avenue can be better imagined than described ; the soldiers whom I pass immediately commence yelling at their comrades ahead to call their attention, while epauletted officers forget for the moment their military dignity and reserve as they turn their affrighted chargers around and gaze after me, stupefied with astonishment ; perhaps they are wondering whether I am not some supernatural being connected in some way with the celebration of the Sultan's birthday—a winged messenger, perhaps, from the Prophet.

Upon reaching the city I repair at once to the large custom-house caravanserai and engage a room for the night. The proprietor of the rooms seems a sensible fellow, with nothing of the inordinate inquisitiveness of the average native about him, and instead of throwing the weight of his influence and his persuasive powers on the side of the importuning crowd, he authoritatively bids them "*haidy!*" locks the bicycle in my room, and gives me the key.   The Erzingan caravanserai—and all these caravanserais are essentially similar—is a square court-yard surrounded by the four sides of a two-storied brick building ; the ground-floor is occupied by the offices of the importers of foreign goods and the custom-house authorities ; the upper floor is divided into small rooms for the accommodation of travellers and caravan men arriving with goods from Trebizond.   Sallying forth in search of supper, I am taken in tow by a couple of Armenians, who volunteer the welcome information that there is an "*Americanish hakim*" in the city ; this intelligence is an agreeable surprise, for Erzeroum is the nearest place in which I have been expecting to find an English-speaking person.   While searching about for the *hakim*, we pass near the *zaptieh* headquarters ; the officers are enjoying their nargileh in the cool evening air outside the building, and seeing an Englishman, beckon us over.   They desire to examine my *teskeri*, the first occasion on which it has been officially demanded since landing at Ismidt, although I have voluntarily produced it on previous occasions, and at Sivas requested the Vali to attach his seal and signature ; this is owing to the proximity of Erzingan to the Russian frontier, and the suspicions that any stranger may be a subject of the Czar, visiting the military centres for sinister reasons. They send an officer with me to hunt up the resident pasha ; that worthy and enlightened personage is found busily engaged in playing a game of chess with a military officer, and barely takes the trouble to glance at the proffered passport: "It is *viséd* by the Sivas Vali," he says, and lackadaisically waves us adieu.   Upon returning to the *zaptieh* station, a quiet, unassuming American comes forward and introduces himself as Dr. Van Nordan, a physician formerly connected with the Persian mission.   The doctor is a spare-built and not over-robust man, and would perhaps be considered by most people as a trifle eccentric ; instead of being connected with any missionary organization, he nowadays wanders hither and thither, acquiring knowledge and seeking whom he can persuade from the error of their ways, meanwhile supporting himself by the practice of his profession.   Among other interesting things spoken of, he tells me something of his recent journey to Khiva (the doctor pronounces it "Heevah") ; he was surprised, he says, at finding the Khivans a mild-mannered and harmless sort of people, among whom the carrying of weapons is as much the ex-

ception as it is the rule in Asiatic Turkey.   Doubtless the fact of
Khiva being under the Russian Government has something to do
with the latter otherwise unaccountable fact.

   After supper we sit down on a newly arrived bale of Manchester
calico in the caravanserai court, cross one knee and whittle chips
like Michigan grangers at a cross-roads post-office, and spend two
hours conversing on different topics.   The good doctor's mind

The Pasha was Playing Chess.

wanders as naturally into serious channels as water gravitates to its
level ; when I inquire if he has heard anything of the whereabout
of Mahmoud Ali and his gang lately, the pious doctor replies
chiefly by hinting what a glorious thing it is to feel prepared to
yield up the ghost at any moment ; and when I recount something
of my experiences on the journey, instead of giving me credit for
pluck, like other people, he merely inquires if I don't recognize
the protecting hand of Providence ; native modesty prevents me

telling the doctor of my valuable missionary work at Sivas. After the doctor's departure I wander forth into the bazaar to see what it looks like after dark ; many of the stalls are closed for the day, the principal places remaining open being *kahvay-khans* and Armenian wine-shops, and before these petroleum lamps are kept burning ; the remainder of the bazaar is in darkness. I have not strolled about many minutes before I am corralled as usual by Armenians ; they straightway send off for a youthful compatriot of theirs who has been to the missionary's school at Kaizareah and can speak a smattering of English. After the usual programme of questions, they suggest :

" Being an Englishman, you are of course a Christian," by which they mean that I am not a Mussulman.

" Certainly," I reply ; whereupon they lug me into one of their wine-shops and tender me a glass of *raki* (a corruption of "arrack" —raw, fiery spirits of the kind known among the English soldiers in India by the suggestive pseudonym of "fixed bayonets"). Smelling the *raki,* I make a wry face and shove it away ; they look surprised and order the waiter to bring cognac ; to save the waiter the trouble, I make another wry face, indicative of disapproval, and suggest that he bring vishner-*su.*

"Vishner-*su !* " two or three of them sing out in a chorus of blank amazement ; "Ingilis ? Christi-an ? vishner-*su !* " they exclaim, as though such a preposterous and unaccountable thing as a Christian partaking of a non-intoxicating beverage like vishner-*su* is altogether beyond their comprehension. The youth who has been to the Kaizareah school then explains to the others that the American missionaries never indulge in intoxicating beverages ; this seems to clear away the clouds of their mystification to some extent, and they order vishner-*su,* eying me critically, however, as I taste it, as though expecting to observe me make yet another wry countenance and acknowledge that in refusing the fiery, throat-blistering *raki* I had made a mistake.

Nothing in the way of bedding or furniture is provided in the caravanserai rooms, but the proprietor gets me plenty of quilts, and I pass a reasonably comfortable night. In the morning I obtain breakfast and manage to escape from town without attracting a crowd of more than a couple of hundred people ; a remarkable occurrence in its way, since Erzingan contains a population of about twenty thousand. The road eastward from Erzingan is level, but heavy with dust, leading through a low portion of the valley that earlier in the season is swampy, and gives the city an unenviable reputation for malarial fevers. To prevent the travellers drinking the unwholesome water in this part of the valley, some benevolent Mussulman or public-spirited pasha has erected at intervals, by the road side, compact mud huts, and placed there in huge earthenware vessels, holding perhaps fifty gallons each ; these are kept supplied with pure spring-water and provided with a wooden drinking-scoop.

Fourteen miles from Erzingan, at the entrance to a ravine whence flows the boisterous stream that supplies a goodly proportion of the irrigating water for the valley, is situated a military outpost station. My road runs within two hundred yards of the building, and the officers, seeing me evidently intending to pass without stopping, motion for me to halt. I know well enough they want to examine my passport, and also to satisfy their curiosity concerning the bi- cycle, but determine upon spurting ahead and escaping their bother altogether. This movement at once arouses the official suspicion as to my being in the country without proper authority, and causes them to attach some mysterious significance to my strange vehicle, and several soldiers forthwith receive racing orders to intercept me. Unfortunately, my spurting receives a prompt check at the stream, which is not bridged, and here the doughty warriors intercept my progress, taking me into custody with broad grins of satisfaction, as though pretty certain of having made an important capture. Since there is no escaping, I conclude to have a little quiet amuse- ment out of the affair, anyway, so I refuse point-blank to accom- pany my captors to their officer, knowing full well that any show of reluctance will have the very natural effect of arousing their sus- picions still further.

The bland and childlike soldiers of the Crescent receive this show of obstinacy quite complacently, their swarthy countenances wreathed in knowing smiles ; but they make no attempt at com- pulsion, satisfying themselves with addressing me deferentially as "Effendi," and trying to coax me to accompany them. Seeing that there is some difficulty about bringing me, the two officers come down, and I at once affect righteous indignation of a mild order, and desire to know what they mean by arresting my prog- ress. They demand my *teskeri* in a manner that plainly shows their doubts of my having one. The *teskeri* is produced. One of the officers then whispers something to the other, and they both glance knowingly mysterious at the bicycle, apologize for having detained me, and want to shake hands. Having read the passport, and satisfied themselves of my nationality, they attach some deep mysterious significance to my journey in this incomprehensible manner up in this particular quarter ; but they no longer wish to offer any impediment to my progress, but rather to render me assistance. Poor fellows! how suspicious they are of their great overgrown neighbor to the north. What good-humored fellows these Turkish soldiers are! what simple-hearted, overgrown children! What a pity that they are the victims of a criminally incompetent government that neither pays, feeds, nor clothes them a quarter as well as they deserve! In the fearful winters of Erzeroum, they have been known to have no clothing to wear but the linen suits provided for the hot weather. Their pay, insignificant though it be, is as uncertain as gambling ; but they never raise a murmur. Being by nature and religion fatalists, they cheerfully accept these undeserved hardships as the will of Allah.

To-day is the hottest I have experienced in Asia Minor, and soon after leaving the outpost I once more encounter the everlasting mountains, following now the Trebizond and Erzingan caravan trail. Once again I get benighted in the mountains, and push ahead for some time after dark. I am beginning to think of camping out supperless again when I hear the creaking of a buffalo *araba* some distance ahead. Soon I overtake it, and, following it for half a mile off the trail, I find myself before an enclosure of several acres, surrounded by a high stone wall with quite imposing gateways. It is the walled village of Houssenbegkhan, one of those places built especially for the accommodation of the Trebizond caravans in the winter. I am conducted into a large apartment, which appears to be set apart for the hospitable accommodation of travellers. The apartment is found already occupied by three travellers, who, from their outward appearance, might well be taken for cutthroats of the worst description ; and the villagers swarming in, I am soon surrounded by the usual ragged, flea-bitten congregation. There are various arms and warlike accoutrements hanging on the wall, enough of one kind or other to arm a small company. They all belong to the three travellers, however ; my modest little revolver seems really nothing compared with the warlike display of swords, daggers, pistols and guns hanging around ; the place looks like a small armory. The first question is—as is usual of late—"Russ or Ingilis?" Some of the younger and less experienced men essay to doubt my word, and, on their own supposition that I am a Russian, begin to take unwarrantable liberties with my person ; one of them steals up behind and commences playing a tattoo on my helmet with two sticks of wood, by way of bravado, and showing his contempt for a subject of the Czar. Turning round, I take one of the sticks away and chastise him with it until he howls for Allah to protect him, and then, without attempting any sort of explanation to the others, resume my seat ; one of the travellers then solemnly places his forefingers together and announces himself as *kardash* (my brother), at the same time

"A Russian, am I?"

pointing significantly to his choice assortment of ancient weapons. I shake hands with him and remind him that I am somewhat hungry ; whereupon he orders a villager to for thwith contribute six eggs, another butter to fry them in, and a third bread ; a *tezek* fire is already burning, and with his own hands he fries the eggs, and makes my ragged audience stand at a respectful distance while I eat ; if I were to ask him, he would probably clear the room of them *instanter*. About ten o'clock my *impromptu* friend and his companion order their horses, and buckle their arms and accoutrements about them to depart ; my "brother" stands before me and loads up his flintlock rifle ; it is a fearful and wonderful process ; it takes him at least two minutes ; he does not seem to know on which particular part of his wonderful paraphernalia to find the slugs, the powder, or the patching, and he finishes by tearing a piece of rag off a by-standing villager to place over the powder in the pan. While he is doing all this, and especially when ramming home the bullet, he looks at me as though expecting me to come and pat him approvingly on the shoulder.

When they are gone, the third traveller, who is going to remain over night, edges up beside me, and pointing to his own imposing armory, likewise announces himself as my brother; thus do I unexpectedly acquire two brothers within the brief space of an evening. The villagers scatter to their respective quarters; quilts are provided for me, and a ghostly light is maintained by means of a cup of grease and a twisted rag. In one corner of the room is a paunchy youngster of ten or twelve summers, whom I noticed during the evening as being without a single garment to cover his nakedness; he has partly inserted himself into a large, coarse, nose-bag, and lies curled up in that ridiculous position, probably imagining himself in quite comfortable quarters. " Oh, wretched youth!" I mentally exclaim, "what will you do when that nose-bag has petered out?" and soon afterward I fall asleep, in happy consciousness of perfect security beneath the protecting shadow of brother number two and his formidable armament of ancient weapons.

Ten miles of good ridable road from Houssenbegkhan, and I again descend into the valley of the west fork of the Euphrates, crossing the river on an ancient stone bridge; I left Houssenbegkhan without breakfasting, preferring to make my customary early start and trust to luck. I am beginning to doubt the propriety of having done so, and find myself casting involuntary glances toward a Koordish camp that is visible some miles to the north of my route, when, upon rounding a mountain-spur jutting out into the valley, I descry the minaret of Mamakhatoun in the distance ahead. A minaret hereabout is a sure indication of a town of sufficient importance to support a public eating-*khan*, where, if not a very elegant, at least a substantial meal is to be obtained. I obtain an acceptable breakfast of kabobs and boiled sheeps'-trotters; killing two birds with one stone by satisfying my own appetite and at the same time giving a first-class entertainment to a *khan*-ful of wondering-eyed people, by eating with the *khan-jee's* carving-knife and fork in preference to my fingers. Here, as at Houssenbegkhan, there is a splendid, large caravanserai; here it is built chiefly of hewn stone, and almost massive enough for a fortress; this is a mountainous, elevated region, where the winters are stormy and severe, and these commodious and substantial retreats are absolutely necessary for the safety of Erzingan and Trebizond caravans during the winter.

The country now continues hilly rather than mountainous. The road is generally too heavy with sand and dust, churned up by the Erzingan mule-caravans, to admit of riding wherever the grade is unfavorable; but much good wheeling surface is encountered on long, gentle declivities and comparatively level stretches. During the forenoon I meet a company of three splendidly armed and mounted Circassians; they remain speechless with astonishment until I have passed beyond their hearing; they then conclude among themselves that I am something needing investigation; they come galloping after me, and having caught up, their spokesman gravely delivers himself of the solitary monosyllable, "Russ?" "Ingilis," I reply, and they resume the even tenor of their way without questioning me further. Later in the day the hilly country develops into a mountainous region, where the trail intersects numerous deep ravines whose sides are all but perpendicular. Between the ravines the riding is ofttimes quite excellent, the composition being soft shale, that packs down hard and smooth beneath the animals' feet. Deliciously cool streams flow at the bottom of these ravines. At one crossing I find an old man washing his feet, and mournfully surveying sundry holes in the bottom of his sandals; the day is hot, and I likewise halt a few minutes to cool my pedal extremities in the crystal water. With that childlike simplicity I have so often mentioned, and which is nowhere encountered as in the Asiatic Turk, the old fellow blandly asks me to exchange my comparatively sound moccasins for his worn-out sandals, at the same time ruefully pointing out the dilapidated condition of the latter, and looking as dejected as though it were the only pair of sandals in the world.

This afternoon I am passing along the same road where Mahmoud Ali's gang robbed a large party of Armenian harvesters who had been south to help harvest the wheat, and were returning home in a body with the wages earned during the summer. This happened but a few days before, and notwithstanding the well-known saying that lightning never strikes twice in the same place, one is scarcely so unimpressionable as not to find himself involuntarily scanning his surroundings, half expecting to be attacked. Nothing startling turns up, however, and at five o'clock I come to a village which is enveloped in clouds of wheat chaff; being a breezy evening, winnowing is going briskly forward on several threshing-floors. After duly *binning*, I am taken under the protecting wing of a prominent

villager, who is walking about with his hand in a sling, the reason
whereof is a crushed finger ; he is a sensible, intelligent fellow, and
accepts my reply that I am not a crushed-finger *hakim* with all
reasonableness ; he provides a substantial supper of bread and
yaort, and then installs me in a small, windowless, unventilated
apartment adjoining the buffalo-stall, provides me with quilts,
lights a primitive grease-lamp, and retires. During the evening
the entire female population visit my dimly-lighted quarters, to satisfy their feminine curiosity by taking a timid peep at their neighbor's strange guest and his wonderful *araba*. They imagine I am
asleep and come on tiptoe part way across the room, craning their
necks to obtain a view in the semi-darkness.

An hour's journey from this village brings me yet again into
the West Euphrates Valley. Just where I enter the valley the river
spreads itself over a wide stony bed, coursing along in the form of
several comparatively small streams. There is, of course, no bridge
here, and in the chilly, almost frosty, morning I have to disrobe and
carry clothes and bicycle across the several channels. Once across,
I find myself on the great Trebizond and Persian caravan route, and
in a few minutes am partaking of breakfast at a village thirty-five
miles from Erzeroum, where I learn with no little satisfaction that
my course follows along the Euphrates Valley, with an artificial
wagon-road, the whole distance to the city. Not far from the village the Euphrates is recrossed on a new stone bridge. Just beyond the bridge is the camp of a road-engineer's party, who are
putting the finishing touches to the bridge. A person issues from
one of the tents as I approach and begins chattering away at me
in French. The face and voice indicates a female, but the costume
consists of jack-boots, tight-fitting broadcloth pantaloons, an ordinary pilot-jacket, and a fez. Notwithstanding the masculine
apparel, however, it turns out not only to be a woman, but a Pari-
sienne, the better half of the Erzeroum road engineer, a Frenchman, who now appears upon the scene. They are both astonished
and delighted at seeing a "velocipede," a reminder of their own
far-off France, on the Persian caravan trail, and they urge me to remain and partake of coffee.

I now encounter the first really great camel caravans, *en route*
to Persia with tea and sugar and general European merchandise ;
they are all camped for the day alongside the road, and the camels
scattered about the neighboring hills in search of giant thistles

and other outlandish vegetation, for which the patient ship of the desert entertains a partiality. Camel caravans travel entirely at night during the summer. Contrary to what, I think, is a common belief in the Occident, they can endure any amount of cold weather, but are comparatively distressed by the heat; still, this may not characterize all breeds of camels any more than the different breeds of other domesticated animals. During the summer, when the camels are required to find their own sustenance along the road, a large caravan travels but a wretched eight miles a day, the remainder of the time being occupied in filling his capacious thistle and camel-thorn receptacle; this comes of the scarcity of good grazing along the route, compared with the number of camels, and the consequent necessity of wandering far and wide in search of pasturage, rather than because of the camel's absorptive capacity, for he is a comparatively abstemious animal. In the winter they are fed on balls of barley flour, called *nawalla*; on this they keep fat and strong, and travel three times the distance. The average load of a full-grown camel is about seven hundred pounds.

Before reaching Erzeroum I have a narrow escape from what might have proved a serious accident. I meet a buffalo *araba* carrying a long projecting stick of timber; the sleepy buffaloes pay no heed to the bicycle until I arrive opposite their heads, when they give a sudden lurch sidewise, swinging the stick of timber across my path; fortunately the road happens to be of good width, and by a very quick swerve I avoid a collision, but the tail end of the timber just brushes the rear wheel as I wheel past. Soon after noon I roll into Erzeroum, or rather, up to the Trebizond gate, and dismount. Erzeroum is a fortified city of considerable importance, both from a commercial and a military point of view; it is surrounded by earthwork fortifications, from the parapets of which large siege guns frown forth upon the surrounding country, and forts are erected in several commanding positions round about, like watch-dogs stationed outside to guard the city. Patches of snow linger on the Palantokan Mountains, a few miles to the south; the Deve Boyun Hills, a spur of the greater Palantokans, look down on the city from the east; the broad valley of the West Euphrates stretches away westward and northward, terminating at the north in another mountain range.

Repairing to the English consulate, I am gratified at finding

several letters awaiting me, and furthermore by the cordial hospitality extended by Yusuph Effendi, an Assyrian gentleman, the
chargé d'affaires of the consulate for the time being, Colonel E——,
the consul, having left recently for Trebizond and England, in consequence of numerous sword-wounds received at the hands of a
desperado who invaded the consulate for plunder at midnight. The
Colonel was a general favorite in Erzeroum, and is being tenderly
carried (Thursday, September 3, 1885) to Trebizond on a stretcher
by relays of willing natives, no less than forty accompanying him
on the road. Yusuph Effendi tells me the story of the whole lamentable affair, pausing at intervals to heap imprecations on the
head of the malefactor, and to bestow eulogies on the wounded
consul's character.

It seems that the door-keeper of the consulate, a native of a
neighboring Armenian village, was awakened at midnight by an
acquaintance from the same village, who begged to be allowed to
share his quarters till morning. No sooner had the servant admitted him to his room than he attacked him with his sword, intending—as it afterward leaked out—to murder the whole family,
rob the house, and escape. The servant's cries for assistance awakened Colonel E——, who came to his rescue without taking the
trouble to provide himself with a weapon. The man, infuriated
at the detection and the prospect of being captured and brought
to justice, turned savagely on the consul, inflicting several severe
wounds on the head, hands, and face. The consul closed with him
and threw him down, and called for his wife to bring his revolver.
The wretch now begged so piteously for his life, and made such
specious promises, that the consul magnanimously let him up, neg-
lecting—doubtless owing to his own dazed condition from the
scalp wounds—to disarm him. Immediately he found himself released he commenced the attack again, cutting and slashing like
a demon, knocking the revolver from the consul's already badly
wounded hand while he yet hesitated to pull the trigger and take
his treacherous assailant's life. The revolver went off as it struck
the floor and wounded the consul himself in the leg—broke it?
The servant now rallied sufficiently to come to his assistance, and
together they succeeded in disarming the robber, who, however,
escaped and bolted up-stairs, followed by the servant with the
sword. The consul's wife, with praiseworthy presence of mind,
now appeared with a second revolver, which her husband grasped

27

in his left hand, the right being almost hacked to pieces. Dazed and faint with the loss of blood, and, moreover, blinded by the blood flowing from the scalp-wounds, it was only by sheer strength of will that he could keep from falling. At this juncture the servant unfortunately appeared on the stairs, returning from an unsuccessful pursuit of the robber. Mistaking the servant with the sword in his hand for the desperado returning to the attack, and realizing his own helpless condition, the consul fired two shots at him, wounding him with both shots. The would-be murderer is now (September 3, 1885), captured and in durance vile ; the servant lies here in a critical condition, and the consul and his sorrowing family are *en route* to England.

Having determined upon resting here until Monday, I spend a good part of Friday looking about the city. The population is a mixture of Turks, Armenians, Russians, Persians, and Jews. Here I first make the acquaintance of a Persian *tchai-khan* (tea-drinking shop). With the exception of the difference in the beverages, there is little difference between a *tchai-khan* and a *kahvay-khan*, although in the case of a swell establishment, the *tchai-khan* blossoms forth quite gaudily with scores of colored lamps. The tea is served scalding hot in tiny glasses, which are first half-filled with loaf-sugar. If the proprietor is desirous of honoring or pleasing a new or distinguished customer, he drops in lumps of sugar until it protrudes above the glass. The tea is made in a samovar—a brass vessel, holding perhaps a gallon of water, with a hollow receptacle in the centre for a charcoal fire. Strong tea is made in an ordinary queen's-ware teapot that fits into the hollow ; a small portion of this is poured into the glass, which is then filled up with hot water from a tap in the samovar.

There is a regular Persian quarter in Erzeroum, and I am not suffered to stroll through it without being initiated into the fundamental difference between the character of the Persians and the Turks. When an Osmanli is desirous of seeing me ride the bicycle, he goes honestly and straightforwardly to work at coaxing and worrying ; except in very rare instances they have seemed incapable of resorting to deceit or sharp practice to gain their object. Not so childlike and honest, however, are our new acquaintances, the Persians. Several merchants gather round me, and pretty soon they cunningly begin asking me how much I will sell the bicycle for. " Fifty liras," I reply, seeing the deep, deep scheme

hidden beneath the superficial fairness of their observations, and
thinking this will quash all further commercial negotiations. But
the wily Persians are not so easily disposed of as this. "Bring it
round and let us see how it is ridden," they say, "and if we like it
we will purchase it for fifty liras, and perhaps make you a present
besides." A Persian would rather try to gain an end by deceit
than by honest and above-board methods, even if the former were
more trouble. Lying, cheating, and deception is the universal
rule among them ; honesty and straightforwardness are unknown
virtues. Anyone whom they detect telling the truth or acting
honestly they consider a simpleton unfit to transact business.

The missionaries and their families are at present tenting out,
five miles south of the city, in a romantic little ravine called Kirkdagheman, or the place of the forty mills ; and on Saturday morning I receive a pressing invitation to become their guest during the
remainder of my stay. The Erzeroum mission is represented by
Mr. Chambers, his brother—now absent on a tour—their respective families, and Miss Powers. Yusuph Effendi accompanies us
out to the camp on a spendid Arab steed, that curvets gracefully
the whole way. Myself and the—other missionary people (bicycle
work at Sivas, and again at Erzeroum) ride more sober and deco-
ous animals. Kirkdagheman is found to be near the entrance to
a pass over the Palantokan Mountains. Half a dozen small tents
are pitched beneath the only grove of trees for many a mile around.
A dancing stream of crystal water furnishes the camp with an
abundance of that necessary, as also a lavish supply of such music
as babbling brooks coursing madly over pebbly beds are wont to
furnish. To this particular section of the little stream legendary
lore has attached a story which gives the locality its name, Kirkdagheman :

"Once upon a time, a worthy widow found herself the happy
possessor of no less than forty small grist-mills strung along this
stream. Soon after her husband's death, the lady's amiable quali-
ties—and not unlikely her forty mills into the bargain—attracted
the admiration of a certain wealthy owner of flocks in the neighborhood, and he sought her hand in marriage. 'No,' said the
lady, who, being a widow, had perhaps acquired wisdom ; 'no ; I
have forty sons, each one faithfully laboring and contributing
cheerfully toward my support ; therefore, I have no use for a husband.' 'I will kill your forty sons, and compel you to become my

wife,' replied the suitor, in a huff at being rejected. And he went and sheared all his sheep, and, with the multitudinous fleeces, dammed up the stream, caused the water to flow into other channels, and thereby rendered the widow's forty mills useless and unproductive. With nothing but ruination before her, and seeing no alternative, the widow's heart finally softened, and she suffered herself to be wooed and won. The fleeces were removed, the stream returned to its proper channel, and the merry whir of the forty mills henceforth mingled harmoniously with the bleating of the sheep."

Two days are spent at the quiet missionary camp, and thoroughly enjoyed. It seems like an oasis of home life in the surrounding desert of uncongenial social conditions. I eagerly devour the contents of several American newspapers, and embrace the opportunities of the occasion, even to the extent of nursing the babies (missionaries seem rare folks for babies), of which there are three in camp. The altitude of Erzeroum is between six thousand and seven thousand feet ; the September nights are delightfully cool, and there are no blood-thirsty mosquitoes. I am assigned a sleeping-tent close alongside a small waterfall, whose splashing music is a soporific that holds me in the bondage of beneficial repose until breakfast is announced both mornings ; and on Monday morning I feel as though the hunger, the irregular sleep, and the rough-and-tumble dues generally of the past four weeks were but a troubled dream. Again the bicycle contributes its curiosity-quickening and question-exciting powers for the benefit of the sluggish-minded pupils of the mission school. The Persian consul and his sons come to see me ride ; he is highly interested upon learning that I am travelling on the wheel to the Persian capital, and he visés my passport and gives me a letter of introduction to the Pasha Khan of Ovahjik, the first village I shall come to beyond the frontier.

It is nearly 3 p.m., September 7th, when I bid farewell to everybody, and wheel out through the Persian Gate, accompanied by Mr. Chambers on horseback, who rides part way to the Deve Boyun (camel's neck) Pass. On the way out he tells me that he has been intending taking a journey through the Caucasus this autumn, but the difficulties of obtaining permission, on account of his being a clergyman, are so great—a special permission having to be obtained from St. Petersburg—that he has about relinquished the idea for the present season.

Deve Boyun Pass leads over a comparatively low range of hills. It was here where the Turkish army, in November, 1877, made their last gallant attempt to stem the tide of disaster that had, by the fortunes of war and the incompetency of their commanders, set in irresistibly against them, before taking refuge inside the walls of the city. An hour after parting from Mr. Chambers I am wheeling briskly down the same road on the eastern slope of the pass where Mukhtar Pasha's ill-fated column was drawn into the fatal ambuscade that suddenly turned the fortunes of the day against them. While rapidly gliding down the gentle gradient, I fancy I can see the Cossack regiments, advancing toward the Turkish position, the unwary and over-confident Osmanlis leaping from their intrenchments to advance along the road and drive them back ; now I come to the Nabi Tchai ravines, where the concealed masses of Russian infantry suddenly sprang up and cut off their retreat ; I fancy I can see—chug ! wh-u-u-p ! thud !—stars, and see them pretty distinctly, too, for while gazing curiously about, locating the Russian ambushment, the bicycle strikes a sand-hole, and I am favored with the worst header I have experienced for many a day. I am—or rather was, a minute ago—bowling along quite briskly ; the header treats me to a fearful shaking up ; I am sore all over the next morning, and present a sort of a stiff-necked, woe-begone appearance for the next four days. A bent handle-bar and a slightly twisted rear wheel fork likewise forcibly remind me that, while I am beyond the reach of repair shops, it will be Solomon-like wisdom on my part to henceforth survey battle-fields with a larger margin of regard for things more immediately interesting.

From the pass, my road descends into the broad and cultivated valley of the Passin Su ; the road is mostly ridable, though heavy with dust. Part way to Hassen Kaleh I am compelled to use considerable tact to avoid trouble with a gang of riotous *katir-jees* whom I overtake ; as I attempt to wheel past, one of them wantonly essays to thrust his stick into the wheel ; as I spring from the saddle for sheer self-protection, they think I have dismounted to attack him, and his comrades rush forward to his protection, brandishing their sticks and swords in a menacing manner. Seeing himself reinforced, as it were, the bold aggressor raises his stick as though to strike me, and peremptorily orders me to *bin* and *haidi !* Very naturally I refuse to remount the bicycle while surrounded by this evidently mischievous crew ; there are about twenty of them, and it re-

quires much self-control to prevent a conflict, in which, I am persuaded, somebody would have been hurt ; however, I finally manage
to escape their undesirable company and ride off amid a fusillade of
stones.

This incident reminds me of Yusuph Effendi's warning, that
even though I had come thus far without a *zaptieh* escort, I should
require one now, owing to the more lawless disposition of the people near the frontier.   Near dark I reach Hassan Kaleh, a large
village nestling under the shadow of its former importance as a
fortified town, and seek the accommodation of a Persian *tchai-khan ;*

Wantonly Assaulted.

it is not very elaborate or luxurious accommodation, consisting
solely of tiny glasses of sweetened tea in the public room and a
shake-down in a rough, unfurnished apartment over the stable ;
eatables have to be obtained elsewhere, but it matters little so long
as they are obtainable somewhere.   During the evening a Persian
troubadour and story-teller entertains the patrons of the *tchai-khan*
by singing ribaldish songs, twanging a tambourine-like instrument,
and telling stories in a sing-song tone of voice.   In deference to
the mixed nationality of his audience, the sagacious troubadour
wears a Turkish fez, a Persian coat, and a Russian metallic-faced
belt ; the burden of his songs are of Erzeroum, Erzingan, and Is-

pahan ; the Russians, it would appear, are too few and unpopular to justify risking the displeasure of the Turks by singing any Russian songs. So far as my comprehension goes, the stories are chiefly of intrigue and love affairs among pashas, and would quickly bring the righteous retribution of the Lord Chamberlain down about his ears, were he telling them to an English audience.

I have no small difficulty in getting the bicycle up the narrow

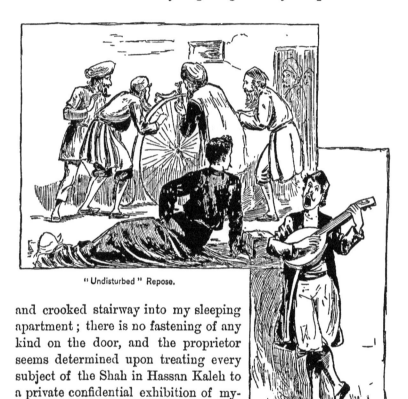

" Undisturbed " Repose.

and crooked stairway into my sleeping apartment ; there is no fastening of any kind on the door, and the proprietor seems determined upon treating every subject of the Shah in Hassan Kaleh to a private confidential exhibition of myself and bicycle, after I have retired to bed. It must be near midnight, I think, when I am again awakened from my uneasy, oft-disturbed slumbers by murmuring voices and the shuffling of feet ; examining the bicycle by the feeble glimmer of a classic lamp are a dozen meddlesome Persians. Annoyed at their unseemly midnight intrusion, and at being repeatedly awakened, I rise up and sing out at them rather authoratively ; I have exhibited the *marifet* of my Smith & Wesson during the evening, and these intruders seem really afraid I might be going to practise on them with it. The Persians are apparently timid mortals; they evidently regard me as a strange being of unknown temperament, who might possibly break loose and encompass their destruction on the slightest provocation, and the proprietor and another equally intrepid individual hurriedly come to my couch, and pat me soothingly on the shoulders, after which they all retire, and I am disturbed no more till morning.

The "rocky road to Dublin" is nothing compared to the road leading eastward from Hassan Kaleh for the first few miles, but afterward it improves into very fair wheeling. Eleven miles down the Passin Su Valley brings me to the Armenian village of Kuipri Kui. Having breakfasted before starting I wheel on without halting, crossing the Araxes River at the junction of the Passin Su, on a very ancient stone bridge known as the *Tchebankerpi*, or the bridge of pastures, said to be over a thousand years old. Nearing Dele Baba Pass, a notorious place for robbers, I pass through a village of sedentary Koords. Soon after leaving the village a wild-looking Koord, mounted on an angular sorrel, overtakes me and wants me to employ him as a guard while going through the pass, backing up the offer of his presumably valuable services by unsheathing a semi-rusty sword and waving it valiantly aloft. He intimates, by tragically graphic pantomime, that unless I traverse the pass under the protecting shadow of his ancient and rusty blade, I will be likely to pay the penalty of my rashness by having my throat cut. Yusuph Effendi and the Erzeroum missionaries have thoughtfully warned me against venturing through the Dele Baba Pass alone, advising me to wait and go through with a Persian caravan; but this Koord looks like anything but a protector; on the contrary, I am inclined to regard him as a suspicious character himself, interviewing me, perhaps, with ulterior ideas of a more objectionable character than that of faithfully guarding me through the Dele Baba Pass. Showing him the shell-extracting mechanism of my revolver, and explaining the rapidity with which it can be fired, I give him to understand that I feel quite capable of guarding myself, consequently have no earthly use for his services. A tea caravan of some two hundred camels are resting near the approach to the pass, affording me an excellent opportunity of having company through by waiting and journeying with them in the night; but warnings of danger have been repeated so often of late, and they have proved themselves groundless so invariably that I should feel the taunts of self-reproach were I to find myself hesitating to proceed on their account.

Passing over a mountain spur, I descend into a rocky cañon, with perpendicular walls of rock towering skyward like giant battlements, inclosing a space not over fifty yards wide; through this runs my road, and alongside it babbles the Dele Baba Su. The cañon is a wild, lonely-looking spot, and looks quite appropriate to the reputation it bears. Professor Vambery, a recognized authority on Asiatic matters, and whose party encountered a gang of marauders here, says the Dele Baba Pass bore the same

A Suspicious Offer of Protection.

unsavory reputation that it bears to-day as far back as the time of Herodotus. However, suffice it to say, that I get through without molestation; mounted men, armed to the teeth, like almost everybody else hereabouts, are encountered in the pass; they invariably halt and look back after me as though endeavoring to comprehend who and what I am, but that is all. Emerging from the cañon, I follow in a general course the tortuous windings of the Dele Baba Su through another ravine-riven battle-field of the late war, and up toward its source in a still more mountainous and elevated region beyond.

CHAPTER XVIII.

MOUNT ARARAT AND KOORDISTAN.

The shades of evening are beginning to settle down over the wild mountainous country round about. It is growing uncomfortably chilly for this early in the evening, and the prospects look favorable for a supperless and most disagreeable night, when I descry a village perched in an opening among the mountains a mile or thereabouts off to the right. Repairing thither, I find it to be a Koordish village, where the hovels are more excavations than buildings; buffaloes, horses, goats, chickens, and human beings all find shelter under the same roof; their respective quarters are nothing but a mere railing of rough poles, and as the question of ventilation is never even thought of, the effect upon one's olfactory nerves upon entering is anything but reassuring. The filth and rags of these people is something abominable; on account of the chilliness of the evening they have donned their heavier raiment; these have evidently had rags patched on top of other rags for years past until they have gradually developed into thick-quilted garments, in the innumerable seams of which the most disgusting entomological specimens, bred and engendered by their wretched mode of existence, live and perpetuate their kind. However, repulsive as the outlook most assuredly is, I have no alternative but to cast my lot among them till morning.

I am conducted into the Sheikh's apartment, a small room partitioned off with a pole from a stable-full of horses and buffaloes, and where darkness is made visible by the sickly glimmer of a grease lamp. The Sheikh, a thin, sallow-faced man of about forty years, is reclining on a mattress in one corner smoking cigarettes; a dozen ill-conditioned ragamuffins are squatting about in various attitudes, while the rag, tag, and bobtail of the population crowd into the buffalo-stable and survey me and the bicycle from outside the partition-pole.

A circular wooden tray containing an abundance of bread, a bowl of *yaort*, and a small quantity of peculiar stringy cheese that resembles chunks of dried codfish, warped and twisted in the drying, is brought in and placed in the middle of the floor. Everybody in the room at once gather round it and begin eating with as little formality as so many wild animals; the Sheikh silently motions for me to do the same. The *yaort* bowl contains one solitary wooden spoon, with which they take turns at eating mouthfuls. One is compelled to draw the line somewhere, even under the most uncompromising circumstances, and I naturally draw it against eating *yaort* with this same wooden spoon; making small scoops with pieces of bread, I dip up *yaort* and eat scoop and all together. These particular Koords seem absolutely ignorant of anything in the shape of mannerliness, or of consideration for each other at the table. When the *yaort* has been dipped into twice or thrice all round, the Sheikh coolly confiscates the bowl, eats part of what is left, pours water into the remainder, stirs it up with his hand, and deliberately drinks it all up; one or two others seize all the cheese, utterly regardless of the fact that nothing remains for myself and their companions, who, by the by, seem to regard it as a perfectly natural proceeding.

After supper they return to their squatting attitudes around the room, and to a resumption of their never-ceasing occupation of scratching themselves. The eminent economist who lamented the wasted energy represented in the wagging of all the dogs' tails in the world, ought to have travelled through Asia on a bicycle and have been compelled to hob-nob with the villagers; he would undoubtedly have wept with sorrow at beholding the amount of this same wasted energy, represented by the above-mentioned occupation of the people. The most loathsome member of this interesting company is a wretched old hypocrite who rolls his eyes about and heaves a deep-drawn sigh of Allah! every few minutes, and then looks furtively at myself and the Sheikh to observe its effects; his sole garment is a round-about mantle that reaches to his knees, and which seems to have been manufactured out of the tattered remnants of other tattered remnants tacked carelessly together without regard to shape, size, color, or previous condition of cleanliness; his thin, scrawny legs are bare, his long black hair is matted and unkempt, his beard is stubby and unlovely to look upon, his small black eyes twinkle in the semi-darkness like ferret's eyes, while soap and water have to all appearances been altogether stricken from the category of his personal requirements.

Probably it is nothing but the lively workings of my own imagination, but this wretch appears to me to entertain a decided preference for my society, constantly insinuating himself as near me as possible, necessitating constant watchfulness on my part to avoid actual contact with him ; eternal vigilance is in this case the price of what it is unnecessary to expatiate upon, further than to say that self-preservation becomes, under such conditions, pre-eminently the first law of Occidental nature. Soon the sallow-faced Sheikh suddenly bethinks himself that he is in the august presence of a *hakim*, and beckoning me to his side, displays an ugly wound on his knee which has degenerated into a running sore, and which he says was done with a sword ; of course he wants me to perform a cure. While examining the Sheikh's knee, another old party comes forward and unbares his arm, also wounded with a sword. This not unnaturally sets me to wondering what sort of company I have gotten into, and how they came by sword wounds in these peaceful times ; but my inquisitiveness is compelled to remain in abeyance to my limited linguistic powers. Having nothing to give them for the wounds, I recommend an application of warm salt water twice a day ; feeling pretty certain, however, that they will be too lazy and trifling to follow the advice. Before dispersing to their respective quarters, the occupants of the room range themselves in a row and go through a religious performance lasting fully half an hour ; they make almost as much noise as howling dervishes, meanwhile exercising themselves quite violently. Having made themselves holier than ever by these exercises, some take their departure, others make up couches on the floor with sheepskins and quilts.

Thin ice covers the still pools of water when I resume my toilsome route over the mountains at daybreak, a raw wind comes whistling from the east, and until the sun begins to warm things up a little, it is necessary to stop and buffet occasionally to prevent benumbed hands. Obtaining some small lumps of wheaten dough cooked crisp in hot grease, like unsweetened doughnuts, from a horseman on the road, I push ahead toward the summit and then down the eastern slope of the mountains ; rounding an abutting hill about 9.30, the glorious snow-crowned peak of Ararat suddenly bursts upon my vision ; it is a good forty leagues away, but even at this distance it dwarfs everything else in sight. Although surrounded by giant mountain chains that traverse the country at every conceivable angle, Ararat stands alone in its solitary grandeur, a glistening white cone rearing its giant height proudly and conspicuously above surrounding eminences ; about mountains that are insignificant only in comparison with the white-robed monarch that has been a beacon-light of sacred history since sacred history has been in existence.

Descending now toward the Alashgird Plain, a prominent theatre of action during the war, I encounter splendid wheeling for some miles ; but once fairly down on the level, cultivated plain, the road becomes heavy with dust. Villages dot the broad, expansive plain in every direction ; conical stacks of *tezek* are observable among the houses, piled high up above the roofs, speaking of commendable forethought for the approaching cold weather. In one of the Armenian villages I am not a little surprised at finding a lone German ; he says he prefers an agricultural life in this country with all its disadvantages, to the hard, grinding struggle for existence, and the compulsory military service of the Fatherland. "Here," he goes on to explain, "there is no foamy lager, no money, no comfort, no amusement of any kind, but there is individual liberty, and it is very easy making a living ; therefore it is for me a better country than Deutschland." "Everybody to their liking," I think, as I continue on across the plain ; but for a European to be living in one of these little agricultural villages comes the nearest to being buried alive of anything I know of. The road improves in hardness as I proceed eastward, but the peculiar disadvantages of being a conspicuous and incomprehensible object on a populous level plain soon becomes manifest. Seeing the bicycle glistening in the sunlight as I ride along, horsemen come wildly galloping from villages miles away. Some of these wonderstricken people endeavor to pilot me along branch trails leading to their villages, but the main caravan trail is now too easily distinguishable for any little deceptions of this kind to succeed. Here, on the Alashgird Plain, I first hear myself addressed as " Hamsherri," a term which now takes the place of Effendi for the next five hundred miles.

Owing to the disgust engendered by my unsavory quarters in the wretched Dele Baba village last night, I have determined upon seeking the friendly shelter of a wheat-shock again to-night, preferring the chances of being frozen out at midnight to the entomological possibilities of village hovels. Accordingly, near sunset, I repair to a village not far from the road, for the purpose of obtaining something to eat before seeking out a rendezvous for the night. It turns out to be the Koordish village of Malosman, and the people are found to be so immeasurably superior in every particular to their kinsfolk of Dele Baba that I forthwith cancel my determination and accept their proffered hospitality. The Malosmanlis are comparatively clean and comfortable ; are reasonably well-dressed, seem well-to-do, and both men and women are, on the average, handsomer than the people of any village I have seen for days past. Almost all possess a conspicuously beautiful set of teeth, pleasant, smiling countenances and good physique ; they also seem to have, somehow, acquired easy, agreeable manners.

The secret of the whole difference, I opine, is that, instead of being located among the inhospitable soil of barren hills they are cultivating the productive soil of the Alashgird Plain, and, being situated on the great Persian caravan trail, they find a ready market for their grain in supplying the caravans in winter. Their Sheikh is a handsome and good-natured young fellow, sporting white clothes trimmed profusely with red braid ; he spends the evening in my company, examining the bicycle, revolver, telescopic pencil-case, L. A. W. badge, etc., and hands me his carved ivory case to select cigarettes from. It would have required considerable inducements to have trusted either my L. A. W. badge or the Smith & Wesson in the custody of any of our unsavory acquaintances of last night, notwithstanding their great outward show of piety. There are no deep-drawn sighs of Allah, nor ostentatious praying among the Malosmanlis, but they bear the stamp of superior trustworthiness plainly on their faces and their bearing. There appears to be far more jocularity than religion among these prosperous villagers, a trait that probably owes its development to their apparent security from want ; it is no newly discovered trait of human character to cease all prayers and supplications whenever the granary is overflowing with plenty, and to commence devotional exercises again whenever the supply runs short. This rule would hold good among the childlike natives here, even more so than it does among our more enlightened selves.

I sally forth into the chilly atmosphere of early morning from Malosman, and wheel eastward over an excellent road for some miles ; an obliging native, *en route* to the harvest field, turns his buffalo *araba* around and carts me over a bridgeless stream, but several others have to be forded ere reaching Kirakhan, where I obtain breakfast. Here I am required to show my *teskeri* to the *mudir*, and the *zaptieh* escorting me thither becomes greatly mystified over the circumstance that I am a Frank and yet am wearing a Mussulman head-band around my helmet (a new one I picked up on the road); this little fact appeals to him as something savoring of an attempt to disguise myself, and he grows amusingly mysterious while whisperingly bringing it to the *mudir's* notice. The habitual serenity and complacency of the corpulent *mudir's* mind, however, is not to be unduly disturbed by trifles, and the untutored *zaptieh's* disposition to attach some significant meaning to it, meets with nothing from his more enlightened superior but the silence of unconcern.

More streams have to be forded ere I finally emerge on to higher ground; all along the Alashgird Plain, Ararat's glistening peak has been peeping over the mountain framework of the plain like a white beacon-light showing above a dark rocky shore; but approaching toward the eastern extremity of the plain, my road hugs the base of the intervening hills and it temporarily disappears from view. In this portion of the country, camels are frequently employed in bringing the harvest from field to village threshing-floor; it is a curious sight to see these awkwardly moving animals walking along beneath tremendous loads of straw, nothing visible but their heads and legs. Sometimes the meandering course of the Euphrates—now the eastern fork, and called the Moorad-Chai—brings it near the mountains, and my road leads over bluffs immediately above it; the historic river seems well supplied with trout hereabouts, I can look down from the bluffs and observe speckled beauties sporting about in its pellucid waters by the score. Toward noon I fool away fifteen minutes trying to beguile one of them into swallowing a grasshopper and a bent pin, but they are not the guileless creatures they seem to be when surveyed from an elevated bluff, so they steadily refuse whatever blandishments I offer. An hour later I reach the village of Daslische, inhabited by a mixed population of Turks and Persians. At a shop kept by one of the latter I obtain some bread and *ghee* (clarified butter), some tea, and a handful of wormy raisins for dessert; for these articles, besides building a fire especially to prepare the tea, the unconscionable Persian charges the awful sum of two piastres (ten cents); whereupon the Turks, who have been interested spectators of the whole nefarious proceeding, commence to abuse him roundly for over-charging a stranger unacquainted with the prices of the locality, calling him the son of a burnt father, and other names that tingle unpleasantly in the Persian ear, as though it was a matter of pounds sterling.

Beyond Daslische, Ararat again becomes visible ; the country immediately around is a ravine-riven plateau, covered with bowld-ers. An hour after leaving Daslische, while climbing the eastern slope of a ravine, four rough-looking footmen appear on the oppo-site side of the slope ; they are following after me, and shouting *"Kardash!"* These people with their old swords and pistols con-spicuously about them, always raise suspicions of brigands and evil characters under such circumstances as these, so I continue on up the slope without heeding their shouting until I observe two of them turn back; I then wait, out of curiosity, to see what they really want. They approach with broad grins of satisfaction at having overtaken me : they have run all the way from Daslische in order to overtake me and see the bicycle, having heard of it after I had left. I am now but a short distance from the Russian fron-tier on the north, and the first Turkish patrol is this afternoon patrolling the road ; he takes a wondering interest in my wheel, but doesn't ask the oft-repeated question, " Russ or Ingiliz? " It is presumed that he is too familiar with the Muscovite " phiz " to make any such question necessary.

About four o'clock I overtake a jack-booted horseman, who straightway proceeds to try and make himself agreeable ; as his flowing remarks are mostly unintelligible, to spare him from wasting the sweetness of his eloquence on the desert air around me, I reply, *" Turkchi binmus."* Instead of checking the impetuous torrent of his remarks at hearing this, he canters companionably alongside, and chatters more persistently than ever. *" T-u-r-k-chi b-i-n-m-u-s !"* I repeat, becoming rather annoyed at his persistent gar-rulousness and his refusal to understand. This has the desired effect of reducing him to silence ; but he canters doggedly behind, and, after a space creeps up alongside again, and, pointing to a large stone building which has now become visible at the base of a mountain on the other side of the Euphrates, timidly ventures upon the explanation that it is the Armenian Gregorian Monastery of Sup Ogwanis (St. John). Finding me more favorably disposed to listen than before, he explains that he himself is an Armenian, is acquainted with the priests of the monastery, and is going to remain there over night; he then proposes that I accompany him thither, and do likewise.

I am, of course, only too pleased at the prospect of experiencing something out of the common, and gladly avail myself of the opportunity; moreover, monasteries and religious institutions in general, have somehow always been pleasantly associated in my thoughts as inseparable accompaniments of orderliness and cleanliness, and I smile serenely to myself at the happy prospect of snowy sheets, and scrupulously clean cooking.

Crossing the Euphrates on a once substantial stone bridge, now in a sadly dilapidated condition, that was doubtless built when Armenian monasteries enjoyed palmier days than the present, we skirt the base of a compact mountain and in a few minutes alight at the monastery village. Exit immediately all visions of cleanliness; the village is in no wise different from any other cluster of mud hovels round about, and the rag-bedecked, flea-bitten objects that come outside to gaze at us, if such a thing were possible, compare unfavorably even with the Dele Baba Koords. There is apparent at once, however, a difference between the respective dispositions of the two peoples: the Koords are inclined to be pig-headed and obtrusive, as though possessed of their full share of the spirit of self-assertion; the Sup Ogwanis people, on the contrary, act like beings utterly destitute of anything of the kind, cowering beneath one's look and shunning immediate contact as though habitually overcome with a sense of their own inferiority. The two priests come out to see the bicycle ridden; they are stout, bushy-whiskered, greasy-looking old jokers, with small twinkling black eyes, whose expression would seem to betoken anything rather than saintliness, and, although the Euphrates flows hard by, they are evidently united in their enmity against soap and water, if in nothing else; in fact, judging from outward appearances, water is about the only thing concerning which they practise abstemiousness. The monastery itself is a massive structure of hewn stone, surrounded by a high wall loop-holed for defence; attached to the wall inside is a long row of small rooms or cells, the habitations of the monks in more prosperous days; a few of them are occupied at present by the older men.

At 5.30 P.M., the bell tolls for evening service, and I accompany my guide into the monastery; it is a large, empty-looking edifice

28

of simple, massive architecture, and appears to have been built
with a secondary purpose of withstanding a siege or an assault,
and as a place of refuge for the people in troublous times ; containing among other secular appliances a large brick oven for baking bread.   During the last war, the place was actually bombarded
by the Russians in an effort to dislodge a body of Koords who had
taken possession of the monastery, and from behind its solid walls,
harassed the Russian troops advancing toward Erzeroum.   The
patched up holes made by the Russians' shots are pointed out, as
also some light earthworks thrown up on the Russian position
across the river.   In these degenerate days one portion of the
building is utilized as a storehouse for grain ; hundreds of pigeons
are cooing and roosting on the crossbeams, making the place their
permanent abode, passing in and out of narrow openings near the
roof ; and the whole interior is in a disgustingly filthy condition.
Rude fresco representations of the different saints in the Gregorian calendar formerly adorned the walls, and bright colored tiles
embellished the approach to the altar.   Nothing is distinguishable
these days but the crumbling and half-obliterated evidences of
past glories ; both priests and people seem hopelessly sunk in the
quagmire of avariciousness and low cunning on the one hand, and
of blind ignorance and superstition on the other.   Clad in greasy
and seedy-looking cowls, the priests go through a few nonsensical
mânoeuvres, consisting chiefly of an ostentatious affectation of reverence toward an altar covered with tattered drapery, by never turning their backs toward it while they walk about, Bible in hand,
mumbling and sighing.   My self-constituted guide and myself
comprise the whole congregation during the " services."   Whenever the priests heave a particularly deep-fetched sigh or fall to
mumbling their prayers on the double quick, they invariably cast
a furtive glance toward me, to ascertain whether I am noticing the
impenetrable depth of their holiness.   They needn't be uneasy on
that score, however ; the most casual observer cannot fail to perceive that it is really and truly impenetrable—so impenetrable, in
fact, that it will never be unearthed, not even at the day of judgment.   In about ten minutes the priests quit mumbling, bestow a
pharisaical kiss on the tattered coverlet of their Bibles, graciously
suffer my jack-booted companion to do likewise, as also two or
three ragamuffins who have come sneaking in seemingly for that
special purpose, and then retreat hastily behind a patch-work cur-

tain ; the next minute they reappear in a cowlless condition, their countenances wearing an expression of intense relief, as though happy at having gotten through with a disagreeable task that had been weighing heavily on their minds all day.

We are invited to take supper with their Reverences in their cell beneath the walls, which they occupy in common. The repast consists of *yaort* and *pillau*, to which is added, by way of compliment to visitors, five salt fishes about the size of sardines. The most greasy-looking of the divines thoughtfully helps himself to a couple of the fishes as though they were a delicacy quite irresistible, leaving one apiece for us others. Having created a thirst with the salty fish, he then seizes what remains of the *yaort*, pours water into it, mixes it thoroughly together with his unwashed hand, and gulps down a full quart of the swill with far greater gusto than mannerliness. Soon the priests commence eructating aloud, which appears to be a well-understood signal that the limit of their respective absorptive capacities are reached, for three hungry-eyed laymen, who have been watching our repast with seemingly begrudging countenances, now carry the wooden tray bodily off into a corner and ravenously devour the remnants. Everything about the cell is abnormally filthy, and I am glad when the inevitable cigarettes are ended and we retire to the quarters assigned us in the village. Here my companion produces from some mysterious corner of his clothing a pinch of tea and a few lumps of sugar. A villager quickly kindles a fire and cooks the tea, performing the services eagerly, in anticipation of coming in for a modest share of what to him is an unwonted luxury. Being rewarded with a tiny glassful of tea and a lump of sugar, he places the sweet morsel in his mouth and sucks the tea through it with noisy satisfaction, prolonging the presumably delightful sensation thereby produced to fully a couple of minutes. During this brief indulgence of his palate, a score of his ragged co-religionists stand around and regard him with mingled envy and covetousness ; but for two whole minutes he occupies his proud eminence in the lap of comparative luxury, and between slow, lingering sucks at the tea, regards their envious attention with studied indifference. One can scarcely conceive of a more utterly wretched people than the monastic community of Sup Ogwanis ; one would not be surprised to find them envying even the pariah curs of the country.

The wind blows raw and chilly from off the snowy slopes of

Ararat next morning, and the shivering, half-clad wretches shuffle off toward the fields and pastures, with blue noses and unwilling faces, humping their backs and shrinking within themselves and wearing most lugubrious countenances; one naturally falls to wondering what they do in the winter. The independent villagers of the surrounding country have a tough enough time of it, worrying through the cheerless winters of a treeless and mountainous country; but they at least have no domestic authority to obey but their own personal and family necessities, and they consume the days huddled together in their unventilated hovels over a smouldering *tezek* fire; but these people seem but helpless dolts under the vassalage of a couple of crafty-looking, coarse-grained priests, who regard them with less consideration than they do the monastery buffaloes.

Eleven miles over a mostly ridable trail brings me to the large village of Dyadin. Dyadin is marked on my map as quite an important place, consequently I approach it with every assurance of obtaining a good breakfast. My inquiries for refreshments are met with importunities of *bin bacalem*, from five hundred of the rag-tag and bob-tail of the frontier, the rowdiest and most inconsiderate mob imaginable. In their eagerness and impatience to see me ride, and their exasperating indifference to my own pressing wants, some of them tell me bluntly there is no bread; others, more considerate, hurry away and bring enough bread to feed a dozen people, and one fellow contributes a couple of onions. Pocketing the onions and some of the bread, I mount and ride away from the madding crowd with whatever despatch is possible, and retire into a secluded dell near the road, a mile from town, to eat my frugal breakfast in peace and quietness. While thus engaged, it is with veritable savage delight that I hear a company of horsemen go furiously galloping past; they are Dyadin people endeavoring to overtake me for the kindly purpose of worrying me out of my senses, and to prevent me even eating a bite of bread unseasoned with their everlasting gabble. Although the road from Dyadin eastward leads steadily upward, they fancy that nothing less than a wild, sweeping gallop will enable them to accomplish their fell purpose; I listen to their clattering hoof-beats dying away in the dreamy distance, with a grin of positively malicious satisfaction, hoping sincerely that they will keep galloping onward for the next twenty miles.

No such happy consummation of my wishes occurs, however ; a couple of miles up the ascent I find them hobnobbing with some Persian caravan men and patiently awaiting my appearance, having learned from the Persians that I had not yet gone past. Mingled with the keen disappointment of overtaking them so quickly, is the pleasure of witnessing the Persians' camels regaling themselves on a patch of juicy thistles of most luxuriant growth ; the avidity with which they attack the great prickly vegetation, and the expression of satisfaction, utter and peculiar, that characterizes a camel while munching a giant thistle stalk that protrudes two feet out of his mouth, is simply indescribable.

From this pass I descend into the Aras Plain, and, behold the gigantic form of Ararat rises up before me, seemingly but a few miles away ; as a matter of fact it is about twenty miles distant, but with nothing intervening between myself and its tremendous proportions but the level plain, the distance is deceptive. No human habitations are visible save the now familiar black tents of Koordish tribesmen away off to the north, and as I ride along I am overtaken by a sensation of being all alone in the company of an overshadowing and awe-inspiring presence. One's attention seems irresistibly attracted toward the mighty snow-crowned monarch, as though the immutable law of attraction were sensibly exerting itself to draw lesser bodies to it, and all other objects around seemed dwarfed into insignificant proportions. One obtains a most comprehensive idea of Ararat's 17,325 feet when viewing it from the Aras Plain, as it rises sheer from the plain, and not from the shoulders of a range that constitutes of itself the greater part of the height, as do many mountain peaks. A few miles to the eastward is Little Ararat, an independent conical peak of 12,800 feet, without snow, but conspicuous and distinct from surrounding mountains ; its proportions are completely dwarfed and overshadowed by the nearness and bulkiness of its big brother. The Aras Plain is lava-strewn and uncultivated for a number of miles ; the spongy, spreading feet of innumerable camels have worn paths in the hard lava deposit that makes the wheeling equal to English roads, except for occasional stationary blocks of lava that the animals have systematically stepped over for centuries, and which not infrequently block the narrow trail and compel a dismount. Evidently Ararat was once a volcano ; the lofty peak which now presents a wintry appearance even in the hottest summer weather,

formerly belched forth lurid flames that lit up the surrounding
country, and poured out fiery torrents of molten lava that stratified
the abutting hills, and spread like an overwhelming flood over the
Aras Plain. Abutting Ararat on the west are stratiform hills, the
strata of which are plainly distinguishable from the Persian trail,
and which, were their inclination continued, would strike Ararat
at or near the summit. This would seem to indicate the layers to
be representations of the mountain's former volcanic overflowings.

I am sitting on a block of lava making an outline sketch of Ararat, when a peasant happens along with a bullock-load of cucum-

bers which he is
taking to the
Koordish camps;
he is pretty badly
scared at finding
himself all alone
on the Aras Plain
with such a non-
descript and dan-
gerous - looking
object as a helmet-
ed wheelman, and
when I halt him
with inquiries
concerning the
nature of his wares
he turns pale and
becomes almost
speechless with
fright. He would
empty his sacks as

Well Guarded at Lunch.

a peace-offering at my feet without venturing upon a remonstrance, were he ordered to do so ; and when I relieve him of but
one solitary cucumber, and pay him more than he would obtain
for it among the Koords, he becomes stupefied with astonishment;
when he continues on his way he hardly knows whether he is on
his head or his feet. An hour later I arrive at Kizil Dizah, the last
village in Turkish territory, and an official station of considerable
importance, where passports, caravan permits, etc., of everybody
passing to or from Persia have to be examined. An officer here

provides me with refreshments, and while generously permitting the population to come in and enjoy the extraordinary spectacle of seeing me fed, he thoughtfully stations a man with a stick to keep them at a respectful distance. A later hour in the afternoon finds me trundling up a long acclivity leading to the summit of a low mountain ridge ; arriving at the summit I stand on the boundary-line between the dominions of the Sultan and the Shah, and I pause a minute to take a brief, retrospective glance.

The cyclometer, affixed to the bicycle at Constantinople, now registers within a fraction of one thousand miles ; it has been on the whole an arduous thousand miles, but those who in the foregoing pages have followed me through the strange and varied experiences of the journey will agree with me when I say that it has proved more interesting than arduous after all. I need not here express any blunt opinions of the different people encountered ; it is enough that my observations concerning them have been jotted down as I have mingled with them and their characteristics from day to day ; almost without exception, they have treated me the best they knew how ; it is only natural that some should know how better than others.

Bidding farewell, then, to the land of the Crescent and the home of the unspeakable Osmanli, I wheel down a gentle slope into a mountain-environed area of cultivated fields, where Persian peasants are busy gathering their harvest. The strange apparition observed descending from the summit of the boundary attracts universal attention ; I can hear them calling out to each other, and can see horsemen come wildly galloping from every direction. In a few minutes the road in my immediate vicinity is alive with twenty prancing steeds ; some are bestrode by men who, from the superior quality of their clothes and the gaudy trappings of their horses, are evidently in good circumstances ; others by wild-looking, bare-legged bipeds, whose horses' trappings consist of nothing but a bridle. The transformation brought about by crossing the mountain ridge is novel and complete ; the fez, so omnipresent throughout the Ottoman dominions, has disappeared, as if by magic ; the better class Persians wear tall, brimless black hats of Astrakan lamb's wool ; some of the peasantry wear an unlovely, close-fitting skull-cap of thick gray felt, that looks wonderfully like a bowl clapped on top of their heads, others sport a huge woolly head-dress like the Roumanians ; this latter imparts to them a fierce, war-like appearance, that the meek-eyed Persian ryot (tiller of the soil) is far from feeling. The national garment is a sort of frock-coat gathered at the waist, and with a skirt of ample fulness, reaching nearly to the knees; among the wealthier class the material of this garment is usually cloth of a solid, dark color, and among the ryots or peasantry, of calico or any cheap fabric they can obtain. Loose-fitting pantaloons of European pattern, and sometimes top-boots, with tops ridiculously ample in their looseness, characterize the nether garments of the better classes; the ryots go mostly bare-legged in summer, and wear loose, slipper-like foot-gear ; the soles of both boots and shoes are frequently pointed, and made to turn up and inwards, after the fashion in England centuries ago.

Nightfall overtakes me as, after travelling several miles of variable road, I commence following a winding trail down into the valley of a tributary of the Arasces toward Ovahjik, where resides the Pasha Khan, to whom I have a letter ; but the crescent-shaped moon sheds abroad a silvery glimmer that exerts a softening influence upon the mountains outlined against the ever-arching dome, from whence here and there a star begins to twinkle. It is one of those beautiful, calm autumn evenings when all nature seems hushed in peaceful slumbers ; when the stars seem to first peep cautiously from the impenetrable depths of their hiding-place, and then to commence blinking benignantly and approvingly upon the world ; and when the moon looks almost as though fair Luna has been especially decorating herself to embellish a scene that without her lovely presence would be incomplete. Such is my first autumn evening beneath the cloudless skies of Persia.

Soon the village of Ovahjik is reached, and some peasants guide me to the residence of the Pasha Khan. The servant who presents my letter of introduction fills the untutored mind of his master with wonderment concerning what the peasants have told him about the bicycle. The Pasha Khan makes his appearance without having taken the trouble to open the envelope. He is a dull-faced, unin-tellectual-looking personage, and without any preliminary palaver he says : " *Bin bacalem*," in a dictatorial tone of voice. " *Bacalem yole lazim, bacalem saba*," I reply, for it is too dark to ride on unknown ground this evening. " *Bin bacalem !* " repeats the Pasha Khan, even more dictatorial than before, ordering a servant to bring' a tallow candle, so that I can have no excuse. There appears to be such a total absence of all consideration for myself that I am not·

disposed to regard very favorably or patiently the obtrusive med-
dlesomeness of two younger men—whom I afterward discover to
be sons of the Pasha Khan—who seem almost inclined to take the
bicycle out of my charge altogether, in their excessive impatience
and inordinate inquisitiveness to examine everything about it. One
of them, thinking the cyclometer to be a watch, puts his ear down
to see if he can hear it tick, and then persists in fingering it about,
to the imminent danger of the tally-pin. After telling him several

The Persistent Son is Shoved into the Water.

times not to meddle with it, and receiving overbearing gestures in
reply, I deliberately throw him backward into an irrigating ditch.
A gleam of intelligence overspreads the stolid countenance of the
Pasha Khan at seeing his offspring floundering about on his back
in the mud and water, and he gives utterance to a chuckle of delight. The discomfited young man betrays nothing of the spirit
of resentment upon recovering himself from the ditch, and the other
son involuntarily retreats as though afraid his turn was coming next.

The servant now arrives with the lighted candle, and the Pasha Kahn leads the way into his garden, where there is a wide brick-paved walk ; the house occupies one side of the garden, the other three sides are inclosed by a high mud wall. After riding a few times along the brick-paved walk, and promising to do better in the morning, I naturally expect to be taken into the house, instead of which the Pasha Khan orders the people to show me the way to the caravanserai. Arriving at the caravanserai, and finding myself thus thrown unexpectedly upon my own resources, I inquire of some bystanders where I can obtain *ekmek ;* some of them want to know how many liras I will give for *ekmek !* When it is reflected that a lira is nearly five dollars, one realizes from this something of the unconscionable possibilities of the Persian commercial mind.

While this question is being mooted, a figure appears in the doorway, toward which the people one and all respectfully salaam and give way. It is the great Pasha Khan ; he has bethought himself to open my letter of introduction, and having perused it and discovered who it was from and all about me, he now comes and squats down in the most friendly manner by my side for a minute, as though to remove any unfavorable impressions his inhospitable action in sending me here might have made, and then bids me accompany him back to his residence. After permitting him to eat a sufficiency of humble pie in the shape of coaxing, to atone for his former incivility, I agree to his proposal and accompany him back. Tea is at once provided, the now very friendly Pasha Khan putting extra lumps of sugar into my glass with his own hands and stirring it up ; bread and cheese comes in with the tea, and under the mistaken impression that this constitutes the Persian evening meal I eat sufficient to satisfy my hunger. While thus partaking freely of the bread and cheese, I do not fail to notice that the others partake very sparingly, and that they seem to be rather astonished because I am not following their example. Being chiefly interested in satisfying my appetite, however, their silent observations have no effect save to further mystify my understanding of the Persian character. The secret of all this soon reveals itself in the form of an ample repast of savory chicken *pillau,* brought in immediately afterward ; and while the Pasha Khan and his two sons proceed to do full justice to this highly acceptable dish, I have to content myself with nibbling at a piece of chicken, and ruminating on the unhappy and ludicrous mistake of having satisfied my hunger with

dry bread and cheese. Thus does one pay the penalty of being unacquainted with the domestic customs of a country when first entering upon its experiences.

There seems to be no material difference between the social
position of the women here and in Turkey ; they eat their meals
by themselves, and occupy entirely separate apartments, which are
unapproachable to members of the opposite sex save their husbands. The Pasha Khan of Ovahjik, however, seems to be a kind,
indulgent husband and father, requesting me next morning to ride
up and down the brick-paved walk for the benefit of his wives and
daughters. In the
seclusion of their
own walled prem-
ises the Persian
females are evi-
dently not so par-
ticular about con-
cealing their feat-
ures, and I ob-
tained a glimpse
of some very pret-
faces ; oval faces
large dreamy
black eyes, and a
flush of warm sun-
set on brownish
cheeks. The in-
door costume of
Persian women is

Riding for the Pasha Khan's Ladies.

but an inconsiderable improvement upon the costume of our an-
cestress in the garden of Eden, and over this they hastily don a
flimsy shawl-like garment to come out and see me ride. They are
always much less concerned about concealing their nether extremities than about their faces, and as they seem but little concerned
about anything on this occasion save the bicycle, after riding for
them I have to congratulate myself that, so far as sight-seeing is
concerned, the ladies leave me rather under obligations than otherwise.

After supper the Pasha Khan's falconer brings in several fine
falcons for my inspection, and in reply to questions concerning one

with his eyelids tied up in what appears to be a cruel manner, I am told that this is the customary way of breaking the spirits of the young falcons and rendering them tractable and submissive ; the eyelids are pierced with a hole, a silk thread is then fastened to each eyelid and the ends tied together over the head, sufficiently tight to prevent them opening their eyes. Falconing is considered the chief out-door sport of the Persian nobility, but the average Persian is altogether too indolent for out-door sport, and the keeping of falcons is fashionable, because regarded as a sign of rank and nobility rather than for sport.

In the morning the Pasha Khan is wonderfully agreeable, and appears anxious to atone as far as possible for the little incivility of yesterday evening, and to remove any unfavorable impressions I may perchance entertain of him on that account before I leave. His two sons and a couple of soldiers accompany me on horseback some distance up the valley. The valley is studded with villages, and at the second one we halt at the residence of a gentleman named Abbas Koola Khan, and partake of tea and light refreshments in his garden. Here I learn that the Pasha Khan has carried his good intentions to the extent of having made arrangements to provide me armed escort from point to point ; how far ahead this well-meaning arrangement is to extend I am unable to understand ; neither do I care to find out, being already pretty well convinced that the escort will prove an insufferable nuisance to be gotten rid of at the first favorable opportunity. Abbas Koola Khan now joins the company until we arrive at the summit of a knoll commanding an extensive view of my road ahead so they can stand and watch me when they all bid me farewell save the soldier who is to accompany me further on. As we shake hands, the young man whom I pushed into the irrigating ditch, points to a similar receptacle near by and shakes his head with amusing solemnity ; whether this is expressive of his sorrow that I should have pushed him in, or that he should have annoyed me to the extent of having deserved it, I cannot say ; probably the latter.

My escort, though a soldier, is dressed but little different from the better-class villagers ; he is an almond-eyed individual, with more of the Tartar cast of countenance than the Persian. Besides the short Persian sword, he is armed with a Martini Henry rifle of the 1862 pattern ; numbers of these rifles having found their way into the hands of Turks, Koords and Persians, since the Russo-

Turkish war.   My predictions concerning his turning out an in-supportable nuisance are not suffered to remain long unverified, for he appears to consider it his chief duty to gallop ahead and notify the villagers of my approach, and to work them up to the highest expectations concerning my marvellous appearance.   The result of all this is a swelling of his own importance at having so wonderful a person under his protection, and my own transforma-tion from an unostentatious traveller to something akin to a free circus for crowds of barelegged ryots.   I soon discover that, with characteristic Persian truthfulness, he has likewise been spreading the interesting report that I am journeying in this extraordinary manner to carry a message from the "Ingilis Shah " to the "Shah in Shah of Iran " (the Persians know their own country as Iran) thereby increasing his own importance and the wonderment of the people concerning myself.   The Persian villages, so far, are little different from the Turkish, but such valuable property as melon-gardens, vineyards, etc., instead of being presided over by a watch-man, are usually surrounded by substantial mud walls ten or twelve feet high.   The villagers themselves, being less improvident and altogether more thoughtful of number one than the Turks, are on the whole, a trifle less ragged ; but that is saying very little indeed, and their condition is anything but enviable.   During the summer they fare comparatively well, needing but little clothing, and they are happy and contented in the absence of actual suffering ; they are perfectly satisfied with a diet of bread and fruit and cucumbers, rarely tasting meat of any kind.   But fuel is as scarce as in Asia Minor, and like the Turks and Armenians, in winter they have re-source to a peculiar and economical arrangement to keep themselves warm ; placing a pan of burning *tezek* beneath a low table, the whole family huddle around it, covering the table and themselves —save of course their heads—up with quilts ; facing each other in this ridiculous manner, they chat and while away the dreary days of winter.

At the third village after leaving the sons of the Pasha Khan, my Tartar-eyed escort, with much garrulous injunction to his suc-cessor, delivers me over to another soldier, himself returning back ; this is my favorable opportunity, and soon after leaving the village I bid my valiant protector return.   The man seems totally un-able to comprehend why I should order him to leave me, and makes an elaborate display of his pantomimic abilities to impress upon me the information that the country ahead is full of very bad Koords, who will kill and rob me if I venture among them unprotected by a soldier. The expressive action of drawing the finger across the throat appears to be the favorite method of signifying personal danger among all these people; but I already understand that the Persians live in deadly fear of the nomad Koords. Consequently his warnings, although evidently sincere, fall on biased ears, and I peremptorily order him to depart. The Tabreez trail is now easily followed without a guide, and with a sense of perfect freedom and unrestraint, that is destroyed by having a horseman cantering alongside one, I push ahead, finding the roads variable, and passing through several villages during the day.

An every-day Occurrence.

The chief concern of the ryots is to detain me until they can bring the resident Khan to see me ride, evidently from a servile desire to cater to his pleasure. They gather around me and prevent my departure until he arrives. An appeal to the revolver will invariably secure my release, but one naturally gets ashamed of threatening people's lives even under the exasperating circumstances of a forcible detention. Once to-day I managed to outwit them beautifully. Pretending acquiescence in their proposition of waiting till the arrival of their Khan, I propose mounting and riding a few yards for their own edification while waiting; in their eagerness to see they readily fall into the trap, and the next minute sees me flying down the road with a swarm of bare-legged ryots in full chase after me, yelling for me to stop. Fortunately, they have no horses handy, but some of these lanky fellows can run like deer almost, and nothing but an excellent piece of road enables me to outdistance my pursuers. Wily as the Persians are, compared to the Osmanlis, one could play this game on them quite frequently, owing to their eagerness to see the bicycle ridden; but it is seldom that the road is sufficiently smooth to justify the attempt. I was gratified to learn from the Persian consul at Erzeroum that my stock of Turkish would answer me as far as Teheran, the people west of the capital speaking a dialect known as Tabreez Turkish; still, I find quite a difference. Almost every Persian points to the bicycle

Politeness in a Koordish Tent.

and says: " Boo; námi nádder? " ("This; what is it?") and it is several days ere I have an opportunity of finding out exactly what they mean. They are also exceedingly prolific in using the endearing term of kardash when accosting me. The distance is now reckoned by farsakhs (roughly, four miles) instead of hours; but, although the farsakh is a more tangible and comprehensive measurement than the Turkish hour, in reality it is almost as unreliable to go by.

Towards evening I ascend into a more mountainous region, inhabited exclusively by nomad Koords; from points of vantage their tents are observable clustered here and there at the bases of the mountains. Descending into a grassy valley or depression, I find myself in close proximity to several different camps, and eagerly avail myself of the opportunity to pass a night among them. I am now in the heart of Northern Koordistan, which embraces both Persian and Turkish territory, and the occasion is most opportune for seeing something of these wild nomads in their own mountain pastures. The greensward is ridable, and I dismount before the Sheikh's tent in the presence of a highly interested and interesting audience. The half-wild dogs make themselves equally interesting in another and a less desirable sense as I approach, but the men pelt them with stones, and when I dismount they conduct me and the bicycle at once into the tent of their chieftain. The Sheikh's tent is capacious enough to shelter a regiment almost, and it is divided into compartments similar to a previous description ; the Sheikh is a big, burly fellow, of about forty-five, wearing a turban the size of a half-bushel measure, and dressed pretty much like a well-to-do Turk ; as a matter of fact, the Koords admire the Osmanlis and despise the Persians. The bicycle is reclined against a carpet partition, and after the customary interchange of questions, a splendid fellow, who must be six feet six inches tall, and broad-shouldered in proportion, squats himself cross-legged beside me, and proceeds to make himself agreeable, rolling me cigarettes, asking questions, and curiously investigating anything about me that strikes him as peculiar. I show them, among other things, a cabinet photograph of myself in all the glory of needle-pointed mustache and dress-parade apparel ; after a critical examination and a brief conference among themselves they pronounce me an " English Pasha." I then hand the Sheikh a set of sketches, but they are not sufficiently civilized to appreciate the sketches ; they hold them upside down and sidewise ; and not being able to make anything out of them, the Sheikh holds them in his hand and looks quite embarrassed, like a person in possession of something he doesn't know what to do with.

Noticing that the women are regarding these proceedings with much interest from behind a low partition, and not having yet become reconciled to the Mohammedan idea of women being habitually ignored and overlooked, I venture upon taking the photograph to them ; they seem much confused at finding themselves the object of direct attention, and they appear several degrees wilder than the men, so far as comprehending such a product of civilization as a photograph is an indication. It requires more material objects than sketches and photos to meet the appreciation of these semi-civilized children of the desert. They bring me their guns and spears to look at and pronounce upon, and then my stalwart entertainer grows inquisitive about my revolver. First extracting the cartridges to prevent accident, I hand it to him, and he takes it for the Sheikh's inspection. The Sheikh examines the handsome little Smith & Wesson long and wistfully, and then toys with it several minutes, apparently reluctant about having to return it; finally he asks me to give him a cartridge and let him go out and test its accuracy. I am getting a trifle uneasy at his evident covetousness of the revolver, and in this request I see my opportunity of giving him to understand that it would be a useless weapon for him to possess, by telling him I have but a few cartridges and that others are not procurable in Koordistan or neighboring countries. Recognizing immediately its uselessness to him under such circumstances, he then returns it without remark ; whether he would have confiscated it without this timely explanation, it is difficult to say.

Shortly after the evening meal, an incident occurs which causes considerable amusement. Everything being unusually quiet, one sharp-eared youth happens to hear the obtrusive ticking of my Waterbury, and strikes a listening attitude, at which everybody else likewise begins listening ; the tick, tick is plainly discernible to everybody in the compartment and they become highly interested and amused, and commence looking at me for an explanation. With a view to humoring the spirit of amusement thus awakened, I likewise smile, but affect ignorance and innocence concerning the origin of the mysterious ticking, and strike a listening attitude as well as the others. Presuming upon our interchange of familiarity, our six-foot-sixer then commences searching about my clothing for the watch, but being hidden away in a pantaloon fob, and minus a chain, it proves beyond his power of discovery. Nevertheless, by bending his head down and listening, he ascertains and announces it to be somewhere about my person ; the Waterbury is then produced, and the loudness of its ticking awakes the wonder and admiration of the Koords, even to a greater extent than the Turks.

During the evening, the inevitable question of Russ, Osmanli, and English crops up, and I win unanimous murmurs of approval

29

by laying my forefingers together and stating that the English and
the Osmanlis are *kardash*.   I show them my Turkish *teskeri*, upon
which several of them bestow fervent kisses, and when, by means
of placing several stones here and there I explained to them how
in 1877, the hated Muscov occupied different Mussulman cities one
after the other, and was ·prevented by the English from occupying
their dearly beloved Stamboul itself, their admiration knows no
bounds.   Along the trail, not over a mile from camp, a large Persian caravan has been halting during the day ; late in the evening

Explaining England's Friendly Offices.

loud shouting and firing of guns announces them as prepared to
start on their night's journey.   It is customary when going through
this part of Koordistan for the caravan men to fire guns and make
as much noise as possible, in order to impress the Koords with exaggerated ideas concerning their strength and number ; everybody
in the Sheikh's tent thoroughly understands the meaning of the
noisy demonstration, and the men exchange significant smiles.   The
firing and the. shouting produce a truly magical effect upon a
blood-thirsty youngster of ten or twelve summers ; he becomes

wildly hilarious, gamboling about the tent, and rolling over and kicking up his heels. He then goes to the Sheikh, points to me, and draws his finger across his throat, intimating that he would like the privilege of cutting somebody's throat, and why not let him cut mine? The Sheikh and others laugh at this, but instead of chiding him for his tragical demonstration, they favor him with the same admiring glances that grown people bestow upon precocious youngsters the world over. Under these circumstances of abject fear on the one hand, and inbred propensity for violence and plunder on the other, it is really surprising to find the Koords in Persian territory behaving themselves as well as they do.

Quilts are provided for me, and I occupy this same compartment of the tent, in common with several of the younger men. In the morning, before departing, I am regaled with bread and rich, new cream, and when leaving the tent I pause a minute to watch the busy scene in the female department. Some are churning butter in sheep-skin churns which are suspended from poles and jerked back and forth ; others are weaving carpets, preparing curds for cheese, baking bread, and otherwise industriously employed. I depart from the Koordish camp thoroughly satisfied with my experience of their hospitality, but the cerulean waist-scarf bestowed upon me by our Hungarian friend Igali, at Belgrade, no longer adds its embellishments to my personal adornments. Whenever a favorable opportunity presents, certain young men belonging to the noble army of hangers-on about the Sheikh's apartments invariably glide inside, and importune the guest from Frangistan for any article of his clothing that excites the admiration of their semi-civilized minds. This scarf, they were doubtless penetrating enough to observe, formed no necessary part of my wardrobe, and a dozen times in the evening, and again in the morning, I was worried to part with it, so I finally presented it to one of them. He hastily hid it away among his clothes and disappeared, as though fearful, either that the Sheikh might see it and make him return it, or that one of the chieftain's favorites might take a fancy to it and summarily appropriate it to his own use.

Not more than five miles eastward from the camp, while trundling over a stretch of stony ground, I am accosted by a couple of Koordish shepherds ; but as the country immediately around is wild and unfrequented, save by Koords, and knowing something of their little weaknesses toward travellers under tempting, one-sided conditions, I deem it advisable to pay as little heed to them as possible. Seeing that I have no intention of halting, they come running up, and undertake to forcibly detain me by seizing hold of the bicycle, at the same time making no pretence of concealing their eager curiosity concerning the probable contents of my luggage. Naturally disapproving of this arbitrary conduct, I push them roughly away. With a growl more like the voice of a wild animal than of human beings, one draws his sword and the other picks up a thick knobbed stick that he had dropped in order to the better pinch and sound my packages. Without giving them

Koordish Highwaymen.

time to reveal whether they seriously intend attacking me, or only to try intimidation, I have them nicely covered with the Smith & Wesson. They seem to comprehend in a moment that I have them at a disadvantage, and they hurriedly retreat a short distance, executing a series of gyral antics, as though expecting me to fire at their legs.

They are accompanied by two dogs, tawny-coated monsters, larger than the largest mastiffs, who now proceed to make things lively and interesting around myself and the bicycle. Keeping the revolver in my hand, and threatening to shoot their dogs if they don't call them away, I continue my progress toward where · the stony ground terminates in favor of smooth camel-paths, about a hundred yards farther on. At this juncture I notice several other "gentle shepherds" coming racing down from the adjacent knolls; but whether to assist their comrades in catching and robbing me, or to prevent a conflict between us, will always remain an uncertainty. I am afraid, however, that with the advantage on their side, the Koordish herdsmen rarely trouble themselves about any such uncongenial task as peace-making. Reaching the smooth ground before any of the new-comers overtake me, I mount and speed away, followed by wild yells from a dozen Koordish throats, and chased by a dozen of their dogs. Upon sober second thought, when well away from the vicinity, I conclude this to have been a rather ticklish incident; had they attacked me in the absence of anything else to defend myself with, I should have been compelled to shoot them; the nearest Persian village is about ten miles distant; the absence of anything like continuously ridable road would have made it impossible to out-distance their horsemen, and a Persian village would have afforded small security against a party of enraged Koords, after all.

The first village I arrive at to-day, I again attempt the "skedaddling" dodge on them that proved so successful on one occasion yesterday; but I am foiled by a rocky "jump-off" in the road to-day. The road is not so favorable for spurting as yesterday, and the racing ryots grab me amid much boisterous merriment ere I overcome the obstruction; they take particular care not to give me another chance until the arrival of the Khan. The country hereabouts consists of gravelly, undulating plateaus between the mountains, and well-worn camel-paths afford some excellent wheeling. Near mid-day, while laboriously ascending a long but not altogether unridable ascent, I meet a couple of mounted soldiers; they obstruct my road, and proceed to deliver themselves of voluble Tabreez Turkish, by which I understand · that they are the advance guard of a party in which there is a Ferenghi (the Persian term for an Occidental). While talking with them I am somewhat taken by surprise at seeing a lady on horseback and two children in a *kajaveh* (mule panier) appear over the slope, accompanied by about a dozen Persians.

If I am surprised, the lady herself not unnaturally evinces even greater astonishment at the apparition of a lone wheelman here on the caravan roads of Persia ; of course we are mutually delighted. With the assistance of her servant, the lady alights from the saddle and introduces herself as Mrs. R——, the wife of one of the Persian missionaries ; her husband has lately returned home, and she is on the way to join him. The Persians accompanying her comprise her own servants, some soldiers procured of the Governor of Tabreez by the English consul to escort her as far as the Turkish frontier, and a couple of unattached travellers keeping with the party for company and society. A mule driver has charge of pack-mules carrying boxes containing, among other things, her husband's library. During the course of ten minutes' conversation the lady informs me that she is compelled to travel in this manner the whole distance to Trebizond, owing to the practical impossibility of passing through Russian territory with the library. Were it not for this a comparatively short and easy journey would take them to Tiflis, from which point there would be steam communication with Europe. Ere the poor lady gets to Trebizond she will be likely to reflect that a government so civilized as the Czar's might relax its gloomy laws sufficiently to allow the affixing of official seals to a box of books, and permit its transportation through the country, on condition—if they will—that it should not be opened in transit ; surely there would be no danger of the people's minds being enlightened —not even a little bit—by coming in contact with a library tightly boxed and sealed. At the frontier an escort of Turkish *zaptiehs* will take the place of the Persian soldiers, and at Erzeroum the missionaries will, of course, render her every assistance to Trebizond ; but it is not without feelings of anxiety for the health of a lady travelling in this rough manner unaccompanied by her natural protector, that I reflect on the discomforts she must necessarily put up with between here and Erzeroum. She seems in good spirits, however, and says that meeting me here in this extraordinary manner is the "most romantic" incident in her whole experiences of missionary life in Persia. Like many another, she says, she can scarcely conceive it possible that I am travelling without attendants and without being able to speak the languages. One of the unattached travellers gives me a note of introduction to Mohammed Ali Khan, the Governor of Peri, a suburban village of Khoi, which I expect to reach some time this afternoon.

CHAPTER XIX.

A SHORT trundle to the summit of a sloping pass, and then a winding descent of several miles brings me to a position commanding a view of an extensive valley that looks from this distance as lovely as a dreamy vision of Paradise. An hour later and I am bowling along beneath overhanging peach and mulberry trees, following a volunteer horseman to Mohammed Ali Khan's garden. Before reaching the garden a gang of bare-legged laborers engaged in patching up a mud wall favor me with a fusillade of stones, one of which caresses me on the ankle, and makes me limp like a Greenwich pensioner when I dismount a minute or two afterward. This is their peculiar way of complimenting a lone Ferenghi. Mohammed Ali Khan is found to be rather a moon-faced individual under thirty, who, together with his subordinate officials, are occupying tents in a large garden. Here, during the summer, they dispense justice to applicants for the same within their jurisdiction, and transact such other official business as is brought before them. In Persia the distribution of justice consists chiefly in the officials ruthlessly looting the applicants of everything lootable, and the weightiest task of the officials is intriguing together against the pocket of the luckless wight who ventures upon seeking equity at their hands.

A sorrowful-visaged husbandman is evidently experiencing the easy simplicity of Persian civil justice as I enter the garden ; he wears the mournful expression of a man conscious of being irretrievably doomed, while the festive Kahn and his equally festive *moonshi bashi* (chief secretary) are laying their wicked heads together and whispering mysteriously, fifty paces away from everybody, ever and anon looking suspiciously around as though fearful of the presence of eavesdroppers. After duly *binning*, a young man yclept Abdullah, who seems to be at the beck and call of everybody, brings forth the samovar, and we drink the customary tea of good fellowship, after which they examine such of my modest effects as take their fancy.

The *moonshi bashi*, as becomes a man of education, is quite infatuated with my pocket map of Persia ; the fact that Persia occupies so great a space on the map in comparison with the small portions of adjoining countries visible around the edges makes a powerful appeal to his national vanity, and he regards me with increased affection every time I trace out for him the comprehensive boundary line of his native Iran. After nightfall we repair to the principal tent, and Mohammed Ali Khan and his secretary consume the evening hours in the joyous occupation of alternately smoking the kalian (Persian water-pipe, not unlike the Turkish nargileh, except that it has a straight stem instead of a coiled tube), and swallowing glasses of raw arrack every few minutes ; they furthermore amuse themselves by trying to induce me to follow their noble example, and in poking fun at another young man because his conscientious scruples regarding the Mohammedan injunction against intoxicants forbids him indulging with them. About eight o'clock the Khan becomes a trifle sentimental and very patriotic. Producing a pair of silver-mounted horse-pistols from a corner of the tent, and waving them theatrically about, he proclaims aloud his mighty devotion to the Shah. At nine o'clock Abdullah brings in the supper. The Khan's vertebra has become too limp and willowy to enable him to sit upright, and he has become too indifferent to such coarse, unspiritual things as stewed chicken and musk-melons to care about eating any, while the *moonshi bashi's* affection for me on account of the map has become so overwhelming that he deliberately empties all the chicken on to my sheet of bread, leaving none whatever for himself and the phenomenal young person with the conscientious scruples.

When bedtime arrives it requires the united exertions of Abdullah and the phenomenal young man to partially undress Mohammed Ali Khan and drag him to his couch on the floor, the Kahn being limp as a dish-rag and a moderately bulky person. The *moonshi bashi*, as becomes an individual of lesser rank and superior mental attainments, is not quite so helpless as his official superior, but on retiring he humorously reposes his feet on the pillow and his head on nothing but the bare floor of the tent, and stubbornly refuses to permit Abdullah to alter either his pillow or his position. The phenomenal young man and myself likewise seek our respective pile of quilts, Abdullah removes the lamp, draws a curtain over the entrance of the tent, and retires.

The Persians, as representing the Shiite division of the Mohammedan religion, consider themselves by long odds the holiest people on the earth, far holier than the Turks, whom they religiously despise as Sunnites and unworthy to loose the latchets of their shoes. The Koran strictly enjoins upon them great moderation in the use of intoxicating drinks, yet certain of the Persian nobility are given to drinking this raw intoxicant by the quart daily. When

"Limp as a Dish-rag."

asked why they don't use it in moderation, they reply, "What is the good of drinking arrack unless one drinks enough to become drunk and happy?" Following this brilliant idea, many of them get "drunk and happy" regularly every evening. They likewise frequently consume as much as a pint before each meal to create a false appetite and make themselves feel boozy while eating.

In the morning the *moonshi bashi*, with a soldier for escort, accompanies me on horseback to Khoi, which is but about seven miles distant over a perfectly level road. Sad to say, the *moonshi bashi*, besides his yearning affection for fiery, untamed arrack, is a confirmed opium smoker, and after last night's debauch for supper and "hitting the pipe" this morning for breakfast, he doesn't feel very dashing in the saddle; consequently I have to accommodate myself to his pace. It is the slowest seven miles ever ridden on the road by a wheelman, I think; a funeral procession is a lively, rattling affair, beside our onward progress toward the mud battlements of Khoi, but there is no help for it. Whenever I venture to the fore a little the dreamy-eyed *moonshi bashi* regards me with a gaze of mild reproachfulness, and sings out in a gently-chide-the-erring tone of voice : "*Kardash?* *Kardash?*" meaning "If we are brothers, why do you seem to want to leave me?" Human nature could scarcely be proof against an appeal wherein endearment and reproach are so beautifully and harmoniously blended, and it always brings me back to a level with his horse.

Reaching the suburbs of Khoi, I am initiated into a new departure—new to myself at this time—of Persian sanctimoniousness. Halting at a fountain to obtain a drink, the soldier shapes himself for pouring the water out of the earthenware drinking vessel into my hands; supposing this to be merely an indication of the Persian's own method of drinking, I motion my preference for drinking out of the jar itself. The soldier looks appealingly toward the *moonshi bashi*, who tells him to let me drink, and then orders him to smash the jar. It then dawns upon my unenlightened mind, that being a Ferenghi, I should have known better than to have touched my unhallowed lips to a drinking vessel at a public fountain, defiling it by so doing, so that it must be smashed in order that the sons of the "true prophet" may not unwittingly drink from it afterward and themselves become defiled. The *moonshi bashi* pilots me to the residence of a certain wealthy citizen outside the city walls; this person, a mild-mannered, purring-voiced man, is seated in a room with a couple of seyuds, or descendants of the prophet; they are helping themselves from a large platter of the finest pears, peaches, and egg plums I ever saw anywhere. The room is carpeted with costly rugs and carpets in which one's feet sink perceptibly at every step; the walls and ceiling are artistically stuccoed, and the doors and windows are gay with stained glass.

Abandoning myself to the guidance of the *moonshi bashi*, I ride around the garden-walks, show them the bicycle, revolver, map of Persia, etc. ; like the *moonshi bashi*, they become deeply interested in the map, finding much amusement and satisfaction in having me point out the location of different Persian cities, seemingly regarding my ability to do so as evidence of exceeding cleverness and erudition. The untravelled Persians of the northern provinces regard Teheran as the grand idea of a large and important city ; if

Doing the Agreeable.

there is any place in the whole world larger and more important, they think it may perhaps be Stamboul. The fact that Stamboul is not on my map while Teheran is, they regard as conclusive proof of the superiority of their own capital. The *moonshi bashi's* chief purpose in accompanying me hither has been to introduce me to the attention of the " *hoikim* "; although the pronunciation is a little different from *hakim*, I attribute this to local brogue, and have been surmising this personage to be some doctor, who, perhaps, having graduated at a Frangistan medical college, the *moonshi bashi* thinks will be able to converse with me. After partaking of fruit and tea we continue on our way to the nearest gate-way of the city proper, Khoi being surrounded by a ditch and battlemented mud wall. Arriving at a large, public inclosure, my guide sends in a letter, and shortly afterward delivers me over to some soldiers, who forthwith conduct me into the presence of—not a doctor, but Ali Khan, the Governor of the city, an officer who hereabouts rejoices in the title of the "*hoikim*."

The Governor proves to be a man of superior intelligence ; he has been Persian ambassador to France some time ago, and understands French fairly well ; consequently we manage to understand each other after a fashion. Although he has never before seen a bicycle, his knowledge of the mechanical ingenuity of the Ferenghis causes him to regard it with more intelligence than an untravelled native, and to better comprehend my journey and its object. Assisted by a dozen mollahs (priests) and officials in flowing gowns and henna-tinted beards and finger-nails, the Governor is transacting official business, and he invites me to come into the council chamber and be seated. In a few minutes the noon-tide meal is announced ; the Governor invites me to dine with them, and then leads the way into the dining-room, followed by his counsellors, who form in line behind him according to their rank. The dining-room is a large, airy apartment, opening into an extensive garden ; a bountiful repast is spread on yellow-checkered table-cloths on the carpeted floor ; the Governor squats cross-legged at one end, the stately-looking wiseacres in flowing gowns range themselves along each side in a similar attitude, with much solemnity and show of dignity ; they—at least so I fancy—evidently are anything but rejoiced at the prospect of eating with an infidel Ferenghi.

The Governor, being a far more enlightened and consequently less bigoted personage, looks about him a trifle embarrassed, as if searching for some place where he can seat me in a position of becoming honor without offending the prejudices of his sanctimonious counsellors. Noticing this, I at once come to his relief by taking the position farthest from him, attempting to imitate them in their cross-legged attitude. My unhappy attempt to sit in this uncomfortable attitude—uncomfortable at least to anybody unaccustomed to it—provokes a smile from His Excellency, and he straightway orders an attendant to fetch in a chair and a small table ; the counsellors look on in silence, but they are evidently too deeply im-

PERSIA AND THE TABREEZ CARAVAN TRAIL. 461

pressed with their own dignity and holiness to commit themselves
to any such display of levity as a smile. A portion of each dish is
placed upon my table, together with a travellers' combination knife,
fork and spoon, a relic, doubtless, of the Governor's Parisian experience. His Excellency having waited and kept the counsellors
waiting until these preparations are finished, motions for me to
commence eating, and then begins himself. The repast consists of
boiled mutton, rice *pillau* with curry, mutton chops, hard-boiled
eggs with lettuce, a pastry of sweetened rice-flour, musk-melons,
water-melons, several kinds of fruit, and for beverage glasses of iced
sherbet ; of all the company I alone use knife, fork, and plates.
Before each Persian is laid a broad sheet of bread ; bending their
heads over this they scoop up small handfuls of *pillau*, and toss it
dextrously into their mouths ; scattering particles missing the expectantly opened receptacle fall back on to the bread ; this handy
sheet of bread is used as a plate for placing a chop or anything else
on, as a table-napkin for wiping finger-tips between courses, and
now and then a piece is pulled off and eaten. When the meal is
finished, an attendant waits on each guest with a brazen bowl, an
ewer of water and a towel.

After the meal is over the Governor is no longer handicapped
by the religious prejudices of the mollahs, and leaving them he invites me into the garden to see his two little boys go through their
gymnastic exercises. They are clever little fellows of about seven
and nine, respectively, with large black eyes and clear olive complexions ; all the time we are watching them the Governor's face is
wreathed in a fond, parental smile. The exercises consist chiefly in
climbing a thick rope dangling from a cross-beam. After seeing
me ride the bicycle the Governor wants me to try my hand at gym-
nastics, but being nothing of a gymnast I respectfully beg to be
excused. While thus enjoying a pleasant hour in the garden, a
series of resounding thwacks are heard somewhere near by, and
looking around some intervening shrubs I observe a couple of *farrashes* bastinadoing a culprit ; seeing me more interested in this
novel method of administering justice than in looking at the youngsters trying to climb ropes, the Governor leads the way thither.
The man, evidently a ryot, is lying on his back, his feet are lashed
together and held soles uppermost by means of an horizontal pole,
while the *farrashes* briskly belabor them with willow sticks. The
soles of the ryot's feet are hard and thick as rhinoceros hide almost

from habitually walking barefooted, and under these conditions his punishment is evidently anything but severe. The flagellation goes merrily and uninterruptedly forward until fifty sticks about five feet long and thicker than a person's thumb are broken over his feet without eliciting any signals of distress from the horny-hoofed ryot, except an occasional sorrowful groan of "A-l-l-ah!" He is then loosed and limps painfully away, but it looks like a rather hypocritical limp, after all; fifty sticks, by the by, is a comparatively light punishment, several hundred sometimes being broken at a single punishment. Upon taking my leave the Governor kindly details a couple of soldiers to show me to the best caravanserai, and to remain and protect me from the worry and annoyance of the crowds until my departure from the city.

Arriving at the caravanserai, my valiant protectors undertake to keep the following crowd from entering the courtyard; the crowd refuses to see the justice of this arbitrary proceeding, and a regular pitched battle ensues in the gateway. The caravanserai-jees reinforce the soldiers, and by laying on vigorously with thick sticks, they finally put the rabble to flight. They then close the caravanserai gates until the excitement has subsided. Khoi is a city of perhaps fifty thousand inhabitants, and among them all there is no one able to speak a word of English. Contemplating the surging mass of woolly-hatted Persians from the bala-khana (balcony; our word is taken from the Persian), of the caravanserai, and hearing nothing but unintelligible language, I detect myself unconsciously recalling the lines : "Oh it was pitiful; in a whole city full——." It is the first large city I have visited without finding somebody capable of speaking at least a few words of my own language. Locking the bicycle up, I repair to the bazaar, my watchful and zealous attendants making the dust fly from the shoulders of such unlucky wights whose eager inquisitiveness to obtain a good close look brings them within the reach of their handy staves. We are followed by immense crowds, a Ferenghi being a rara avis in Khoi, and the fame of the wonderful asp-i-awhan (horse of iron) has spread like wild-fire through the city. In the bazaar I obtain Russian silver money, which is the chief currency of the country as far east as Zendjan. Partly to escape from the worrying crowds, and partly to ascertain the way out next morning, as I intend making an early start, I get the soldiers to take me outside the city wall and show me the Tabreez road.

A new caravanserai is in process of construction just outside the Tabreez gate, and I become an interested spectator of the Persian mode of building the walls of a house ; these of the new caravanserai are nearly four feet thick.   Parallel walls of mud bricks are built up, leaving an interspace of two feet or thereabouts ; this is filled with stiff, well-worked mud, which is dumped in by bucketsful and continually tramped by barefooted laborers ; harder bricks are used for the doorways and windows.   The bricklayer uses mud for mortar and his hands for a trowel ; he works without either level or plumb-line, and keeps up a doleful, melan-

Taking a Drink.

choly chant from morning to night.   The mortar is handed to him by an assistant by handsful ; every workman is smeared and spattered with mud from head to foot, as though glorying in covering themselves with the trade-mark of their calling.

Strolling away from the busy builders we encounter a man— the " wather bhoy av the ghang "—bringing a three-gallon pitcher of water from a spring half a mile away.   Being thirsty, the soldiers shout for him to bring the pitcher.   Scarcely conceiving it possible that these humble mud-daubers would be so wretchedly sanctimonious, I drink from the jar, much to the disgust of the poor water-carrier, who forthwith empties the remainder away and returns with hurried trot to the spring for a fresh supply ; he would doubtless have smashed the vessel had it been smaller and of lesser value.   Naturally I feel a trifle conscience-stricken at having caused him so much trouble, for he is rather an elderly man, but the soldiers display no sympathy for him whatever, apparently regarding an humble water-carrier as a person of small consequence anyhow, and they laugh heartily at seeing him trotting briskly back half a mile for another load.   Had he taken the first water after a Ferenghi had drank from it and allowed his fellow-workmen to unwittingly partake of the same, it would probably have fared badly with the old fellow had they found it out afterward.

Returning cityward we meet our friend, the *moonshi bashi*, looking me up ; he is accompanied by a dozen better-class Persians, scattering friends and acquaintances of his, whom he has collected during the day chiefly to show them my map of Persia ; the mechanical beauty of the bicycle and the apparent victory over the laws of equilibrium in riding it being, in the opinion of the scholarly *moonshi bashi*, quite overshadowed by a map which shows Teheran and Khoi, and doesn't show Stamboul, and which shows the whole broad expanse of Persia, and only small portions of other countries.   This latter fact seems to have made a very deep impression upon the *moonshi bashi's* mind ; it appears to have filled him with the unalterable conviction that all other countries are insignificant compared with Persia ; in his own mind this patriotic person has always believed this to be the case, but he is overjoyed at finding his belief verified—as he fondly imagines—by the map of a Ferenghi.   Returning to the caravanserai, we find the courtyard crowded with people, attracted by the fame of the bicycle.   The *moonshi bashi* straightway ascends to the *bala-khana*, tenderly unfolds my map, and displays it for the inspection of the gaping multitude below ; while five hundred pairs of eyes gaze wonderingly upon it, without having the slightest conception of what they are looking at, he proudly traces with his finger the outlines of Persia.   It is one of the most amusing scenes imaginable ; the *moonshi bashi* and myself, surrounded by his little company of friends, occupying the *bala-khana*, proudly displaying to a mixed crowd of fully five hundred people a shilling map as a thing to be wondered at and admired.

After the departure of the *moonshi bashi* and his friends, by invitation I pay a visit of curiosity to a company of dervishes (they themselves pronounce it "darwish") occupying one of the caravanserai rooms.   There are eight of them lolling about in one small room ; their appearance is disgusting and yet interesting ; they are all but naked in deference to the hot weather and to obtain a little relief from the lively tenants of their clothing.   Prominent among their effects are panther or leopard skins which they use as cloaks, small steel battle-axes, and huge spiked clubs.   Their whole appearance is most striking and extraordinary ; their long black hair is dangling about their naked shoulders ; they have the wild, haggard countenances of men whose lives are being spent in debauch-

The Patriotic Moonshi-Bashi.

ery and excesses ; nevertheless, most of them have a decidedly intellectual expression.   The Persian dervishes are a strange and interesting people ; they spend their whole lives in wandering from one end of the country to another, subsisting entirely by mendicancy ; yet their cry, instead of a beggar's supplication for charity, is "huk, huk" (my right, my right) ; they affect the most wildly picturesque and eccentric costumes, often wearing nothing whatever but white cotton drawers and a leopard or panther skin thrown carelessly about their shoulders, besides which they carry a huge spiked club or steel battle-axe and an alms-receiver ; this latter is usually made of an oval gourd, polished and suspended on small brass chains.   Sometimes they wear an embroidered conical cap

30

decorated with verses from the Koran, but often they wear no head-gear save the covering provided by nature. The better-class Persians have little respect for these wandering fakirs; but their wild, eccentric appearance makes a deep impression upon the simple-hearted villagers, and the dervishes, whose wits are sharpened by constant knocking about, live mostly by imposing on their good nature and credulity. A couple of these worthies, arriving at a small village, affect their wildest and most grotesque appearance and proceed to walk with stately, majestic tread through the streets, gracefully brandishing their clubs or battle-axes, gazing fixedly at vacancy and reciting aloud from the Koran with a peculiar and impressive intonation; they then walk about the village holding out their alms-receiver and shouting " *huk yah huk! huk yah huk!* " Half afraid of incurring their displeasure, few of the villagers refuse to contribute a copper or portable cooked provisions.

Most dervishes are addicted to the intemperate use of opium, *bhang* (a preparation of Indian hemp), arrack, and other baleful intoxicants, generally indulging to excess whenever they have collected sufficient money ; they are likewise credited with all manner of debauchery ; it is this that accounts for their pale, haggard appearance. The following quotation from "In the Land of the Lion and Sun," and which is translated from the Persian, is eloquently descriptive of the general appearance of the dervish :

> The dervish had the dullard air,
> The maddened look, the vacant stare,
> That *bhang* and contemplation give.
> He moved, but did not seem to live ;
> His gaze was savage, and yet sad ;
> What we should call stark, staring mad.
> All down his back, his tangled hair
> Flowed wild, unkempt; his head was bare ;
> A leopard's skin was o'er him flung ;
> Around his neck huge beads were hung,
> And in his hand—ah! there's the rub—
> He carried a portentous club.

After visiting the dervishes I spend an hour in an adjacent *tchai-khan* drinking tea with my escort and treating them to sundry well-deserved kalians. Among the rabble collected about the doorway is a half-witted youngster of about ten or twelve summers with a suit of clothes consisting of a waist-string and a piece of rag about the size of an ordinary pen-wiper. He is the unfortunate possessor of a stomach disproportionately large and which intrudes itself upon other people's notice like a prize pumpkin at an agricultural fair.

This youth's chief occupation appears to be feeding melon-rinds to a pet sheep belonging to the *tchai-khan* and playing a resonant tattoo on his abnormally obtrusive paunch with the palms of his hands. This produces a hollow, echoing sound like striking an inflated bladder with a stuffed club; and considering that the youth also introduces a novel and peculiar squint into the performance, it is a remarkably edifying spectacle. Supper-time coming round, the soldiers show the way to an eating place, where

A Yankee Artist's Idea of Dervishes.

we sup off delicious bazaar-kabobs, one of the most tasteful preparations of mutton one could well imagine. The mutton is minced to the consistency of paste and properly seasoned; it is then spread over flat iron skewers and grilled over a glowing char-

coal fire; when nicely browned they are laid on a broad pliable
sheet of bread in lieu of a plate, and the skewers withdrawn, leaving before the customer a dozen long flat fingers of nicely browned
kabobs reposing side by side on the cake of wheaten bread—a
most appetizing and digestible dish.

Returning to the caravanserai, I dismiss my faithful soldiers
with a suitable present, for which they loudly implore the blessings
of Allah on my head, and for the third or fourth time impress upon
the caravanserai-*jee* the necessity of making my comfort for the
night his special consideration. They fill that humble individual's
mind with grandiloquent ideas of my personal importance by
dwelling impressively on the circumstance of my having eaten with
the Governor, a fact they likewise have lost no opportunity of
heralding throughout the bazaar during the afternoon. The cara-
vanserai-*jee* spreads quilts and a pillow for me on the open *bala-
khana*, and I at once prepare for sleep. A gentle-eyed and youthful seyud wearing an enormous white turban and a flowing gown
glides up to my couch and begins plying me with questions. The
soldiers noticing this as they are about leaving the court-yard
favor him with a torrent of imprecations for venturing to disturb
my repose; a score of others yell fiercely at him in emulation of
the soldiers, causing the dreamy-eyed youth to hastily scuttle away
again. Nothing is now to be heard all around but the evening
prayers of the caravanserai guests; listening to the multitudinous
cries of Allah-il-Allah around me, I fall asleep. About midnight I
happen to wake again; everything is quiet, the stars are shining
brightly down into the court-yard, and a small grease lamp is
flickering on the floor near my head, placed there by the caravan-
serai-*jee* after I had fallen asleep. The past day has been one full
of interesting experiences; from the time of leaving the garden of
Mohammed Ali Khan this morning in company with the *moonshi
bashi*, until lulled to sleep three hours ago by the deep-voiced
prayers of fanatical Mohammedans the day has proved a series of
surprises, and I seem more than ever before to have been the sport
and plaything of fortune; however, if the fickle goddess never
used anybody worse than she has used me to-day there would be
little cause for complaining.

As though to belie their general reputation of sanctimoniousness, a tall, stately seyud voluntarily poses as my guide and protector *en route* through the awakening bazaar toward the Tabreez

gate next morning, cuffing obtrusive youngsters right and left, and chiding grown-up people whenever their inordinate curiosity appeals to him as being aggressive and impolite ; one can only account for this strange condescension on the part of this holy man by attributing it to the marvellous civilizing and levelling influence of the bicycle. Arriving outside the gate, the crowd of followers are well repaid for their trouble by watching my progress for a couple of miles down a broad straight roadway admirably kept and shaded with thrifty chenars or plane-trees. Wheeling down this pleasant avenue I encounter mule-trains, the animals festooned with strings of merrily jingling bells, and camels gayly caparisoned, with huge, nodding tassels on their heads and pack-saddles, and deep-toned bells of sheet iron swinging at their throats and sides ; likewise the omnipresent donkey heavily laden with all manner of village produce for the Khoi market.

My road after leaving the avenue winds around the end of projecting hills, and for a dozen miles traverses a gravelly plain that ascends with a scarcely perceptible gradient to the summit of a ridge ; it then descends by a precipitous trail into the valley of Lake Ooroomiah. Following along the northern shore of the lake I find fairly level roads, but nothing approaching continuous wheeling, owing to wash-outs and small streams leading from a range of mountains near by to the left, between which and the briny waters of the lake my route leads. Lake Ooroomiah is somewhere near the size of Salt Lake, Utah, and its waters are so heavily impregnated with saline matter that one can lie down on the surface and indulge in a quiet, comfortable snooze ; at least, this is what I am told by a missionary at Tabreez who says he has tried it himself ; and even allowing for the fact that missionaries are but human after all and this gentleman hails originally from somewhere out West, there is no reason for supposing the statement at all exaggerated. Had I heard of this beforehand I should certainly have gone far enough out of my course to try the experiment of being literally rocked on the cradle of the deep.

Near midday I make a short circuit to the north, to investigate the edible possibilities of a village nestling in a *cul-de-sac* of the mountain foot-hills. The resident Khan turns out to be a regular jovial blade, sadly partial to the flowing bowl. When I arrive he is perseveringly working himself up to the proper pitch of booziness for enjoying his noontide repast by means of copious potations of arrack ; he introduces himself as Hassan Khan, offers me arrack, and cordially invites me to dine with him.   After dinner, when examining my revolver, map, etc., the Khan greatly admires a photograph of myself as a peculiar proof of Ferenghi skill in producing a person's physiognomy, and blandly asks me to "make him one of himself," doubtless thinking that a person capable of riding on a wheel is likewise possessed of miraculous all 'round abilities.

Hassan Khan takes a Lesson.

The Khan consumes not less than a pint of raw arrack during the dinner hour, and, not unnaturally, finds himself at the end a trifle funny and venturesome. When preparing to take my departure he proposes that I give him a ride on the bicycle ; nothing loath to humor him a little in return for his hospitality, I assist him to mount, and wheel him around for a few minutes, to the unconcealed delight of the whole population, who gather about to see the astonishing spectacle of their Khan riding on the Ferenghi's wonderful *asp-i-awhan*.   The Khan being short and pudgy is unable to reach the pedals, and the confidence-inspiring fumes of arrack lead him to announce to the assembled villagers that if his legs were only a little longer he could certainly go it alone, a statement that evidently fills the simple-minded ryots with admiration for the Khan's alleged newly-discovered abilities.

The road continues level but somewhat loose and sandy ; the scenery around becomes strikingly beautiful, calling up thoughts of "Arabian Nights" entertainments, and the genii and troubadours of Persian song.   The bright, blue waters of Lake Ooroomiah stretch away southward to where the dim outlines of mountains, a hundred miles away, mark the southern shore ; rocky islets at a lesser distance, and consequently more pronounced in character and contour, rear their jagged and picturesque forms sheer from the azure surface of the liquid mirror, the face of which is unruffled by a single ripple and unspecked by a single animate or inanimate object ; the beach is thickly incrusted with salt, white and glistening in the sunshine ; the shore land is mingled sand and clay of a deep-red color, thus presenting the striking and beautiful phenomena of a lake shore painted red, white, and blue by the inimitable hand of nature.   A range of rugged gray mountains run parallel with the shore but a few miles away ; crystal streams come bubbling lake-ward over pebble-bedded channels from sources high up the mountain slopes ; villages, hidden amid groves of spreading jujubes and graceful chenars, nestle here and there in the rocky gateways of ravines ; orchards and vineyards are scattered about the plain. They are imprisoned within gloomy mud walls, but, like living creatures struggling for their liberty, the fruit-laden branches extend beyond their prison-walls, and the graceful tendrils of the vines find their way through the sun-cracks and fissures of decay, and trail over the top as though trying to cover with nature's charitable veil the unsightly works of man ; and all is arched over with the cloudless Persian sky.

Roaming the roads of this picturesque region in search of victims is a most persistent and pugnacious species of fly ; rollicking as the blue-bottle, and the veritable double of the green-head horse-fly of the Western prairies, he combines the dash and impetuosity of the one with the ferocity and persistency of the other ; but he is happily possessed of one redeeming feature not possessed by either of the above-mentioned and well-known insects of the Western world.   When either of these settles himself affectionately on the end of a person's nose, and the person, smarting under the indignity, hits himself viciously on that helpless and unoffending portion of his person, as a general thing it doesn't hurt the fly, simply because the fly doesn't wait long enough to be hurt ; but the Lake Ooroomiah fly is a comparatively guileless insect, and quietly remains where he alights until it suits one's convenience to forcibly remove him ; for this redeeming quality I bespeak for him the warmest encomiums of fly-harassed humans everywhere.

Dusk is settling down over the broad expanse of lake, plain, and mountain when I encounter a number of villagers taking donkey-loads of fruit and almonds from an orchard to their village. They cordially invite me to accompany them and accept their hospitality for the night. They are travelling toward a large area of walled orchards but a short distance to the north, and I naturally expect to find their village located among them ; so, not knowing how far ahead the next village may be, I gladly accept their kindly invitation, and follow along behind. It gets dusky, then duskier, then dark ; the stars come peeping out thicker and thicker, and still I am trundling with these people slowly along up the dry and stone-strewn channel of spring-time freshets, expecting every minute to reach their village, only to be as often disappointed, for over an hour, during which we travel out of my proper course perhaps four miles. Finally, after crossing several little streams, or rather, one stream several times, we arrive at our destination, and I am installed, as the guest of a leading villager, beneath a sort of open porch attached to the house. Here, as usual, I quickly become the centre of attraction for a wondering and admiring audience of half-naked villagers. The villager whose guest I become brings forth bread and cheese, some bring me grapes, others newly gathered almonds, and then they squat around in the dim religious light of primitive grease-lamps and watch me feed, with the same wondering interest and the same unconcealed delight with which youthful Londoners at the Zoological Gardens regard a pet monkey devouring their offerings of nuts and ginger-snaps.

I scarcely know what to make of these particular villagers ; they seem strangely childlike and unsophisticated, and moreover, perfectly delighted at my unexpected presence in their midst. It is doubtful whether their unimportant little village among the foot-hills was ever before visited by a Ferenghi ; consequently I am to them a *rara avis* to be petted and admired. I am inclined to think them a village of Yezeeds or devil-worshippers ; the Yezeeds believe that Allah, being by nature kind and merciful, would not injure anybody under any circumstances, consequently there is nothing to be gained by worshipping him. Sheitan (Satan), on the contrary, has both the power and the inclination to do people harm, therefore they think it politic to cultivate his good-will and to pursue a policy of conciliation toward him by worshipping him and revering his name. Thus they treat the name of Satan with even greater reverence than Christians and Mohammedans treat the name of God. Independent of their hospitable treatment of myself, these villagers seem but little advanced in their personal habits above mere animals; the women are half-naked, and seem possessed of little more sense of shame than our original ancestors before the fall. There is great talk of *kardash* among them in reference to myself. They are advocating hospitality of a nature altogether too profound for the consideration of a modest and discriminating Ferenghi—hospitable intentions that I deem it advisable to dissipate at once by affecting deep, dense ignorance of what they are discussing.

. In the morning they search the village over to find the wherewithal to prepare me some tea before my departure. Eight miles from the village I discover that four miles forward yesterday evening, instead of backward, would have brought me to a village containing a caravanserai. I naturally feel a trifle chagrined at the mistake of having journeyed eight unnecessary miles, but am, perhaps, amply repaid by learning something of the utter simplicity of the villagers before their character becomes influenced by intercourse with more enlightened people.

My course now leads over a stony plain. The wheeling is reasonably good, and I gradually draw away from the shore of Lake Ooroomiah. Melon-gardens and vineyards are frequently found here and there across the plain; the only entrance to the garden is a hole about three feet by four in the high mud wall, and this is closed by a wooden door; an arm-hole is generally found in the wall to enable the owner to reach the fastening from the outside. Investigating one of these fastenings at a certain vineyard I discover a lock so primitive that it must have been invented by prehistoric man. A flat, wooden bar or bolt is drawn into a mortise-like receptacle of the wall, open at the top; the man then daubs a handful of wet clay over it; in a few minutes the clay hardens and the door is fast. This is not a burglar-proof lock, certainly, and is only depended upon for a fastening during the temporary absence of the owner in the day-time. During the summer the owner and family not infrequently live in the garden altogether.

During the forenoon the bicycle is the innocent cause of two people being thrown from the backs of their respective steeds. One is a man carelessly sitting sidewise on his donkey; the meek-eyed jackass suddenly makes a pivot of his hind feet and wheels round, and the rider's legs as suddenly shoot upward. He frantically grips his fiery, untamed steed around the neck as he finds himself over-balanced, and comes up with a broad grin and an irrepressible chuckle of merriment over the unwonted spirit displayed by his meek and humble charger, that probably had never scared at anything before in all its life. The other case is unfortunately a lady whose horse literally springs from beneath her, treating her to a clean tumble. The poor lady sings out "Allah!" rather snappishly at finding herself on the ground, so snappishly that it leaves little room for doubt of its being an imprecation; but her rude, unsympathetic attendants laugh right merrily at seeing her floundering about in the sand; fortunately, she is uninjured. Although Turkish and Persian ladies ride *à la Amazon*, a position that is popularly supposed to be several times more secure than side-saddles, it is a noticeable fact that they seem perfectly helpless, and come to grief the moment their steed shies at anything or commences capering about with anything like violence.

On a portion of road that is unridable from sand I am captured by a rowdyish company of donkey-drivers, returning with empty fruit-baskets from Tabreez. They will not be convinced that the road is unsuitable, and absolutely refuse to let me go without seeing the bicycle ridden. After detaining me until patience on my part ceases to be a virtue, and apparently as determined for their purpose as ever, I am finally compelled to produce the convincing argument with five chambers and rifled barrel. These crowds of donkey-men seem inclined to be rather lawless, and scarcely a day passes lately but what this same eloquent argument has to be advanced in the interest of individual liberty. Fortunately the mere sight of a revolver in the hands of a Ferenghi has the magical effect of transforming the roughest and most overbearing gang of ryots into peaceful, retiring citizens. The plain I am now traversing is a broad, gray-looking area surrounded by mountains, and stretching away eastward from Lake Ooroomiah for seventy-five miles. It presents the same peculiar aspect of Persian scenery nearly everywhere—a general verdureless and unproductive country, with the barren surface here and there relieved by small oases of cultivated fields and orchards. The villages being built solely of mud, and consequently of the same color as the general surface, are undistinguishable from a distance, unless rendered conspicuous by trees.

Laboring under a slightly mistaken impression concerning the distance to Tabreez, I push ahead in the expectation of reaching there to-night; the plain becomes more generally cultivated; the caravan routes from different directions come to a focus on broad trails leading into the largest city in Persia, and which is the great centre of distribution for European goods arriving by caravan to Trebizond. Coming to a large, scattering village, some time in the afternoon, I trundle leisurely through the lanes inclosed between lofty and unsightly mud walls thinking I have reached the sub-urbs of Tabreez; finding my mistake upon emerging on the open plain again, I am yet again deceived by another spreading village, and about six o'clock find myself wheeling eastward across an un-cultivated stretch of uncertain dimensions. The broad caravan trail is worn by the traffic of centuries considerably below the level of the general surface, and consists of a number of narrow, parallel trails, along which swarms of donkeys laden with produce from tributary villages daily plod, besides the mule and camel car-avans from a greater distance. These narrow beaten paths afford excellent wheeling, and I bowl along quite briskly. As one ap-proaches Tabreez, the country is found traversed by an intricate network of irrigating ditches, some of them works of considerable magnitude; the embankments on either side of the road are fre-quently high enough to obscure a horseman. These works are al-most as old as the hills themselves, for the cultivation of the Tab-reez plain has remained practically an unchanged system for three thousand years, as though, like the ancient laws of the Medes and Persians, it also were made unchangeable.

About dusk I fall in with another riotous crowd of homeward-bound fruit-carriers, who, not satisfied at seeing me ride past, want to stop me; one of them rushes up behind, grabs my package at-tached to the rear baggage-carrier, and nearly causes an overthrow; frightening him off, I spurt ahead, barely escaping two or three donkey cudgels hurled at me in pure wantonness, born of the courage inspired by a majority of twenty to one. There is no remedy for these unpleasant occurrences except travelling under escort, and the avoiding serious trouble or accident becomes a matter for every-day congratulation. At eighteen miles from the last village it becomes too dark to remain in the saddle without danger of headers, and a short trundle brings me, not to Tabreez even now, but to another village eight miles nearer. Here there is

a large caravanserai. Near the entrance is a hole-in-the-wall sort of
a shop wherein I espy a man presiding over a tempting assortment
of cantaloupes, grapes, and pears. The whirligig of fortune has
favored me to-day with tea, blotting-paper *ekmek*, and grapes for

The Maivah-jee Surprised.

breakfast; later on two small watermelons, and at 2 P.M. blotting-
paper *ekmek* and an infinitesimal quantity of *yaort* (now called *mast*).
It is unnecessary to add that I arrive in this village with an appetite
that will countenance no unnecessary delay. Two splendid ripe

cantaloupes, several fine bunches of grapes, and some pears are devoured immediately, with a reckless disregard of consequences, justifiable only on the grounds of semi-starvation and a temporary barbarism born of surrounding circumstances. After this savage attack on the *maivah-jee's* stock, I learn that the village contains a small *tchai-khan;* repairing thither I stretch myself on the divan for an hour's repose, and afterward partake of tea, bread, and peaches. At bed-time the *khan-jee* makes me up a couch on the

The Khan-jee Escapes through the Window.

divan, locks the door inside, blows out the light, and then, afraid to occupy the same building with such a dangerous-looking individual as myself, climbs to the roof through a hole in the wall.

Eager villagers carry both myself and wheel across a bridgeless stream upon resuming my journey to Tabreez next morning ; the road is level and ridable, though a trifle deep with dust and sand, and in an hour I am threading the suburban lanes of the city. Along these eight miles I certainly pass not less than five hundred pack-donkeys *en route* to the Tabreez market with everything, from baskets of the choicest fruit in the world to huge bundles of prickly camel-thorn and sacks of *tezek* for fuel. No animals in all the world, I should think, stand in more urgent need of the kindly offices of the Society for the Prevention of Cruelty to Animals than the thousands of miserable donkeys engaged in supplying Tabreez with fuel ; their brutal drivers seem utterly callous and indifferent to the pitiful sufferings of these patient toilers. Numbers of instances are observed this morning where the rough, ill-fitting breech-straps and ropes have literally see-sawed their way through the skin and deep into the flesh, and are still rasping deeper and deeper every day, no attempt whatever being made to remedy this evil ; on the contrary, their pitiless drivers urge them on by prodding the raw sores with sharpened sticks, and by belaboring them unceasingly with an instrument of torture in the shape of whips with six inches of ordinary trace-chain for a lash.

As if the noble army of Persian donkey drivers were not satisfied with the refinement of physical cruelty to which they have attained, they add insult to injury by talking constantly to their donkeys while driving them along, and accusing them of all the crimes in the calendar and of every kind of disreputable action. Fancy the bitter sense of humiliation that must overcome the proud, haughty spirit of a mouse-colored jackass at being prodded in an open wound with a sharp stick and hearing himself at the same time thus insultingly addressed : "Oh, thou son of a burnt father and murderer of thine own mother, would that I myself had died rather than my father should have lived to see me drive such a brute as thou art ! " yet this sort of talk is habitually indulged in by the barbarous drivers. While young, the donkeys' nostrils are slit open clear up to the bridge-bone ; this is popularly supposed among the Persians to be an improvement upon nature in that it gives them greater freedom of respiration. Instead of the well known clucking sound used among ourselves as a persuasive, the Persian makes a sound not unlike the bleating of a sheep ; a stranger, being within hearing and out of sight of a gang of donkey drivers in a hurry to reach their destination, would be more likely to imagine himself in the vicinity of a flock of sheep than anything else.

As is usually the case, a volunteer guide bobs serenely up immediately I enter the city, and I follow confidently along, thinking he is piloting me to the English consulate, as I have requested ;

"Take the Horse and leave the Bicycle."

instead of this he steers me into the custom-house and turns me over to the officials. These worthy gentlemen, after asking me to ride around the custom-house yard, pretend to become altogether mystified about what they ought to do with the bicycle, and in the absence of any precedent to govern themselves by, finally conclude among themselves that the proper thing would be to confiscate it. Obtaining a guide to show me to the residence of Mr. Abbott, the English consul-general, that energetic representative of Her Majesty's government smiles audibly at the thoughts of their mystification, and then writes them a letter couched in terms of humorous reproachfulness, asking them what in the name of Allah and the Prophet they mean by confiscating a traveller's horse, his carriage, his camel, his everything on legs and wheels consolidated into the beautiful vehicle with which he is journeying to Teheran to see the Shah, and all around the world to see everybody and everything?—ending by telling them that he never in all his consular experiences heard of a proceeding so utterly atrocious. He sends the letter by the consulate dragoman, who accompanies me back to the custom-house. The officers at once see and acknowledge their mistake; but meanwhile they have been examining the bicycle, and some of them appear to have fallen violently in love with it; they yield it up, but it is with apparent reluctance, and one of the leading officials takes me into the stable, and showing me several splendid horses begs me to take my choice from among them and leave the bicycle behind.

Mr. and Mrs. Abbott cordially invite me to become their guest while staying at Tabreez. To-day is Thursday, and although my original purpose was only to remain here a couple of days, the innovation from roughing it on the road, to roast duck for dinner, and breakfast in one's own room of a morning, coupled with warnings against travelling on the Sabbath and invitations to dinner from the American missionaries, proves a sufficient inducement for me to conclude to stay till Monday, satisfied at the prospect of reaching Teheran in good season. It is now something less than four hundred miles to Teheran, with the assurance of better roads than I have yet had in Persia, for the greater portion of the distance; besides this, the route is now a regular post route with *chapar-khanas* (posthouses) at distances of four to five farsakhs apart. On Friday night Tabreez experienced two slight shocks of an earthquake, and in the morning Mr. Abbott points out several fissures in the masonry of

the consulate, caused by previous visitations of the same undesirable nature ; the earthquakes here seem to resemble the earthquakes
of California in that they come reasonably mild and often. The
place likewise awakens memories of the Golden State in another and
more appreciative particular : nowhere, save perhaps in California,
does one find such delicious grapes, peaches, and pears as at ancient
Taurus, a specialty for which it has been justly celebrated from
time immemorial. On Saturday I take dinner with Mr. Oldfather,
one of the missionaries, and in the evening we all pay a visit to Mr.
Whipple and family, the consulate link-boy lighting the way before us with a huge cylindrical lantern of transparent oiled muslin
called a *farnooze*.

These lanterns are always carried after night before people of
wealth or social consequence, varying in size according to the person's idea of their own social importance. The size of the *farnooze*
is supposed to be an index of the social position of the person or
family, so that one can judge something of what sort of people are
coming down the street, even on the darkest night, whenever the
attendant link-boy heaves in sight with the *farnooze*. Some of these
social indicators are the size of a Portland cement barrel, even in
Persia ; it is rather a smile-provoking thought to think what tremendous *farnoozes* would be seen lighting up the streets on gloomy
evenings, were this same custom prevalent among ourselves ; few
of us but what could call to memory people whose *farnoozes* would
be little smaller than brewery mash-tubs, and which would have to
be carried between six-foot link-boys on a pole.

Ameer-i-Nazan, the *Valiat* or heir apparent to the throne, and
at present nominal governor of Tabreez, has seen a tricycle in
Teheran, one having been imported some time ago by an English
gentleman in the Shah's service ; but the fame of the bicycle excites his curiosity and he sends an officer around to the consulate
to examine and report upon the difference between bicycle and
tricycle, and also to discover and explain the *modus operandi* of
maintaining one's balance on two wheels. The officer returns with
the report that my machine won't even stand up, without somebody holding it, and that nobody but a Ferenghi who is in league
with Sheitan, could possibly hope to ride it. Perhaps it is this
alarming report, and the fear of exciting the prejudices of the
mollahs and fanatics about him, by having anything to do with a
person reported on trustworthy authority to be in league with His

31

Satanic Majesty, that prevents the Prince from requesting me to
ride before him in Tabreez; but I have the pleasure of meeting
him at Hadji Agha on the evening of the first day out. Mr. Whipple kindly makes out an itinerary of the villages and *chapar-khanas*
I shall pass on the journey to Teheran; the superintendent of the
Tabreez station of the Indo-European Telegraph Company voluntarily telegraphs to the agents at Miana and Zendjan when to expect me, and also to Teheran; Mrs. Abbott fills my coat pockets
with roast chicken, and thus equipped and prepared, at nine o'clock
on Monday morning I am ready for the home-stretch of the season,
before going into winter quarters.

The Turkish consul-general, a corpulent gentleman whose avoirdupois I mentally jot down at four hundred pounds, comes around
with several others to see me take a farewell spin on the bricked
pavements of the consulate garden. Like all persons of four hundred pounds weight, the Effendi is a good-natured, jocose individual, and causes no end of merriment by pretending to be anxious
to take a spin on the bicycle himself, whereas it requires no inconsiderable exertion on his part to waddle from his own residence
hard by into the consulate. Three soldiers are detailed from the
consulate staff to escort me through the city; *en route* through the
streets the pressure of the rabble forces one unlucky individual
into one of the dangerous narrow holes that abound in the streets,
up to his neck; the crowd yell with delight at seeing him tumble
in, and nobody stops to render him any assistance or to ascertain
whether he is seriously hurt. Soon a poor old ryot on a donkey,
happens amid the confusion to cross immediately in front of the
bicycle; whack! whack! whack! come the ready staves of the zealous and vigilant soldiers across the shoulders of the offender; the
crowd howls with renewed delight at this, and several hilarious
hobble-de-hoys endeavor to shove one of their companions in the
place vacated by the belabored ryot, in the hope that he likewise
will come in for the visitation of the soldiers' o'er-willing staves.

The broad suburban road, where the people have been fondly
expecting to see the bicycle light out in earnest for Teheran at a
marvellous rate of speed, is found to be nothing less than a bed of
loose sand and stones, churned up by the narrow hoofs of multitudinous donkeys. Quite a number of better class Persians accompany me some distance further on horseback; when taking their
departure, a gentleman on a splendid Arab charger, shakes hands

and says: "Good-by, my dear," which apparently is all the English he knows. He has evidently kept his eyes and ears open when happening about the English consulate, and the happy thought striking him at the moment, he repeats, parrot-like, this term of endearment, all unsuspicious of the ridiculousness of its application in the present case.

For several miles the road winds tortuously over a range of low, stony hills, the surface being generally loose and unridable. The water-supply of Tabreez is conducted from these hills by an ancient system of kanaats or underground water-ditches ; occasionally one comes to a sloping cavern leading down to the water ; on descending to the depth of from twenty to forty feet, a small, rapidly-coursing stream of delicious cold water is found, well rewarding the thirsty traveller for his trouble ; sometimes these cavernous openings are simply sloping, bricked archways, provided with steps.   The course of these subterranean water-ways can always be traced their entire length by uniform mounds of earth, piled up at short intervals on the surface ; each mound represents the excavations from a perpendicular shaft, at the bottom of which the crystal water can be seen coursing along toward the city ; they are merely man-holes for the purpose of readily cleaning out the channel of the kanaat.   The water is conducted underground, chiefly to avoid the waste by evaporation and absorption in surface ditches.   These kanaats are very extensive affairs in many places ; the long rows of surface mounds are visible, stretching for mile after mile across the plain as far as eye can penetrate, or until losing themselves among the foot-hills of some distant mountain chain ; they were excavated in the palmy days of the Persian Empire to bring pure mountain streams to the city fountains and to irrigate the thirsty plain ; it is in the interest of self-preservation that the Persians now keep them from falling into decay.

At noon, while seated on a grassy knoll discussing the before-mentioned contents of my pockets, I am favored with a free exhibition of what a physical misunderstanding is like among the Persian ryots.   Two companies of *katir-jees* happen to get into an altercation about something, and from words it gradually develops into blows ; not blows of the fist, for they know nothing of fisticuffs, but they belabor each other vigorously with their long, thick donkey persuaders, sticks that are anything but small and willowy ; it is an amusing spectacle, and seated on the commanding knoll

nibbling " drum-sticks" and wish-bones, I can almost fancy myself
a Roman of old, eating peanuts and watching a gladiatorial contest
in the amphitheatre. The similitude, however, is not at all striking, for thick as are their quarter-staffs the Persian ryots don't
punish each other very severely. Whenever one of them works
himself up to a fighting-pitch, he commences belaboring one of the
others on the back, apparently always striking so that the blow
produces a maximum of noise with a minimum of punishment; the

Persian Katir-jees Differ.

person thus attacked never ventures to strike back, but retreats
under the blows until his assailant's rage becomes spent and he
desists. Meanwhile the war of words goes merrily forward; perchance in a few minutes the person recently attacked suddenly becomes possessed of a certain amount of rage-inspired courage, and
he in turn commences a vigorous assault upon somebody, probably
his late assailant; this worthy, having become a little cooler, has
mysteriously lost his late pugnacity, and now likewise retreats
without once attempting to raise his own stick in self-defence. The

lower and commercial class Persians are pretty quarrelsome among themselves, but they quarrel chiefly with their tongues ; when they fight without sticks it is an ear-pulling, clothes-tugging, wrestling sort of a scuffle, which continues without greater injury than a torn garment until they become exhausted if pretty evenly matched, or until separated by bystanders ; they never, never hurt each other unless they are intoxicated, when they sometimes use their short swords ; there is no intoxication, except in private drinking-parties.

CHAPTER XX.

THE wheeling improves in the afternoon, and alongside my road runs a bit of civilization in the shape of the splendid iron poles of the Indo-European Telegraph Company. Half a dozen times this afternoon I become the imaginary enemy of a couple of cavalry-men travelling in the same direction as myself ; they swoop down upon me from the rear at a charging gallop, valiantly whooping and brandishing their Martini-Henrys ; when they arrive within a few yards of my rear wheel they swerve off on either side and rein their fiery chargers up, allowing me to forge ahead ; they amuse themselves by repeating this interesting performance over and over again. Being usually a good rider, the dash and courage of the Persian cavalryman is something extraordinary in time of peace ; no more brilliant and intrepid cavalry charge on a small scale could be well imagined than I have witnessed several times this afternoon. But upon the outbreak of serious hostilities the average warrior in the Shah's service suddenly becomes filled with a wild, pathetic yearning after the peaceful and honorable calling of a *katir-jee*, an uncontrollable desire to become a humble, con-tented tiller of the soil, or handy-man about a *tchai-khan*, anything, in fact, of a strictly peaceful character. Were I a hostile trooper with a red jacket, and a general warlike appearance, and the bi-cycle a machine gun, though our whooping, charging cavalrymen were twenty instead of two, they would only charge once, and that would be with their horses' crimson-dyed tails streaming in the breeze toward me. The Shah's soldiers are gentle, unwarlike creatures at heart ; there are probably no soldiers in the whole world that would acquit themselves less creditably in a pitched battle ; they are, nevertheless, not without certain soldierly quali-ties, well adapted to their country ; the cavalrymen are very good riders, and although the infantry does not present a very encourag-ing appearance on the parade-ground, they would meander across five hundred miles of country on half rations of blotting-paper *ek-mek* without any vigorous remonstrance, and wait uncomplainingly for their pay until the middle of next year.

About five o'clock I arrive at Hadji Agha, a large village forty miles from Tabreez ; here, as soon as it is ascertained that I intend remaining over night, I am actually beset by rival *khan-jees*, who commence jabbering and gesticulating about the merits of their

They Swoop Down on Me from the Rear.

respective establishments, like hotel-runners in the United States ; of course they are several degrees less rude and boisterous, and more considerate of one's personal inclinations than their proto-types in America, but they furnish yet another proof that there is nothing new under the sun. Hadji Agha is a village of seyuds, or descendants of the Prophet, these and the mollahs being the most bigoted class in Persia ; when I drop into the *tchai-khan* for a glass or two of tea, the sanctimonious old joker with henna-tinted beard and finger-nails, presiding over the samovar, rolls up his eyes in holy horror at the thoughts of waiting upon an unhallowed Ferenghi, and it requires considerable pressure from the younger and less fanatical men to overcome his disinclination ; he probably breaks the glass I drank from after my departure.

About dusk the *Valiat* and his courtiers arrive on horseback from Tabreez ; the Prince immediately seeks my quarters at the *khan*, and, after examining the bicycle, wants me to take it out and ride ; it is getting rather dark, however, so I put him off till morning ; he remains and smokes cigarettes with me for half an hour, and then retires to the residence of the local Khan for the night. The Prince seems an amiable, easy-going sort of a person ; while in my company his countenance is wreathed in a pleasant smile continually, and I fancy he habitually wears that same expression. His youthful courtiers seem frivolous young bloods, putting in most of the half-hour in showing me their accomplishments in the way of making floating rings of their cigarette smoke. Later in the evening I stroll around to the *tchai-khan* again ; it is the gossiping-place of the village, and I find our sanctimonious seyuds indulging in uncomplimentary comments regarding the *Valiat's* conduct in hobnobbing with the Ferenghi ; how bigoted these Persians are, and yet how utterly destitute of principle and moral character !

In the morning the Prince sends me an invitation to come and drink tea with them before starting out ; he bears the same perennial smile as yesterday evening. Although he is generally understood to be completely under the influence of the fanatical and bigoted seyuds and mollahs, who are strictly opposed to the Ferenghi and the Ferenghi's ideas of progress and civilization, he seems withal an amiable, well-disposed young man, whom one could scarce help liking personally, and feeling sorry at the troubles in store for him ahead. He has an elder brother, the Zil-es-Sultan, now governor of the Southern Provinces ; but not being the son of a royal princess, the Shah has nominated Ameer-i-Nazan as his successor to the throne. The Zil-es-Sultan, although of a somewhat cruel disposition, has proved himself a far more capable and energetic person than the *Valiat*, and makes no secret of the fact that he intends disputing the succession with his brother, by force of arms if necessary, at the Shah's demise. He has, so at least it is currently reported, had his sword-blade engraved with

The Valiat gives Me a Race.

the grim inscription, "This is for the *Valiat's* head," and has jocularly notified his inoffensive brother of the fact. The Zil-es-Sultan belongs to the party of progress; recks little of the opinions of priests and fanatics, is fond of Englishmen and European improvements, and keeps a kennel of English bull dogs. Should he become Shah of Persia, Baron Reuter's grand scheme of railways and commercial regeneration, which was foiled by the fanaticism of the seyuds and mollahs soon after the Shah's visit to England, may yet come to something, and the railroad rails now rusting in the swamps of the Caspian littoral may, after all, form part of a railway between the seaboard and the capital.

The road for a short distance east of Hadji Agha is splendid wheeling, and the Prince and his courtiers accompany me for some two miles, finding much amusement in racing with me whenever the road permits of spurting. The country now develops into undulating upland, uncultivated and stone-strewn, except where an occasional stream, affording irrigating facilities, has rendered possible the permanent maintenance of a mud village and a circumscribed area of wheat-fields, melon-gardens, and vineyards.

No sooner does one find himself launched upon the comparatively well-travelled post-route than a difference becomes manifest in the character of the people. Commercially speaking, the Persian is considerably more of a Jew than the Jew himself, and along a route frequented by travellers, the person possessing some little knowledge of the thievish ways of the country and of current prices, besides having plenty of small change, finds these advantages a matter for congratulation almost every hour of the day. The proprietor of a wretched little mud hovel, solemnly presiding over a few thin sheets of bread, a jar of rancid, hirsute butter, and a dozen half-ripe melons, affects a glum, sorrowful expression to think that he should happen to be without small change, and consequently obliged to accept the Hamsherri's fifty kopec piece for provisions of one-tenth the value; but the mysterious frequency of this same state of affairs and accompanying sorrowful expression, taken in connection with the actual plenitude of small change in Persia, awakens suspicions even in the mind of the most confiding and uninitiated person. A peculiar system of commercial mendicancy obtains among the proprietors of melon and cucumber gardens alongside the road of this particular part of the country; observing a likely-looking traveller approaching, they come running

to him with a melon or cucumber that they know to be utterly
worthless, and beg the traveller to accept it as a present; delighted,
perhaps with their apparent simple-hearted hospitality, and, moreover, sufficiently thirsty to appreciate the gift of a melon, the unsuspecting wayfarer tenders the crafty proprietor of the garden a
suitable present of money in return and accepts the proffered
gift; upon cutting it open he finds the melon unfit for anything,
and it gradually dawns upon him that he has just grown a trifle
wiser concerning the inbred cunningness and utter dishonesty of
the Persians than he was before.   Ere the day is ended the same
game will probably be attempted a dozen times.

In addition to these artful customers, one occasionally comes
across small colonies of lepers, who, being compelled to isolate
themselves from their fellows, have taken up their abode in rude
hovels or caves by the road-side, and sally forth in all their hide-
ousness to beset the traveller with piteous cries for assistance.
Some of these poor lepers are loathsome in appearance to the last
degree; their scanty coverings of rags and tatters conceals nothing of the ravages of their dread disease; some sit at the entrance
to their hovels, stretching out their hands and piteously appealing
for alms; others drop down exhausted in the road while endeavoring to run and overtake the passer-by; there is nothing deceptive
about these wretched outcasts, their condition is only too glaringly
apparent.

Toward sundown I arrive at Turcomanchai, a large village,
where in 1828, was drawn up the Treaty of Peace between Persia
and Russia, which transferred the remaining Persian territory of
the Caucasus into the capacious maw of the Northern Bear.   It is
currently reported that after depriving the Persians of their
rights to the navigation of the Caspian Sea the Czar coolly gave
his amiable friend the Shah a practical lesson concerning the irony
of fortune by presenting him with a yacht.   Seeking the guidance
of a native to the caravanserai, this quick-witted individual leads
the way through tortuous alleyways to the other end of the village
and pilots me to the camp of a tea caravan, pitched on the outskirts, thinking I had requested to be guided to a caravan; the
caravan men direct me to the *chapar-khana*, where accommodations
of the usual rude nature are provided.   Sending into the village for eggs, sugar, and tea, the *chapar-khana* keeper and stablemen produce a battered samovar, and after frying my supper,

they prepare tea; they are poor, ragged fellows, but they seem light-hearted and contented; the siren song of the steaming samovar seems to awaken in their semi-civilized breasts a sympathetic response, and they fall to singing and making merry over tiny glasses of sweetened tea quite as naturally as sailors in a seaport groggery, or Germans over a keg of lager. Jolly, happy-go-lucky fellows though they outwardly appear, they prove no exception, however, to the general run of their countrymen in the matter of petty dishonesty; although I gave them money enough to purchase twice the quantity of provisions they brought back, besides promising them the customary small present before leaving, in the morning they make a further attempt on my purse under pretence of purchasing more butter to cook the remainder of the eggs. These are trifling matters to discuss, but they serve to show the wide difference between the character of the peasant classes in Persia and Turkey. The *chapar-khana* usually consists of a walled enclosure containing stabling for a large number of horses and quarters for the stablemen and station-keeper. The quickest mode of travelling in Persia is by *chapar*, or, in other words, on horseback, obtaining fresh horses at each *chapar-khana*.

The country east of Turcomanchai consists of rough, uninteresting upland, with nothing to vary the monotony of the journey, until noon, when after wheeling five farsakhs I reach the town of Miana, celebrated throughout the Shah's dominions for a certain poisonous bug which inhabits the mud walls of the houses, and is reputed to bite the inhabitants while they are sleeping. The bite is said to produce violent and prolonged fever, and to be even dangerous to life. It is customary to warn travellers against remaining over night at Miana, and, of course, I have not by any means been forgotten. Like most of these alleged dreadful things, it is found upon close investigation to be a big bogey with just sufficient truthfulness about it to play upon the imaginative minds of the people. The "Miana bug-bear" would, I think, be a more appropriate name than Miana bug. The people here seem inclined to be rather rowdyish in their reception of a Ferenghi without an escort. While trundling through the bazaar toward the telegraph station I become the unhappy target for covertly thrown melon-rinds and other unwelcome missiles, for which there appears no remedy except the friendly shelter of the station. This is just outside the town, and before the gate is reached, stones are exchanged for melon-rinds, but fortunately without any serious damage being done.

Mr. F——, a young German operator, has charge of the control-station here, and welcomes me most cordially to share his comfortable quarters, urging me to remain with him several days. I gladly accept his hospitality till to-morrow morning. Mr. F—— has a brother who has recently become a Mussulman, and married a couple of Persian wives ; he is also residing temporarily at Miana. He soon comes around to the telegraph station, and turns out to be a wild harum-skarum sort of a person, who regards his transformation into a Mussulman and the setting up of a harem of his own as anything but a serious affair. As a reward for embracing the Mohammedan religion and becoming a Persian subject the Shah has given him a sum of money and a position in the Tabreez mint, besides bestowing upon him the sounding title of Mirza Abdul Kärim Khan. It seems that inducements of a like substantial nature are held out to any Ferenghi of known respectability who formally embraces the Shiite branch of the Mohammedan religion, and becomes a Persian subject—a rare chance for chronic ne'er-do-wells among ourselves, one would think.

This novel and festive convert to Islam readily gives me a mental peep behind the scenes of Persian domestic life, and would unhesitatingly have granted me a peep in person had such a thing been possible. Imagine the ordinary costume of an opera-bouffe artist, shorn of all regard for the difference between real indecency and the suggestiveness of indelicacy permissible behind the footlights, and we have the every-day costume of the Persian harem. In the dreamy eventide the lord of the harem usually betakes himself to that characteristic institution of the East and proceeds to drive dull care away by smoking the kalian and watching an exhibition of the terpsichorean talent of his wives or slaves. This does not consist of dancing, such as we are accustomed to understand the art, but of graceful posturing and bodily contortions, spinning round like a coryphée, with hand aloft, and snapping their fingers or clashing tiny brass cymbals ; standing with feet motionless and wriggling the joints, or bending backward until their loose, flowing tresses touch the ground. Persians able to afford the luxury have their womens' apartment walled with mirrors, placed at appropriate angles, so that when enjoying these exhibitions of his wives' abilities he finds himself not merely in the presence of three or six

wives, as the case may be, but surrounded on all sides by scores of
airy-fairy nymphs, and amid the dreamy fumes and soothing hub-
ble-bubbling of his kalian can imagine himself the happy—or one
would naturally think, unhappy—possessor of a hundred.   The effect of this mirror-work arrangement can be better imagined than
described.

"You haven't got one of those mirrored rooms, have you?" I
inquire, beginning to get a trifle inquisitive, and perhaps rather
impertinent.   "You couldn't manage to smuggle a fellow inside,
disguised as a seyud or——"   "*Nicht*," replies Mirza Abdul Kärim
Khan, laughing, "I have not bothered about a mirror chamber yet,
because I only remain here for another month ; but if you happen
to come to Tabreez any time after I get settled down there, look
me up, and I'll—hello! here comes Prince Assabdulla to see your
velocipede!"

Fatteh-Ali Shah, the grandfather of the present monarch, had
some seventy-two sons, besides no lack of daughters.   As the son
of a prince inherits his father's title in Persia, the numerous descendants of Fatteh-Ali Shah are scattered all over the empire, and
royal princes bob serenely up in every town of any consequence in
the country.   They are frequently found occupying some snug,
but not always lucrative, post under the Government.   Prince Assabdulla has learned telegraphy, and has charge of the government control-station here, drawing a salary considerably less than the agent
of the English company's line.   The Persian Government telegraph
line consists of one wire strung on tumble-down wooden poles.   It
is erected alongside the splendid English line of triple wires and
substantial iron poles, and the control-stations are built adjacent
to the English stations, as though the Persians were rather timid
about their own abilities as telegraphists, and preferred to nestle,
as it were, under the protecting shadow of the English line.
Prince Assabdulla has an elder brother who is Governor of Miana,
and who comes around to see the bicycle during the afternoon ;
they both seem pleasant and agreeable fellows.   When the heat of
the day has given place to cooler eventide, and the moon comes
peeping over the lofty Koflan Koo Mountains, near-by to the eastward, we proceed to a large fruit-garden on the outskirts of the
town, and, sitting on the roof of a building, indulge in luscious
purple grapes as large as walnuts, and pears that melt away in the
mouth.   Mirza Abdul Kärim Khan plays a German accordeon, and

Like a Coryphée with Hand Aloft.

Prince Assabdulla sings a Persian love-song ; the leafy branches of poplar groves are whispering in response to a gentle breeze, and playing hide-and-seek across the golden face of the moon, and the mountains have assumed a shadowy, indistinct appearance. It is a scene of transcendental loveliness, characteristic of a Persian moonlight night.

Afterward we repair to Mirza Abdul Kärim Khan's house to smoke the kalian and drink tea. His favorite wife, whom he has taught to respond to the purely Frangistan name of "Rosie," replenishes and lights the kalian—giving it a few preliminary puffs herself by way of getting it under headway before handing it to her husband—and then serves us with glasses of sweetened tea from the samovar. In deference to her Ferenghi brother-in-law and myself, Rosie has donned a gauzy shroud over the above-mentioned in-door costume of the Persian female. "She is a beautiful dancer," says her husband, admiringly, "I wish it were possible for you to see her dance this evening ; but it isn't ; Rosie herself wouldn't mind, but it would be pretty certain to leak out, and Miana being a rather fanatical place, my life wouldn't be worth that much," and the Khan carelessly snapped his fingers. Supper is brought up to the telegraph station. Prince Assabdulla is invited, and comes round with his servant bearing a number of cucumbers and a bottle of arrack ; the Prince, being a genuine Mohammedan, is forbidden by his religion to indulge ; consequently he consumes the fiery arrack in preference to some light and harmless native wine ; such is the perversity of human nature.

Two princes and a khan are cantering (not khan-tering) alongside the bicycle as I pull out eastward from Miana. They accompany me to the foot-hills approaching the Koflan Koo Pass, and wishing me a pleasant journey, turn their horses' heads homeward again. Reaching the pass proper, I find it to be an exceedingly steep trundle, but quite easy climbing compared with a score of mountain passes in Asia Minor, for the surface is reasonably smooth, and toward the summit is an ancient stone causeway. A new and delightful experience awaits me upon the summit of the pass ; the view to the westward is a revelation of mountain scenery altogether new and novel in my experience, which can now scarcely be called unvaried. I seem to be elevated entirely above the surface of the earth, and gazing down through transparent, ethereal depths upon a scene of everchanging beauty. Fleecy cloudlets are floating lazily over the valley far below my position, producing on the landscape a panoramic scene of constantly changing shadows; through the ethery depths, so wonderfully transparent, the billowy gray foot-hills, the meandering streams fringed with green, and Miana with its blue-domed mosques and emerald gardens, present a phantasmagorical appearance, as though they themselves were floating about in the lower strata of space, and undergoing constant transformation. Perched on an apparently inaccessible crag to the north is an ancient robber stronghold commanding the pass; it is a natural fortress, requiring but a few finishing touches by man to render it impregnable in the days when the maintenance of robber strongholds were possible. Owing to its walls and battlements being chiefly erected by nature, the Persian peasantry call it the Perii-Kasr, believing it to have been built by fairies. While descending the eastern slope, I surprise a gray lizard almost as large as a rabbit, basking in the sunbeams; he briskly scuttles off into the rocks upon being disturbed.

Crossing the Sefid Rûd on a dilapidated brickwork bridge, I cross another range of low hills, among which I notice an abundance of mica cropping above the surface, and then descend on to a broad, level plain, extending eastward without any lofty elevation as far as eye can reach. On this shelterless plain I am overtaken by a furious equinoctial gale; it comes howling suddenly from the west, obscuring the recently vacated Koflan Koo Mountains behind an inky veil, filling the air with clouds of dust, and for some minutes rendering it necessary to lie down and fairly hang on to the ground to prevent being blown about. First it begins to rain, then to hail; heaven's artillery echoes and reverberates in the Koflan Koo Mountains, and rolls above the plain, seeming to shake the hailstones down like fruit from the branches of the clouds, and soon I am enveloped in a pelting, pitiless downpour of hailstones, plenty large enough to make themselves felt wherever they strike. To pitch my tent would have been impossible, owing to the wind and the suddenness of its appearance. In thirty minutes or less it is all over; the sun shines out warmly and dissipates the clouds, and converts the ground into an evaporator that envelops everything in steam. In an hour after it quits raining, the road is dry again, and across the plain it is for the most part excellent wheeling.

About four o'clock the considerable village of Sercham is reached; here, as at Hadji Aghi, I at once become the bone of con-

32

tention between rival *khan-jees* wanting to secure me for a guest, on the supposition that I am going to remain over night. Their anxiety is all unnecessary, however, for away off on the eastern horizon can be observed clusters of familiar black dots that awaken agreeable reflections of the night spent in the Koordish camp between Ovahjik and Khoi. I remain in Sercham long enough to eat a watermelon, ride, against my will, over rough ground to appease

The Bridgeless Streams of Asia.

the crowd, and then pull out toward the Koordish camps which are evidently situated near my proper course.

It seems to have rained heavily in the mountains and not rained at all east of Sercham, for during the next hour I am compelled to disrobe, and ford several freshets coursing down ravines over beds that before the storm were inches deep in dust, the approaching slopes being still dusty ; this little diversion causes me to thank fortune that I have been enabled to keep in advance of the regular rainy season, which commences a little later. Striking a Koordish camp adjacent to the trail I trundle toward one of the tents ; before reaching it I am overhauled by a shepherd who hands me a handful of dried peaches from a wallet suspended from his waist. The evening air is cool with a suggestion of frostiness, and the occupants of the tent are found crouching around a smoking *tezek* fire ; they are ragged and of rather unprepossessing appearance, but being instinctively hospitable, they shuffle around to make me welcome at the fire ; at first I almost fancy myself mistaken in thinking them Koords, for there is nothing of the neatness and cleanliness of our late acquaintances about them ; on the contrary, they are almost as repulsive as their sedentary relatives of Dele Baba—but a little questioning removes all doubt of their being Koords. They are simply an ill-conditioned tribe, without any idea whatever of thrift or good management. They have evidently been to Tabreez or somewhere lately, and invested most of the proceeds of the season's shearing in three-year-old dried peaches that are hard enough to rattle like pebbles ; sacksful of these edibles are scattered all over the tent serving for seats, pillows, and general utility articles for the youngsters to roll about on, jump over, and throw around ; everybody in the camp seems to be chewing these peaches and throwing them about in sheer wantonness because they are plentiful ; every sack contains finger-holes from which one and all help themselves *ad libitum* in wanton disregard of the future.

Nearly everybody seems to be suffering from ophthalmia, which is aggravated by crouching over the densely smoking *tezek ;* and one miserable looking old character is groaning and writhing with the pain of a severe stomach-ache. By loafing lazily about the tent all day, and chewing these flinty dried peaches, this hopeful old joker has well-nigh brought himself to the unhappy condition of the Yosemite valley mule, who broke into the tent and consumed half a bushel of dried peaches ; when the hunters returned to camp and were wondering what marauder had visited their tent and stolen the peaches, they heard a loud explosion behind the tent ; hastily going out they discover the remnants of the luckless mule scattered about in all directions. Of course I am appealed to for a remedy, and I am not sorry to have at last come across an applicant for my services as a *hakim,* for whose ailment I can prescribe with some degree of confidence ; to make assurance doubly sure I give the sufferer a double dose, and in the morning have the satisfaction of finding him entirely relieved from his misery. There seems to be

no order or sense of good manners whatever among these people ;
we have bread and half-stewed peaches for supper, and while they
are cooking, ill-mannered youngsters are constantly fishing them
from the kettles with weed-stalks, meeting with no sort of reproof
from their elders for so doing ; when bedtime arrives, everybody
seizes quilts, peach-sacks, etc., and crawls wherever they can for
warmth and comfort ; three men, two women, and several children

Midnight Intruders.

occupy the same compartment as myself, and gaunt dogs are nosing
hungrily about among us.

About midnight there is a general hallooballoo among the dogs,
and the clatter of horses' hoofs is heard outside the tent ; the occupants of the tent, including myself, spring up, wondering what the
disturbance is all about. A group of horsemen are visible in the
bright moonlight outside, and one of them has dismounted, and

under the guidance of a shepherd, is about entering the tent ; seeing me spring up, and being afraid lest perchance I might misinterpret their intentions and act accordingly, he sings out in a soothing voice, " *Kardash, Hamsherri ; Kardash, Kardash !* " thus assuring me of their peaceful intentions. These midnight visitors turn out to be a party of Persian travellers from Miana, from which it would appear they have less fear of the Koords here than in Koordistan near the frontier ; having, somehow, found out my whereabouts, they have come to try and persuade me to leave the camp and join their company to Zenjan. Although my own unfavorable impressions of my entertainers are seconded by the visitors' reiterated assurances that these Koords are bad people, I decline to accompany them, knowing the folly of attempting to bicycle over these roads by moonlight in the company of horsemen who would be continually worrying me to ride, no matter what the condition of the road ; after remaining in camp half an hour they take their departure.

In the morning I discover that my mussulman hat-band has mysteriously disappeared, and when preparing to depart, a miscellaneous collection of females gather about me, seize the bicycle, and with much boisterous hilarity refuse to let me depart until I have given each one of them some money ; their behavior is on the whole so outrageous, that I appeal to my patient of yesterday evening, in whose bosom I fancy I may perchance have kindled a spark of gratitude ; but the old reprobate no longer has the stomach-ache, and he regards my unavailing efforts to break away from my hoidenish tormentors with supreme indifference, as though there were nothing extraordinary in their conduct. The demeanor of these wild-eyed Koordish females on this occasion fully convinces me that the stories concerning their barbarous conduct toward travellers captured on the road is not an exaggeration, for while preventing my departure they seem to take a rude, boisterous delight in worrying me on all sides, like a gang of puppies barking and harassing anything they fancy powerless to do them harm. After I have finally bribed my freedom from the women, the men seize me and attempt to further detain me until they can send for their Sheikh to come from another camp miles away, to see me ride. After waiting a reasonable time, out of respect for their having accommodated me with quarters for the night, and no signs of the Sheikh appearing, I determine to submit to their impudence no

longer; they gather around me as before, but presenting my revolver and assuming an angry expression, I threaten instant destruction to the next one laying hands on either myself or the bicycle; they then give way with lowering brows and sullen growls
of displeasure. My rough treatment on this occasion compared
with my former visit to a Koordish camp, proves that there is as
much difference between the several tribes of nomad Koords, as
between their sedentary relatives of Dele Baba and Malosman respectively; for their general reputation, it were better that I had
spent the night in Sercham.

A few miles from the camp, I am overtaken by four horsemen
followed by several dogs and a pig; it proves to be the tardy Sheikh
and his retainers, who have galloped several miles to catch me up;
the Sheikh is a pleasant, intelligent fellow of thirty or thereabouts,
and astonishes me by addressing me as "Monsieur;" they canter
alongside for a mile or so, highly delighted, when the Sheikh cheer-
ily sings out "Adieu, monsieur!" and they wheel about and return;
had their Sheikh been in the camp I stayed at, my treatment would
undoubtedly have been different. I am at the time rather puzzled
to account for so strange a sight as a pig galloping briskly behind
the horses, taking no notice of the dogs which continually gambol
about him; but I afterward discover that a pet pig, trained to
follow horses, is not an unusual thing among the Persians and Persian Koords; they are thin, wiry animals of a sandy color, and
quite capable of following a horse for hours; they live in the stable
with their equine companions, finding congenial occupation in
rooting around for stray grains of barley; the horses and pig are
said to become very much attached to each other; when on the
road the pig is wont to signify its disapproval of a too rapid pace,
by appealing squeaks and grunts, whereupon the horse responsively
slacks its speed to a more accommodating speed for its porcine
companion. The road now winds tortuously along the base of
some low gravel hills, and the wheeling perceptibly improves; beyond Nikbey it strikes across the hilly country, and more trundling
becomes necessary. At Nikbey I manage to leave the inhabitants
in a profound puzzle by replying that I am not a Ferenghi, but an
Englishman; this seems to mystify them not a little, and they commence inquiring among themselves for an explanation of the difference; they are probably inquiring yet.

Fifty-eight miles are covered from the Koordish camp, and at three o'clock the blue-tiled domes of the Zendjan mosques appear in sight ; these blue-tiled domes are more characteristic of Persian mosques, which are usually built of bricks, and have no lofty tapering minarets as in Turkey ; the summons to prayers are called from the top of a wall or roof. When approaching the city gate, a half-crazy man becomes wildly excited at the spectacle of a man .on a wheel, and, rushing up, seizes hold of the handle; as I spring from the saddle he rapidly takes to his heels ; finding that I am not pursuing him, he plucks up courage, and timidly approaching, begs me to let him see me ride again. Zendjan is celebrated for the manufacture of copper vessels, and the rat-a-tat-tat of the workmen beating them out in the coppersmiths' quarters is heard fully a mile outside the gate ; the hammering is sometimes deafening while trundling through these quarters, and my progress through it is indicated by what might perhaps be termed a sympathetic wave of silence following me along, the din ceasing at my approach and commencing again with renewed vigor after I have passed.

Mr. F——, a Levantine gentleman in charge of the station here, fairly outdoes himself in the practical interpretation of genuine old-fashioned hospitality, which brooks no sort of interference with the comfort of his guest ; understanding the perpetual worry a person travelling in so extraordinary a manner must be subject to among an excessively inquisitive people like the Persians, he kindly takes upon himself the duty of protecting me from anything of the kind during the day I remain over as his guest, and so manages to secure me much appreciated rest and quiet. The Governor of the city sends an officer around saying that himself and several prominent dignitaries would like very much to see the bicycle. "Very good," replies Mr. F——, "the bicycle is here, and Mr. Stevens will doubtless be pleased to receive His Excellency and the leading officials of Zendjan any time it suits their convenience to call, and will probably have no objections to showing them the bicycle." It is, perhaps, needless to explain that the Governor doesn't turn up ; I, however, have an interesting visitor in the person of the Sheikh-ul-Islam (head of religious affairs in Zendjan), a venerable-looking old party in flowing gown and monster turban, whose hands and flowing beard are dyed to a ruddy yellow with henna. The Sheikh-ul-Islam is considered the holiest personage in Zendjan, and his appearance and demeanor does not in the least belie his reputation ; whatever may be his private opinion of himself, he makes far less display of sanctimoniousness than many of the common seyuds, who usually gather their garments about them whenever they pass a Ferenghi in the bazaar, for fear their clothing should become defiled by brushing against him. The Sheikh-ul-Islam fulfils one's idea of a gentle-bred, worthy-minded old patriarch; he examines the bicycle and listens to the account of my journey with much curiosity and interest, and bestows a flattering mead of praise on the wonderful ingenuity of the Ferenghis as exemplified in my wheel.

From Zendjan eastward the road gradually improves, and after a dozen miles develops into the finest wheeling yet encountered in Asia; the country is a gravelly plain between a mountain chain on the left and a range of lesser hills to the right. Near noon I pass through Sultaneah, formerly a favorite country resort of the Persian monarchs; on the broad, grassy plain, during the autumn, the Shah was wont to find amusement in manœuvring his cavalry regiments, and for several months an encampment near Sultaneah became the head-quarters of that arm of the service. The Shah's palace and the blue dome of a large mosque, now rapidly crumbling to decay, are visible many miles before reaching the village.

The presence of the Shah and his court doesn't seem to have exerted much of a refining or civilizing influence on the common villagers; otherwise they have retrograded sadly toward barbarism again since Sultaneah has ceased to be a favorite resort. They appear to regard the spectacle of a lone Ferenghi meandering through their wretched village on a wheel, as an opportunity of doing something aggressive for the cause of Islam not to be overlooked; I am followed by a hooting mob of bare-legged wretches, who forthwith proceed to make things lively and interesting, by pelting me with stones and clods of dirt. One of these wantonly aimed missiles catches me square between the shoulders, with a force that, had it struck me fairly on the back of the neck, would in all probability have knocked me clean out of the saddle; unfortunately, several irrigating ditches crossing the road immediately ahead prevent escape by a spurt, and nothing remains but to dismount and proceed to make the best of it.

There are only about fifty of them actively interested, and part of these being mere boys, they are anything but a formidable crowd of belligerents if one could only get in among them with a stuffed club; they seem but little more than human vermin in their rags and nakedness, and like vermin, the greatest difficulty is to get hold of them. Seeing me dismount, they immediately take to their heels, only to turn and commence throwing stones again at finding themselves unpursued ; while I am retreating and actively dodging the showers of missiles, they gradually venture closer and closer, until things becoming too warm and dangerous, I drop the bicycle, and make a feint toward them ; they then take to their heels, to return to the attack again as before, when I again com-

Firing over their Heads.

mence retreating. Finally I try the experiment of a shot in the air, by way of notifying them of my ability to do them serious injury ; this has the effect of keeping them at a more respectful distance, but they seem to understand that I am not intending serious shooting, and the more expert throwers manage to annoy me considerably until ridable ground is reached ; seeing me mount, they all come racing pell-mell after me, hurling stones, and howling insulting epithets after me as a Ferenghi, but with smooth road ahead I am, of course, quickly beyond their reach.

The villages east of Sultaneah are observed to be, almost without exception, surrounded by a high mud wall, a characteristic giving them the appearance of fortifications rather than mere agricultural villages; the original object of this was, doubtless, to secure themselves against surprises from wandering tribes; and as the Persians seldom think of changing anything, the custom is still maintained. Bushes are now occasionally observed near the roadside, from every twig of which a strip of rag is fluttering in the breeze; it is an ancient custom still kept up among the Persian peasantry when approaching any place they regard with reverence, as the ruined mosque and imperial palace at Sultaneah, to tear a strip of rag from their clothing and fasten it to some roadside bush; this is supposed to bring them good luck in their undertakings, and the bushes are literally covered with the variegated offerings of the superstitious ryots; where no bushes are handy, heaps of small stones are indicative of the same belief; every time he approaches the well-known heap, the peasant picks up a pebble, and adds it to the pile.

Owing to a late start and a prevailing head-wind, but forty-six miles are covered to-day, when about sundown I seek the accommodation of the *chapar-khana*, at Heeya; but, providing the road continues good, I promise myself to polish off the sixty miles between here and Kasveen, to-morrow. The *chapar-khana* sleeping apartments at Heeya contain whitewashed walls and reed matting, and presents an appearance of neatness and cleanliness altogether foreign to these institutions previously patronized; here, also, first occurs the innovation from "Hamsherri" to "Sahib," when addressing me in a respectful manner; it will be Sahib, from this point clear to, through and beyond India; my various titles through the different countries thus far traversed have been; Monsieur, Herr, Effendi, Hamsherri, and now Sahib; one naturally wonders what new surprises are in store ahead.

A bountiful supper of scrambled eggs (*toke-mi-morgue*) is obtained here, and the customary shake-down on the floor. After getting rid of the crowd I seek my rude couch, and am soon in the land of unconsciousness; an hour afterward I am awakened by the busy hum of conversation; and, behold, in the dim light of a primitive lamp, I become conscious of several pairs of eyes immediately above me, peering with scrutinizing inquisitiveness into my face; others are examining the bicycle standing against the wall at my head. Rising up, I find the *chapar-khana* crowded with caravan teamsters, who, going past with a large camel caravan from the Caspian seaport of Resht, have heard of the bicycle, and come flocking to my room; I can hear the unmelodious clanging of the big sheet-iron bells as their long string of camels file slowly past the building.

Daylight finds me again on the road, determined to make the best of early morning, ere the stiff easterly wind, which seems in-

Passing a Camel Caravan.

clined to prevail of late, commences blowing great guns against me. A short distance out, I meet a string of some three hundred laden camels that have not yet halted after the night's march; scores of large camel caravans have been encountered since leaving Erzeroum, but they have invariably been halting for the day; these camels regard the bicycle with a timid reserve, merely swerving a step or two off their course as I wheel past; they all seem about equally startled, so that my progress down the ranks simply causes a sort of a gentle ripple along the line, as though each successive

camel were playing a game of follow-my-leader. The road this
morning is nearly perfect for wheeling, consisting of well-trodden
camel-paths over a hard gravelled surface that of itself naturally
makes excellent surface for cycling ; there is no wind, and twenty-
five miles are duly registered by the cyclometer when I halt to eat
the breakfast of bread and a portion of yesterday evening's scrambled eggs which I have brought along.

On past Seyudoon and approaching Kasveen, the plain widens
to a considerable extent and becomes perfectly level ; apparent
distances become deceptive, and objects at a distance assume weird,
fantastic shapes ; beautiful mirages hold out their allurements from
all directions ; the sombre walls of villages present the appearance
of battlemented fortresses rising up from the mirror-like surface
of silvery lakes, and orchards and groves seem shadowy, undefinable objects floating motionless above the earth. The telegraph
poles traversing the plain in a long, straight line until lost to view
in the hazy distance, appear to be suspended in mid-air ; camels,
horses, and all moving objects more than a mile away, present the
strange optical illusion of animals walking· through the air many
feet above the surface of the earth. Long rows of kanaat mounds
traverse the plain in every direction, leading from the numerous
villages to distant mountain chains. Descending one of the sloping cavernous entrances before mentioned, for a drink, I am rather
surprised at observing numerous fishes disporting themselves in
the water, which, on the comparatively level plain, flows but slowly ;
perhaps they are an eyeless variety similar to those found in the
Mammoth Cave of Kentucky ; still they get a glimmering light
from the numerous perpendicular shafts. Flocks of wild pigeons
also frequent these underground water-courses, and the peasantry
sometimes capture them by the hundred with nets placed over the
shafts ; the kanaats are not bricked archways, but merely tunnels
burrowed through the ground.

Three miles of loose sand and stones have to be trundled
through before reaching Kasvoen ; nevertheless my promised sixty
miles are overcome, and I enter the city gate at 2 P.M. A trundle
through several narrow, crooked streets brings me to an inner
gateway emerging upon a broad, smooth avenue ; a short ride down
this brings me to a large enclosure containing the custom-house
offices and a fine brick caravanserai. Yet another prince appears
here in the person of a custom-house official ; I readily grant the

requested privilege of seeing me ride, but the title of a Persian
prince is no longer associated in my mind with greatness and importance ; princes in Persia are as plentiful as counts in Italy or
barons in Germany, yet it rather shocks one's dreams of the splen-

Persian " Lutis," or Buffoons.

dor of Oriental royalty to find princes manipulating the keys of a
one wire telegraph control-station at a salary of about forty dollars a month (25 tomans), or attending to the prosy duties of a
small custom-house.

Kasveen is important as being the half-way station between Teheran and the Caspian port of Resht, and on the highway of travel and commerce between Northern Persia and Europe ; added importance is likewise derived from its being the terminus of a broad level road from the capital, and where travellers and the mail from Teheran have to be transferred from wheeled vehicles to the backs of horses for the passage over the rugged passes of the Elburz mountains leading to the Caspian slope, or *vice versa* when going the other way. Locking the bicycle up in a room of the caravanserai, I take a strolling peep at the nearest streets ; a couple of *lutis* or professional buffoons, seeing me strolling leisurely about, come hurrying up ; one is leading a baboon by a string around the neck, and the other is carrying a gourd drum. Reaching me, the man with the baboon commences making the most ludicrous grimaces and causes the baboon to caper wildly about by jerking the string, while the drummer proceeds to belabor the head of his drum, apparently with the single object of extracting as much noise from it as possible. Putting my fingers to my ears I turn away ; ten minutes afterward I observe another similar combination making a bee-line for my person ; waving them off I continue on down the street ; soon afterward yet a third party attempts to secure me for an audience. It is the custom for these strolling buffoons to thus present themselves before persons on the street, and to visit houses whenever there is occasion for rejoicing, as at a wedding, or the birth of a son ; the *lutis* are to the Persians what Italian organ-grinders are among ourselves ; I fancy people give them money chiefly to get rid of their noise and annoyance, as we do to save ourselves from the soul-harrowing tones of a wheezy crank organ beneath the window.

Among the novel conveyances observed in the courtyard of the caravanserai is the *takhtrowan*, a large sedan chair provided with shafts at either end, and carried between two mules or horses ; another is the before-mentioned *kajaveh*, an arrangement not unlike a pair of canvas-covered dog kennels strapped across the back of an animal ; these latter contrivances are chiefly used for carrying women and children.

After riding around the courtyard several different times for crowds continually coming, I finally conclude that there must be a limit to this sort of thing anyhow, and refuse to ride again ; the new-comers linger around, however, until evening, in the hopes that an opportunity of seeing me ride may present itself. A number of them then contribute a handful of coppers, which they give to the proprietor of a tributary *tchai-khan* to offer me as an inducement to ride again. The wily Persians know full well that while a Ferenghi would scorn to accept their handful of coppers, he would probably be sufficiently amused at the circumstance to reward their persistence by riding for nothing; telling the grinning *khan-jee* to pocket the coppers, I favor them with "positively the last entertainment this evening." An hour later the *khan-jee* meets me going toward the bazaar in search of something for supper; inquiring the object of my search, he takes me back to his *tchai-khan*, points significantly to an iron kettle simmering on a small charcoal fire, and bids me be seated; after waiting on a customer or two, and supplying me with tea, he quietly beckons me to the fire, removes the cover and reveals a savory dish of stewed chicken and onions; this he generously shares with me a few minutes later, refusing to accept any payment. As there are exceptions to every rule, so it seems there are individuals, even among the Persian commercial classes, capable of generous and worthy impulses; true the *khan-jee* obtained more than the value of the supper in the handful of coppers—but gratitude is generally understood to be an unknown commodity among the subjects of the Shah.

Soon the obstreperous cries of "Ali Akbar, la-al-lah-il-allah" from the throats of numbers of the faithful perched upon the caravanserai steps, stable-roof, and other conspicuous soul-inspiring places, announces the approach of bedtime. My room is actually found to contain a towel and an old tooth-brush; the towel has evidently not been laundried for some time and a public tooth-brush is hardly a joy inspiring object to contemplate; nevertheless they are evidences that the proprietor of the caravanserai is possessed of vague, shadowy ideas of a Ferenghi's requirements. After a person has dried his face with the slanting sunbeams of early morning, or with his pocket-handkerchief for weeks, the bare possibility of soap, towels, etc., awakens agreeable reflections of coming comforts.

At seven o'clock on the following morning I pull out toward Teheran, now but six *chapar*-stations distant. Running parallel with the road is the Elburz range of mountains, a lofty chain, separating the elevated plateau of Central Persia from the moist and wooded slopes of the Caspian Sea; south of this great dividing ridge the country is an arid and barren waste, a desert, in fact, save where irrigation redeems here and there a circumscribed area, and the mountain slopes are gray and rocky. Crossing over to the northern side of the divide, one immediately finds himself in a moist climate, and a country green almost as 'the British Isles, with dense box-wood forests covering the slopes of the mountains and hiding the foot-hills beneath an impenetrable mantle of green. The Elburz Mountains are a portion of the great water-shed of Central Asia, extending from the Himalayas up through Afghanistan and Persia into the Caucasus, and they perform very much the same office for the Caspian slope of Persia, as the Sierra Nevadas do for the Pacific slope of California, inasmuch as they cause the moisture-laden clouds rolling in from the sea to empty their burthens on the seaward slopes instead of penetrating farther into the interior.

The road continues fair wheeling, but nothing compared with the road between Zendjan and Kasveen ; it is more of an artificial highway ; the Persian government has been tinkering with it, improving it considerably in some respects, but leaving it somewhat lumpy and unfinished generally, and in places it is unridable from sand and loose material on the surface ; it has the appreciable merit of levelness, however, and, for Persia, is a very creditable highway indeed. At four farsakhs from Kasveen I reach the *chapar-khana* of Cawanda, where a breakfast is obtained of eggs and tea ; these two things are among the most readily obtained refreshments in Persia. The country this morning is monotonous and uninteresting, being for the most part a stony, level plain, sparsely covered with gray camel-thorn shrubs. Occasionally one sees in the distance a camp of Eliauts, one of the wandering tribes of Persia; their tents are smaller and of an entirely different shape from the Koordish tents, partaking more of the nature of square-built movable huts than tents ; these camps are too far off my road to justify paying them a visit, especially as I shall probably have abundant opportunities before leaving the Shah's dominions; but I intercept a straggling party of them crossing the road. They have a more docile look about them than the Koords, have more the general appearance of gypsies, and they dress but little different from the ryots of surrounding villages.

At Kishlock, where I obtain a dinner of bread and grapes, I find the cyclometre has registered a gain of thirty-two miles from Kasveen ; it has scarcely been an easy thirty-two miles, for I am again confronted by a discouraging head breeze.

Keaching the Shah Abbas caravanserai of Yeng-Imam (all first-class caravanserais are called Shah Abbas caravanserais, in deference to so many having been built throughout Persia by that monarch) about five o'clock, I conclude to remain here over night, having wheeled fifty-three miles. Yeng-Imam is a splendid large brick serai, the finest I have yet seen in Persia ; many travellers are putting up here, and the place presents quite a lively appearance. In the centre of the court-yard is a large covered spring ; around this is a garden of rose-bushes, pomegranate trees, and flowers ; surrounding the garden is a brick walk, and forming yet a larger square is the caravanserai building itself, consisting of a one-storied brick edifice, partitioned off into small rooms. The building is only one room deep, and each room opens upon a sort of covered porch containing a fireplace where a fire can be made and provisions cooked. Attached to the caravanserai, usually beneath the massive and roomy arched gateway, is a tchai-khan and a small store where bread, eggs, butter, fruit, charcoal, etc., are to be obtained. The traveller hires a room which is destitute of all furniture ; provides his own bedding and cooking utensils, purchases provisions and a sufficiency of charcoal, and proceeds to make himself comfortable. On a pinch one can usually borrow a frying-pan or kettle of some kind, and in such first-class caravanserais as Yeng-Imam there is sometimes one furnished room, carpeted and provided with bedding, reserved for the accommodation of travellers of importance.

After the customary programme of riding to allay the curiosity and excitement of the people, I obtain bread, fruit, eggs, butter to cook them in, and charcoal for a fire, the elements of a very good supper for a hungry traveller. Borrowing a handleless frying-pan, I am setting about preparing my own supper, when a respectable-looking Persian steps out from the crowd of curious on-lookers and voluntarily takes this rather onerous duty out of my hands. Readily obtaining my consent, he quickly kindles a fire, and scrambles and fries the eggs. While my volunteer cook is thus busily engaged, a company of distinguished travellers passing along the road halt at the tchai-khan to smoke a kalian and drink tea. The caravanserai proprietor approaches me, and winking mysteriously, intimates that by going outside and riding for the edification of the

33

new arrivals I will be pretty certain to get a present of a keran
(about twenty cents). As he appears anxious to have me accommodate them, I accordingly go out and favor them with a few turns
on a level piece of ground outside. After they have departed the
proprietor covertly offers me a half-keran piece in a manner so that
everybody can observe him attempting to give me something without seeing the amount. The wily Persian had doubtless solicited
a present from the travellers for me, obtained, perhaps, a couple of
kerans, and watching a favorable opportunity, offers me the half-
keran piece ; the wily ways of these people are several degrees
more ingenious even than the dark ways and vain tricks of Bret
Harte's " Heathen Chinee."

Occupying one of the rooms are two young noblemen travelling
with their mother to visit the Governor of Zendjan ; after I have
eaten my supper, they invite me to their apartments for the evening ; their mother has a samovar under full headway, and a number
of hard boiled eggs. Her two hopeful sons are engaged in a drinking bout of *arrack ;* they are already wildly hilarious and indulging in brotherly embraces and doubtful love-songs. Their fond
mother regards them with approving smiles as they swallow glass
after glass of the raw fiery spirit, and become gradually more intoxicated and hilarious. Instead of checking their tippling, as a
fond and prudent Ferenghi mother would have done, this indulgent parent encourages them rather than otherwise, and the more
deeply intoxicated and hilariously happy the sons become, the happier seems the mother. About nine o'clock they fall to weeping
tears of affection for each other and for myself, and degenerate into
such maudlin sentimentality generally, that I naturally become disgusted, accept a parting glass of tea, and bid them good-evening.

The caravanserai-*jee* assigns me the furnished chamber above
referred to ; the room is found to be well carpeted, contains a mattress and an abundance of flaming red quilts, and on a small table
reposes a well-thumbed copy of the Koran with gilt lettering and
illumined pages ; for these really comfortable quarters I am charged
the trifling sum of one keran.

I am now within fifty miles of Teheran, my destination until
spring-time comes around again and enables me to continue on
eastward toward the Pacific ; the wheeling continues fair, and in
the cool of early morning good headway is made for several miles ;
as the sun peeps over the summit of a mountain spur jutting south-

ward a short distance from the main Elburz Range, a wall of air comes rushing from the east as though the sun were making strenuous exertions to usher in the commencement of another day with a triumphant toot. Multitudes of donkeys are encountered on the road, the omnipresent carriers of the Persian peasantry, taking produce to the Teheran market; the only wheeled vehicle encountered between Kasveen and Teheran is a heavy-wheeled, cumbersome mail wagon, rattling briskly along behind four galloping horses driven abreast, and a newly imported carriage for some notable of the capital being dragged by hand, a distance of two hundred miles from Resht, by a company of soldiers. Pedalling laboriously against a stiff breeze I round the jutting mountain spur about eleven o'clock, and the conical snow-crowned peak of Mount Demavend looms up like a beacon-light from among the lesser heights of the Elburz Range about seventy-five miles ahead. Demavend is a perfect cone, some twenty thousand feet in height, and is reputed to be the highest point of land north of the Himalayas.

From the projecting mountain spur the road makes a bee-line across the intervening plain to the capital; a large willow-fringed irrigating ditch now traverses the stony plain for some distance parallel with the road, supplying the caravanserai of Shahabad and several adjacent villages with water. Teheran itself, being situated on the level plain, and without the tall minarets that render Turkish cities conspicuous from a distance, leaves one undecided as to its precise location until within a few miles of the gate; it occupies a position a dozen or more miles south of the base of the Elburz Mountains, and is flanked on the east by another jutting spur; to the southward is an extensive plain sparsely dotted with villages, and the walled gardens of the wealthier Teheranis.

At one o'clock on the afternoon of September 30th, the sentinels at the Kasveen gate of the Shah's capital gaze with unutterable astonishment at the strange spectacle of a lone Ferenghi riding toward them astride an airy wheel that glints and glitters in the bright Persian sunbeams. They look still more wonder-stricken, and half-inclined to think me some supernatural being, as, without dismounting, I ride beneath the gaudily colored archway and down the suburban streets. A ride of a mile between dead mud walls and along an open business street, and I find myself surrounded by wondering soldiers and citizens in the great central

top-maidan, or artillery square, and shortly afterward am endeavoring to eradicate some of the dust and soil of travel, in a room of a
wretched apology for an hotel, kept by a Frenchman, formerly a
pastry-cook to the Shah.   My cyclometre has registered one thousand five hundred and seventy-six miles from Ismidt; from Liverpool
to Constantinople, where I had no cyclometre, may be roughly esti-
mated at two thousand five hundred, making a total from Liverpool
to Teheran of four thousand and seventy-six miles.   In the evening

Entering the Teheran Gate.

several young Englishmen belonging to the staff of the Indo-
European Telegraph Company came round, and re-echoing my
own above-mentioned sentiments concerning the hotel, generously
invite me to become a member of their comfortable bachelor establishment during my stay in Teheran.   "How far do you reckon it
from London to Teheran by your telegraph line?" I inquire of them
during our after-supper conversation.   "Somewhere in the neighborhood of four thousand miles," is the reply.   "What does your
cyclometre say?"

CHAPTER XXI.

TEHERAN.

THERE is sufficient similarity between the bazaar, the mosques, the residences, the suburban gardens, etc., of one Persian city, and the same features of another, to justify the assertion that the description of one is a description of them all. But the presence of the Shah and his court ; the pomp and circumstance of Eastern royalty ; the foreign ambassadors ; the military ; the improvements introduced from Europe ; the royal palaces of the present sovereign ; the palaces and reminiscences of former kings—all these things combine to effectually elevate Teheran above the somewhat dreary sameness of provincial cities.

A person in the habit of taking daily strolls here and there about the city will scarcely fail of obtaining a glimpse of the Shah, incidentally, every few days. In this respect there is little comparison to be made between him and the Sultan of Turkey, who never emerges from the seclusion of the palace, except to visit the mosque, or on extraordinary occasions ; he is then driven through streets between compact lines of soldiers, so that a glimpse of his imperial person is only to be obtained by taking considerable trouble. Since the Shah's narrow escape from assassination at the hands of the Baabi conspirators in 1867, he has exercised more caution than formerly about his personal safety. Previous to that affair, it was customary for him to ride on horseback well in advance of his body-guard ; but nowadays, he never rides in advance any farther than etiquette requires him to, which is about the length of his horse's neck. When his frequent outings take him beyond the city fortifications, he is generally provided with both saddle-horse and carriage, thus enabling him to change from one to the other at will.

The Shah is evidently not indifferent to the fulsome flattery of the courtiers and sycophants about him, nor insensible of the pomp and vanity of his position ; nevertheless he is not without a fair share of common-sense. Perhaps the worst that can be said of him is, that he seems content to prostitute his own more enlightened and progressive views to the prejudices of a bigoted and fanatical priesthood. He seems to have a generous desire to see the country opened up to the civilizing improvements of the West, and to give the people an opportunity of emancipating themselves from their present deplorable condition ; but the mollahs set their faces firmly against all reform, and the Shah evidently lacks the strength of will to override their opposition. It was owing to this criminal weakness on his part that Baron Reuter's scheme of railways and commercial regeneration for the country proved a failure.

Persia is undoubtedly the worst priest-ridden country in the world ; the mollahs influence everything and everybody, from the monarch downward, to such an extent that no progress is possible. Barring outside interference, Persia will remain in its present wretched condition until the advent of a monarch with sufficient force of character to deliver the people from the incubus of their present power and influence : nothing short of a general massacre, however, will be likely to accomplish complete deliverance.

Without compromising his dignity as "Shah-in-shah," "The Asylum of the Universe," etc., when dealing with his own subjects, Nasr-e-deen Shah has profited by the experiences of his European tour to the extent of recognizing, with becoming toleration, the democratic independence of Ferenghis, whose deportment betrays the fact that they are not dazed by the contemplation of his greatness. The other evening myself and a friend encountered the Shah and his crowd of attendants on one of the streets leading to the winter palace ; he was returning to the palace in state after a visit of ceremony to some dignitary. First came a squad of foot-runners in quaint scarlet coats, knee-breeches, white stockings, and low shoes, and with a most fantastic head-dress, not unlike a peacock's tail on dress-parade ; each runner carried a silver staff; they were clearing the street and shouting their warning for everybody to hide their faces. Behind them came a portion of the Shah's Khajar body-guard, well mounted, and dressed in a gray uniform, braided with black : each of these also carries a silver staff, and besides sword and dagger, has a gun slung at his back in a red baize case. Next came the royal carriage, containing the Shah : the carriage is somewhat like a sheriff's coach of "ye olden tyme," and is drawn by six superb grays ; mounted on the off horses are

three postilions in gorgeous scarlet liveries. Immediately behind
the Shah's carriage came the higher dignitaries on horseback, and
lastly a confused crowd of three or four hundred horsemen. As
the royal procession approached, the Persians—one and all—either
hid themselves, or backed themselves up against the wall, and re-

The Shah's Foot-runners.

mained with heads bowed half-way to the ground until it passed.
Seeing that we had no intention of striking this very submissive
and servile attitude, first the scarlet foot-runners, and then the advance of the Khajar guard, addressed themselves to us personally,
shouting appealingly as though very anxious about it : " Sahib !

Sahib!" and motioned for us to do as the natives were doing. These valiant guardians of the Shah's barbaric gloriousness cling tenaciously to the belief that it is the duty of everybody, whether Ferenghi or native, to prostrate themselves in this manner before him, although the monarch himself has long ceased to expect it, and is very well satisfied if the Ferenghi respectfully doffs his hat as he goes past.

Much of the nonsensical glamour and superstitious awe that formerly surrounded the person of Oriental potentates has been dissipated of late years by the moral influence of European residents and travellers. But a few years ago, it was certain death for any luckless native who failed to immediately scuttle off somewhere out of sight, or to turn his face to the wall, whenever the carriages of the royal ladies passed by ; and Europeans generally turned down a side street to avoid trouble when they heard the attending eunuchs shouting "gitchin, gitchin !" (begone, begone !) down the street. But things may be done with impunity now, that before the Shah's eye-opening visit to Frangistan would have been punished with instant death ; and although the eunuchs shout " gitchin, gitchin !" as lustily as ever, they are now content if people will only avert their faces respectfully as the carriages drive past.

An eccentric Austrian gentleman once saw fit to imitate the natives in turning their faces to the wall, and improved upon the time-honored custom to the extent of making salaams from the back of his head. This singular performance pleased the ladies immensely, and they reported it to the Shah. Sending for the Austrian, the Shah made him repeat the performance in his presence, and was so highly amused that he dismissed him with a handsome present.

Prominent among the improvements that have been introduced in Teheran of late, may be mentioned gas and the electric light. Were one to make this statement and enter into no further explanations, the impression created would doubtless be illusive ; for although the fact remains that these things are in existence here, they could be more appropriately placed under the heading of toys for the gratification of the Shah's desire to gather about him some of the novel and interesting things he had seen in Europe, than improvements made with any idea of benefiting the condition of the city as a whole. Indeed, one might say without exaggeration, that nothing new or beneficial is ever introduced into Persia, except for the personal gratification or glorification of the Shah; hence it is, that, while a few European improvements are to be seen in Teheran, they are found nowhere else in Persia.

Coal of an inferior quality is obtained in the Elburz Mountains, near Kasveen, and brought on the backs of camels to Teheran; and enough gas is manufactured to supply two rows of lamps leading from the *top-maidan* to the palace front, two rows on the east side of the palace, and a dozen more in the *top-maidan* itself. The gas is of the poorest quality, and the lamps glimmer faintly through the gloom of a moonless evening until half-past nine, giving about as much light, or rather making darkness about as visible as would the same number of tallow candles; at this hour they are extinguished, and any Persian found outside of his own house later than this, is liable to be arrested and fined.

The electric light improvements consist of four lights, on ordinary gas-lamp posts, in the *top-maidan*, and a more ornamental and pretentious affair, immediately in front of the palace; these are only used on special occasions. The electric lights are a never-failing source of wonder and mystification to the common people of the city and the peasants coming in from the country. A stroll into the *maidan* any evening when the four electric lights are making the gas-lamps glimmer feebler than ever, reveals a small crowd of natives assembled about each post, gazing wonderingly up at the globe, endeavoring to penetrate the secret of its brightness, and commenting freely among themselves in this wise:

"Mashallah! Abdullah," says one, "where does all the light come from? They put no candles in, no naphtha, no anything; where does it come from?"

"Mashallah!" replies Abdullah, "I don't know; it lights up 'biff!' all of a sudden, without anybody putting matches to it, or going anywhere near it; nobody knows how it comes about except Sheitan (Satan) and Sheitan's children, the Ferenghis."

"Al-lah! it is wonderful!" echoes another, "and our Shah is a wonderful being to give us such things to look at—Allah be praised!"

All these strange innovations and incomprehensible things produce a deep impression on the unenlightened minds of the common Persians, and helps to deify the Shah in their imagination; for although they know these things come from Frangistan, it seems natural for them to sing the praises of the Shah in connection with them. They think these five electric lights in Teheran among the

wonders of the world ; the glimmering gas-lamps and the electric
lights help to rivet their belief that their capital is the most wonderful city in the world, and their Shah the greatest monarch extant.
These extreme ideas are, of course, considerably improved upon
when we leave the ranks of illiteracy ; but the Persians capable of
forming anything like an intelligent comparison between themselves
and a European nation, are confined to the Shah himself, the *corps
diplomatique*, and a few prominent personages who have been abroad.

Always on the lookout for something to please the Shah, the
news of my arrival in Teheran on the bicycle no sooner reaches the
ear of the court officials than the monarch hears of it himself. On
the seventh day after my arrival an officer of the palace calls on
behalf of the Shah, and requests that I favor them all, by following
the soldiers who will be sent to-morrow morning, at eight o'clock,
Ferenghi time, to conduct me to the palace, where it is appointed
that I am to meet the " Shah-in-shah and King of kings, " and ride
with him, on the bicycle, to his summer palace at Doshan Tepe.

"Yes, I shall, of course, be most happy to accommodate ; and to
be the means of introducing to the notice of His Majesty, the wonderful iron horse, the latest wonder from Frangistan," I reply ;
and the officer, after salaaming with more than French politeness,
takes his departure.

Promptly at the hour appointed the soldiers present themselves ;
and after waiting a few minutes for the horses of two young Englishmen who desire to accompany us part way, I mount the ever-ready
bicycle, and together we follow my escort along several fairly ridable
streets to the office of the foreign minister. The soldiers clear the
way of pedestrians, donkeys, camels, and horses, driving them unceremoniously to the right, to the left, into the ditch—anywhere out
of my road ; for am I not for the time being under the Shah's
special protection ? I am as much the Shah's toy and plaything of
the moment, as an electric light, a stop-watch, or as the big Krupp
gun, the concussion of which nearly scared the soldiers out of their
wits, by shaking down the little minars of one of the city gates,
close to which they had unwittingly discharged it on first trial.

The foreign office, like every building of pretension, whether
public or private, in the land of the Lion and the Sun, is a substantial edifice of mud and brick, inclosing a square court-yard or
garden, in which splashing fountains play amid a wealth of vegetation that springs, as if by waft of magician's wand, from the sandy

Soldiers Clearing my Road.

soil of Persia wherever water is abundantly supplied. Tall, slender
poplars are nodding in the morning breeze, the less lofty almond
and pomegranate, sheltered from the breezes by the surrounding
building, rustle never a leaf, but seem to be offering Pomona's choice
products of nuts and rosy pomegranates, with modest mien and
silence ; whilst beds of rare exotics, peculiar to this sunny clime,
imparts to the atmosphere of the cool shaded garden, a pleasing
sense of being perfumed. Here, by means of the Shah's interpreter,
I am introduced to Nasr-i-Mulk, the Persian foreign minister, a
kindly-faced yet business-looking old gentleman, at whose request
I mount and ride with some difficulty around the confined and
quite unsuitable foot-walks of the garden ; a crowd of officials and
farrashes look on in unconcealed wonder and delight. True to their
Persian characteristic of inquisitiveness, Nasr-i-Mulk and the officers
catechise me unmercifully for some time concerning the mechanism
and capabilities of the bicycle, and about the past and future of the
journey around the world.

   In company with the interpreter, I now ride out to the Doshan
Tepe gate, where we are to await the arrival of the Shah. From the
Doshan Tepe gate is some four English miles of fairly good artificial road, leading to one of the royal summer palaces and gardens.
His Majesty goes this morning to the mountains beyond Doshan
Tepe on a shooting excursion, and wishes me to ride out with his
party a few miles, thus giving him a good opportunity of seeing
something of what bicycle travelling is like. The tardy monarch
keeps myself and a large crowd of attendants waiting a full hour at
the gate, ere he puts in an appearance. Among the crowd is the
Shah's chief *shikaree* (hunter), a grizzled old veteran, beneath whose
rifle many a forest prowler of the Caspian slope of Mazanderan has
been laid low. The *shikaree*, upon seeing me ride, and not being
able to comprehend how one can possibly maintain the equilibrium,
exclaims : " Oh, ayab Ingilis ! " (Oh, the wonderful English !)

   Everybody's face is wreathed in smiles at the old *shikaree's* ex-
clamation of wonderment, and when I jokingly advise him that he
ought to do his hunting for the future on a bicycle, and again mount
and ride with hands off handles to demonstrate the possibility of
shooting from the saddle, the delighted crowd of horsemen burst
out in hearty laughter, many of them exclaiming, " Bravo ! bravo ! "
At length the word goes round that the Shah is coming. Everybody dismounts, and as the royal carriage drives up, every Persian

bows his head nearly to the ground, remaining in that highly submissive attitude until the carriage halts and the Shah summons myself and the interpreter to his side.

I am the only Ferenghi in the party, my two English companions
having returned to the city, intending to rejoin me when I separate
from the Shah.

The Shah impresses one as being more intelligent than the
average Persian of the higher class ; and although they are, as a

The Shah Escorts Me to Dohan Tepe.

nation, inordinately inquisitive, no Persian has taken a more lively
interest in the bicycle than His Majesty seems to take, as, through
his interpreter, he plys me with all manner of questions.   Among
other questions he asks if the Koords didn't molest me when coming
through Koordistan without an escort ; and upon hearing the story
of my adventure with the Koordish shepherds between Ovahjik and
Khoi, he seems greatly amused.   Another large party of horsemen
arrived with the Shah, swelling the company to perhaps two hundred attendants.

Pedaling alongside the carriage, in the best position for the Shah to see, we proceed toward Doshan Tepe, the crowd of horsemen following, some behind and others careering over the stony plain through which the Doshan Tepe highway leads. After covering about half a mile, the Shah leaves the carriage and mounts a saddle-horse, in order to the better "put me through some exercises." First he requests me to give him an exhibition of speed; then I have to ride a short distance over the rough stone-strewn plain, to demonstrate the possibility of traversing a rough country, after which he desires to see me ride at the slowest pace possible. All this evidently interests him not a little, and he seems even more amused than interested, laughing quite heartily several times as he rides alongside the bicycle. After awhile he again exchanges for the carriage, and at four miles from the city gate we arrive at the palace garden. Through this garden is a long, smooth walk, and here the Shah again requests an exhibition of my speeding abilities. The garden is traversed with a network of irrigating ditches; but I am assured there is nothing of the kind across the pathway along which he wishes me to ride as fast as possible. Two hundred yards from the spot where this solemn assurance is given, it is only by a lightning-like dismount that I avoid running into the very thing that I was assured did not exist—it was the narrowest possible escape from what might have proved a serious accident.

Riding back toward the advancing party, I point out my good fortune in escaping the tumble. The Shah asks if people ever hurt themselves by falling off bicycles; and the answer that a fall such as I would have experienced by running full speed into the irrigating ditch, might possibly result in broken bones, appeared to strike him as extremely humorous; from the way he laughed I fancy the sending me flying toward the irrigating ditch was one of the practical jokes that he is sometimes not above indulging in. After mounting and forcing my way for a few yards through deep, loose gravel, to satisfy his curiosity as to what could be done in loose ground, I trundle along with him to a small menagerie he keeps at this place. On the way he inquires about the number of wheelmen there are in England and America; whether I am English or American; why they don't use iron tires on bicycles instead of rubber, and many other questions, proving the great interest aroused in him by the advent of the first bicycle to appear in his Capital. The menagerie consists of one cage of monkeys, about a dozen lions, and two or three tigers and leopards. We pass along from cage to cage, and as the keeper coaxes the animals to the bars, the Shah amuses himself by poking them with an umbrella. It was arranged in the original programme that I should accompany them up into their rendezvous in the foot-hills, about a mile beyond the palace, to take breakfast with the party ; but seeing the difficulty of getting up there with the bicycle, and not caring to spoil the favor-

The Shah shows me his Menagerie.

able impression already made, by having to trundle up, I ask per-mission to take my leave at this point. The request is granted, and the interpreter returns with me to the city—thus ends my memor-able bicycle ride with the Shah of Persia.

Soon after my ride with the Shah, the Naib-i-Sultan, the Gov-ernor of Teheran and commander-in-chief of the army, asked me to bring the bicycle down to the military *maidan*, and ride for the edification of himself and officers. Being busy at something or other when the invitation was received, I excused myself and requested that he make another appointment.

I am in the habit of taking a constitutional spin every morning ; by means of which I have figured as an object of interest, and have been stared at in blank amazement by full half the wonder-stricken population of the city. The fame of my journey, the knowledge of my appearance before the Shah, and my frequent appearance upon the streets, has had the effect of making me one of the most conspicuous characters in the Persian Capital ; and the people have bestowed upon me the expressive and distinguishing title of "the asp-i-awhan Sahib" (horse of-iron Sahib).

A few mornings after receiving the Naib-i-Sultan's invitation, I happened to be wheeling past the military maidan, and attracted by the sound of martial music inside, determined to wheel in and investigate. Perhaps in all the world there is no finer military parade ground than in Teheran ; it consists of something over one hundred acres of perfectly level ground, forming a square that is walled completely in by alcoved walls and barracks, with gaily painted bala-khanas over the gates. The delighted guards at the gate make way and present arms, as they see me approaching ; wheeling inside, I am somewhat taken aback at finding a general review of the whole Teheran garrison in progress ; about ten thousand men are manœuvring in squads, companies, and regiments over the ground.

Having, from previous experience on smaller occasions, discovered that my appearance on the incomprehensible "asp-i-awhan" would be pretty certain to temporarily demoralize the troops and create general disorder and inattention, I am for a moment undetermined about whether to advance or retreat. The acclamations of delight and approval from the nearest troopers at seeing me enter the gate, however, determines me to advance ; and I start off at a rattling pace around the square, and then take a zig-zag course through the manœuvring bodies of men.

The sharp-shooters lying prostrate in the dust, mechanically rise up to gaze ; forgetting their discipline, squares of soldiers change into confused companies of inattentive men ; simultaneous confusion takes place in straight lines of marching troops, and the music of the bands degenerates into inharmonious toots and discordant squeaks, from the inattention of the musicians. All along the line the signal runs—not " every Persian is expected to do his duty," but "the *asp-i-awhan* Sahib! the *asp-i-awhan* Sahib!" the whole army is in direful commotion. In the midst of the general confusion, up dashes an orderly, who requests that I accompany him to the presence of the Commander-in-Chief and staff; which, of course, I readily do, though not without certain misgivings as to my probable reception under the circumstances. There is no occasion for misgivings, however; the Naib-i-Sultan, instead of being displeased at the interruption to the review, is as delighted at the appearance of "the *asp-i-auhan*, as is Abdul, the drummer-boy, and he has sent for me to obtain a closer acquaintance. After riding for their edification, and answering their multifarious questions, I suggest to the Commander-in-Chief that he ought to mount the Shah's favorite regiment of Cossacks on bicycles. The suggestion causes a general laugh among the company, and he replies: "Yes, *asp-i-awhan* Cossacks would look very splendid on our dress parade here in the *maidan*; but for scouting over our rough Persian mountains"—and the Naib-i-Sultan finished the sentence with a laugh and a negative shrug of his shoulders.

Two mornings after this I take a spin out on the Doshan Tepe road, and, upon wheeling through the city gate, I find myself in the immediate presence of another grand review, again under the personal inspection of the Naib-i-Sultan. Disturbing two grand reviews within two days is, of course, more than I bargained for, and I would gladly have retreated through the gate; but coming full upon them unexpectedly, I find it impossible to prevent the inevitable result. The troops are drawn up in line about fifty yards from the road, and are for the moment standing at ease, awaiting the arrival of the Shah, while the Commander-in-chief and his staff are indulging in soothing whiffs at the seductive kalian. The cry of "*asp-i-awhan* Sahib!" breaks out all along the line, and scores of soldiers break ranks, and come running helter-skelter toward the road, regardless of the line-officers, who frantically endeavor to wave them back. Dashing ahead, I am soon beyond the lines, congratulating myself that the effects of my disturbing presence is quickly over; but ere long, I discover that there is no other ridable road back, and am consequently compelled to pass before them again on returning. Accordingly, I hasten to return, before the arrival of the Shah. Seeing me returning, the Naib-i-Sultan and his staff advance to the road, with kalians in hand, their oval faces wreathed in smiles of approbation; they extend cordial salu-

34

tations as I wheel past. The Persians seem to do little more than play at soldiering ; perhaps in no other army in the world could a lone cycler demoralize a general review twice within two days, and then be greeted with approving smiles and cordial salutations by the commander-in-chief and his entire staff.

Through November and the early part of December, the weather in Teheran continues, on the whole, quite agreeable, and suitable for short-distance wheeling ; but mindful of the long distance yet before me, and the uncertainty of touching at any point where supplies could be forwarded, I deem it advisable to take my exercise afoot, and save my rubber tires for the more serious work of the journey to the Pacific.

There are no green lanes down which to stroll, nor emerald meads through which to wander about the Persian capital, though what green things there are, retain much of their greenness until the early winter months. The fact of the existence of any green thing whatever—and even to a greater extent, its survival through the scorching summer months—depending almost wholly on irrigation, enables vegetation to retain its pristine freshness almost until suddenly pounced upon and surprised by the frost. There is no springy turf, no velvety greensward in the land of the Lion and the Sun. No sooner does one get beyond the vegetation, called into existence by the moisture of an irrigating ditch or a stream, than the bare, gray surface of the desert crunches beneath one's tread. There is an avenue leading part way from the city to the summer residence of the English Minister at Gulaek, that conjures up memories of an English lane ; but the double row of chenars, poplars, and jujubes are kept alive by irrigation, and all outside is verdureless desert.

Things are valued everywhere for their scarcity, and a patch of greensward large enough to recline on, a shady tree or shrub, and a rippling rivulet are appreciated in Persia at their proper value—appreciated more than broad, green pastures and waving groves of shade-trees in moister climes. Moreover, there is a peculiar charm in these bright emerald gems, set in sombre gray, be they never so small and insignificant in themselves, that is not to be experienced where the contrast is less marked.

Scattered here and there about the stony plain between Teheran and the Elburz foot-hills, are many beautiful gardens—beautiful for Persia—where a pleasant hour can be spent wandering beneath

The Naib-i-Sultan Smiles Approvingly.

the shady avenues and among the fountains. These gardens are simply patches redeemed from the desert plain, supplied with irrigating water, and surrounded with a high mud wall; leading through the garden are gravelled walks, shaded by rows of graceful chenars. The gardens are planted with fig, pomegranate, almond or apricot trees, grape-vines, melons, etc.; they are the property of wealthy Teheranis who derive an income from the sale of the fruit in the Teheran market. The ample space within the city ramparts includes a number of these delightful retreats, some of them presenting the additional charm of historic interest, from having been the property and, peradventure, the favorite summer residence of a former king. Such a one is an extensive garden in the northeast quarter of the city, in which was situated one of the favorite summer palaces of Fatteh-ali Shah, grandfather of Nasr-e-deen.

It was chiefly to satisfy my curiosity as to the truth of the current stories regarding that merry monarch, and his exceedingly novel methods of entertaining himself, that I accepted the invitation of a friend to visit this garden one afternoon. My friend is the owner of a pair of white bull-dogs, who accompany us into the garden. After strolling about a little, we are shown into the summer palace; into the audience room, where we are astonished at the beautiful coloring and marvellously life-like representations in the old Persian frescoing on the walls and ceiling. Depicted in life-size are Fatteh-ali Shah and his courtiers, together with the European ambassadors, painted in the days when the Persian court was a scene of dazzling splendor. The monarch is portrayed as an exceedingly handsome man with a full, black beard, and is covered with a blaze of jewels that are so faithfully pictured as to appear almost like real gems on the walls. It seems strange—almost startling—to come in from contemplating the bare, unlovely mud walls of the city, and find one's self amid the life-like scenes of Fatteh-ali Shah's court; and, amid the scenes to find here and there an English face, an English figure, dressed in the triangular cockade, the long Hessian pigtail, the scarlet coat with fold-back tails, the knee-breeches, the yellow stockings, the low shoes, and the long, slender rapier of a George III. courtier. From here we visit other rooms, glittering rooms, all mirror-work and white stucco. Into rooms we go whose walls consist of myriads of tiny squares of rich stained glass, worked into intricate patterns and geometrical designs, but which are now rapidly falling into decay; and then we go to see the most novel feature of the garden—Fatteh-ali Shah's marble slide, or shute.

Passing along a sloping, arched vault beneath a roof of massive marble, we find ourselves in a small, subterranean court, through which a stream of pure spring water is flowing along a white marble channel, and where the atmosphere must be refreshingly cool even in the middle of summer. In the centre of the little court is a round tank about four feet deep, also of white marble, which can be filled at pleasure with water, clear as crystal, from the running stream. Leading from an upper chamber, and overlapping the tank, is a smooth-worn marble slide or shute, about twenty feet long and four broad, which is pitched at an angle that makes it imperative upon any one trusting themselves to attempt the descent, to slide helplessly into the tank. Here, on summer afternoons, with the chastened daylight peeping through a stained-glass window in the roof, and carpeting the white marble floor with rainbow hues, with the only entrance to the cool and massive marble court, guarded by armed retainers, who while guarding it were conscious of guarding their own precious lives, Fatteh-ali Shah was wont to beguile the hours away by making merry with the bewitching nymphs of his *anderoon*, transforming them for the nonce into naiads.

There are no nymphs nor naiads here now, nothing but the smoothly-worn marble shute to tell the tale of the merry past; but we obtain a realistic idea of their sportive games by taking the bull-dogs to the upper chamber, and giving them a start down the slide. As they clutch and claw, and look scared, and appeal mutely for assistance, only to slide gradually down, down, down, and fall with a splash into the tank at last, we have only to imagine the bull-dogs transformed into Fatteh-ali Shah's naiads, to learn something of the truth of current stories. After we have slid the dogs down a few times, and they begin to realize that they are not sliding hopelessly down to destruction, they enjoy the sport as much as we, or as much as the naiads perhaps did a hundred years ago.

That portion of the Teheran bazaar immediately behind the Shah's winter palace, is visited almost daily by Europeans, and their presence excites little comment or attention from the natives; but I had frequently heard the remark that a Ferenghi couldn't walk through the southern, or more exclusive native quarters, without being insulted. Determined to investigate, I

sallied forth one afternoon alone, entering the bazaar on the east
side of the palace wall, where I had entered it a dozen times before.

The streets outside are sloppy with melting snow, and the
roofed passages of the bazaar, being dry underfoot, are crowded
with people to an unusual extent; albeit they are pretty well
crowded at any time. Most of the dervishes in the city have been
driven, by the inclemency of the weather, to seek shelter in the
bazaar; these, added to the no small number who make the place
their regular foraging ground, render them a greater nuisance than
ever. They are encountered in such numbers, that no matter
which way I turn, I am confronted by a rag-bedecked mendicant,
with a wild, haggard countenance and grotesque costume, thrust-
ing out his gourd alms-receiver, and muttering "huk yah huk!"
each in his own peculiar way.

The mollahs, with their flowing robes, and huge white turbans,
likewise form no inconsiderable proportion of the moving throng;
they are almost without exception scrupulously neat and clean in
appearance, and their priestly costume and Pharisaical deportment
gives them a certain air of stateliness. They wear the placid expression of men so utterly puffed up with the notion of their own
sanctity, that their self-consciousness verily seems to shine through
their skins, and to impart to them a sleek, oily appearance. One
finds himself involuntarily speculating on how they all manage to
make a living; the mollah "toils not, neither does he spin," and
almost every other person one meets is a mollah.

The bazaar is a common thoroughfare for anything and everything that can make its way through. Donkey-riders, horsemen,
and long strings of camels and pack-mules add their disturbing influence to the general confusion; and although hundreds of stalls
are heaped up with every merchantable thing in the city, scores of
donkeys laden with similar products are meandering about among
the crowd, the venders shouting their wares with lusty lungs. In
many places the din is quite deafening, and the odors anything but
agreeable to European nostrils; but the natives are not over fastidious. The steam issuing from the cook-shops, from coppers
of soup, pillau and sheeps'-trotters, and the less objectionable odors
from places where busy men are roasting bazaar-kabobs for hungry customers all day long, mingle with the aromatic contributions from the spice and tobacco shops wedged in between them.

The sleek-looking spice merchant, squatting contentedly beside a
pan of glowing embers, smoking kalian after kalian in dreamy contemplation of his assistant waiting on customers, and also occasionally waiting on him to the extent of replenishing the fire on
the kalian, is undoubtedly the happiest of mortals. With a kabob-
shop on one hand, a sheeps'-trotter-shop on the other, and a
bakery and a fruit-stand opposite, he indulges in tid-bits from
either when he is hungry. With nothing to do but smoke kalians
amid the fragrant aroma of his own spices, and keep a dreamy eye
on what passes on around him, his Persian notions of a desirable
life cause him to regard himself as blest beyond comparison with
those whose avocations necessitate physical exertion. All the
shops are open front places, like small fruit and cigar stands in an
American city, the goods being arranged on boards or shelving,
sloping down to the front, or otherwise exposed to the best advantage, according to the nature of the wares ; the shops have no windows, but are protected at night by wooden shutters.

The piping notes of the flute, or the sing-song voice of the troubadour or story-teller is heard behind the screened entrance of the
tchai-khans, and now and then one happens across groups of angry
men quarrelling violently over some trifling difference in a bargain ;
noise and confusion everywhere reign supreme. Here the road is
blocked up by a crowd of idlers watching a trio of *lutis*, or buffoons,
jerking a careless and indifferent-looking baboon about with a chain
to make him dance ; and a little farther along is another crowd surveying some more *lutis* with a small brown bear. Both the baboon
and the bear look better fed than their owners, the contributions
of the onlookers consisting chiefly of eatables, bestowed upon the
animals for the purpose of seeing them feed.

Half a mile, or thereabouts, from the entrance, an inferior
quarter of the bazaar is reached ; the crowds are less dense, the
noise is not near so deafening, and the character of the shops undergoes a change for the worse. A good many of the shops are
untenanted, and a good many others are occupied by artisans manufacturing the ruder articles of commerce, such as horseshoes, pack-
saddles, and the trappings of camels. Such articles as kalians, *che-
boüks* and other pipes, *geivehs*, slippers and leather shoes, hats,
jewelry, etc., are generally manufactured on the premises in the
better portions of the bazaar, where they are sold. Perched in
among the rude cells of industry are cook-shops and tea-drinking

establishments of an inferior grade ; and the occupants of these places eye me curiously, and call one another's attention to the unusual circumstance of a Ferenghi passing through their quarter. After half a mile of this, my progress is abruptly terminated by a high mud wall, with a narrow passage leading to the right. I am now at the southern extremity of the bazaar, and turn to retrace my footsteps.

So far I have encountered no particular disposition to insult anybody ; only a little additional rudeness and simple inquisitiveness, such as might very naturally have been expected. But ere I have retraced my way three hundred yards, I meet a couple of rowdyish young men of the *charvadar* class ; no sooner have I passed them than one of them wantonly delivers himself of the promised insult—a peculiar noise with the mouth ; they both start off at a run as though expecting to be pursued and punished. As I turn partially round to look, an old pomegranate vender stops his donkey, and with a broad grin of amusement motions me to give chase. When nearing the more respectable quarter again, I stroll up one of the numerous ramifications leading toward what looks like a particularly rough and dingy quarter. Before going many steps I am halted by a friendly-faced sugar merchant, with "*Sahib*," and sundry significant shakes of the head, signifying, if he were me, he wouldn't go up there. And thus it is in the Teheran bazaar ; where a Ferenghi will get insulted once, he will find a dozen ready to interpose with friendly officiousness between him and anything likely to lead to unpleasant consequences. On the whole, a European fares better than a Persian in his national costume would in an Occidental city, in spite of the difference between our excellent police regulations and next to no regulations at all ; he fares better than a Chinaman does in New York.

The Teheran bazaar, though nothing to compare to the world-famous bazaar at Stamboul, is wonderfully extensive. I was under the impression that I had been pretty much all through it at different times ; but a few days after my visit to the "slummy" quarters, I follow a party of corpse-bearers down a passage-way hitherto unexplored, to try and be present at a Persian funeral, and they led the way past at least a mile of shops I had never yet seen. I followed the corpse-bearers through the dark passages and narrow alley-ways of the poorer native quarter, and in spite of the lowering brows of the followers, penetrated even into the house

The Old Pomegranate Vender wants Me to give Chase.

where they washed the corpses before burial; but here the officiating mollahs scowled with such unmistakable displeasure, and refused to proceed in my presence, so that I am forced to beat a retreat. The poorer native quarter of Teheran is a shapeless jumble of mud dwellings, and ruins of the same; the streets are narrow passages describing all manner of crooks and angles in and out among them.

As I emerge from the vaulted bazaar the sun is almost setting, and the musicians in the *bala-khanas* of the palace gates are ushering in the close of another day with discordant blasts from ancient Persian trumpets, and belaboring hemispherical kettle-drums. These musicians are dressed in fantastic scarlet uniforms, not unlike the costume of a fifteen century jester, and every evening at sundown they repair to these *bala-khanas,* and for the space of an hour dispense the most unearthly music imaginable. The trumpets are sounding-tubes of brass about five feet long, which respond to the efforts of a strong-winded person, with a diabolical basso-profundo shriek that puts a Newfoundland fog-horn entirely in the shade. When a dozen of these instruments are in full blast, without any attempt at harmony, it seems to shed a depressing shadow of barbarism over the whole city. This sunset music is, I think, a relic of very old times, and it jars on the nerves like the despairing howl of ancient Persia, protesting against the innovation from the pomp and din and glamour of her old pagan glories, to the present miserable era of mollah rule and feeble dependence for national existence on the forbearance or jealousy of other nations. Beneath the musicians' gate, and I emerge into a small square which is half taken up by a square tank of water; near the tank is a large bronze cannon. It is a huge, unwieldy piece, and a muzzle-loader, utterly useless to such a people as the Persians, except for ornament, or perhaps to help impress the masses with an idea of the Shah's unapproachable greatness.

It is the special hour of prayer, and in every direction may be observed men, halting in whatever they may be doing, and kneeling down on some outer garment taken off for the purpose, repeatedly touch their foreheads to the ground, bending in the direction of Mecca. Passing beneath the second musicians' gate, I reach the artillery square just in time to see a company of army buglers formed in line at one end, and a company of musketeers at the other. As these more modern trumpeters proceed to toot,

the company of musketeers opposite present arms, and then the music of the new buglers, and the hoarse, fog-horn-like blasts of the fantastic tooters on the *bala-khanas* dies away together in a concerted effort that would do credit to a troop of wild elephants.

When the noisy trumpeting ceases, the ordinary noises round about seem like solemn silence in comparison, and above this comparative silence can be heard the voices of men here and there over the city, calling out "Al-lah-il-All-ah; Ali Ak-bar!" (God is greatest; there is no god but one God! etc.) with stentorian voices. The men are perched on the roofs of the mosques, and on noblemen's walls and houses; the Shah has a strong-voiced *muezzin* that can be heard above all the others.

The sun has just set; I can see the snowy cone of Mount Demavend, peeping apparently over the high barrack walls; it has just taken on a distinctive roseate tint, as it oftentimes does at sunset; the reason whereof becomes at once apparent upon turning toward the west, for the whole western sky is aglow with a gorgeous sunset—a sunset that paints the horizon a blood red, and spreads a warm, rich glow over half the heavens.

The moon will be full to-night, and a far lovelier picture even than the glorious sunset and the rose-tinted mountain, awaits anyone curious enough to come out-doors and look. The Persian moonlight seems capable of surrounding the most commonplace objects with a halo of beauty, and of blending things that are nothing in themselves, into scenes of such transcendental loveliness that the mere casual contemplation of them sends a thrill of pleasure coursing through the system. There is no city of the same size (180,-000) in England or America, but can boast of buildings infinitely superior to anything in Teheran; what trees there are in and about the city are nothing compared to what we are used to having about us; and although the gates with their short minars and their gaudy facings are certainly unique, they suffer greatly from a close investigation. Nevertheless, persons happening for the first time in the vicinity of one of these gates on a calm moonlight night, and perchance descrying "fair Luna" through one of the arches or between the minars, will most likely find themselves transfixed with astonishment at the marvellous beauty of the scene presented. By repairing to the artillery square, or to the short street between the square and the palace front, on a moonlight night, one

can experience a new sense of nature's loveliness ; the soft, chas-
tening light of the Persian moon converts the gaudy gates, the
dead mud-walls, the spraggling trees, and the background of snowy
mountains nine miles away, into a picture that will photograph
itself on one's memory forever.

On the way home I meet one of the lady missionaries—which
reminds me that I ought to mention something about the peculiar
position of a Ferenghi lady in these Mohammedan countries, where
it is considered highly improper for a woman to expose her face in
public. The Persian lady on the streets is enveloped in a shroud-
like garment that transforms her into a shapeless and ungraceful-
looking bundle of dark-blue cotton stuff. This garment covers
head and everything except the face ; over the face is worn a white
veil of ordinary sheeting, and opposite the eyes is inserted an oblong peep-hole of open needle-work, resembling a piece of per-
forated card-board. Not even a glimpse of the eye is visible,
unless the lady happens to be handsome and coquettishly inclined ;
she will then manage to grant you a momentary peep at her face ;
but a wise and discreet Persian lady wouldn't let you see her face
on the street—no, not for worlds and worlds !

The European lady with her uncovered face is a conundrum
and an object of intense curiosity, even in Teheran at the present
day ; and in provincial cities, the wife of the lone consul or telegraph employé finds it highly convenient to adopt the native costume, face-covering included, when venturing abroad. Here, in
the capital, the wives and daughters of foreign ministers, European officers and telegraphists, have made uncovered female faces
tolerably familiar to the natives ; but they cannot quite understand but that there is something highly indecorous about it, and
the more unenlightened Persians doubtless regard them as quite
bold and forward creatures. Armenian women conceal their faces
almost as completely as do the Persian, when they walk abroad ;
by so doing they avoid unpleasant criticism, and the rude, inquisitive gaze of the Persian men. Although the Persian readily recognizes the fact that a Sahib's wife or sister must be a superior person
to an Armenian female, she is as much an object of interest to him
when she appears with her face uncovered on the street, as his own
wives in their highly sensational in-door costumes would be to
some of us. In order to establish herself in the estimation of the
average Persian, as all that a woman ought to be, the European

lady would have to conceal her face and cover her shapely, tight-
fitting dress with an inelegant, loose mantle, whenever she ventured outside her own doors.

With something of a *penchant* for undertaking things never
before accomplished, I proposed one morning to take a walk
around the ramparts that encompass the Persian capital. The
question arose as to the distance. Ali Akbar, the head *farrash*,
said it was six farsakhs (about twenty-four miles); Meshedi Abdul said it was more. From the well-known Persian characteristic
of exaggerating things, we concluded from this that perhaps it
might be fifteen miles; and on this basis Mr. Meyrick, of the
Indo-European Telegraph staff, agreed to bear me company. The
ramparts consist of the earth excavated from a ditch some forty
feet wide by twenty deep, banked up on the inner side of the
ditch; and on top of this bank it is our purpose to encompass the
city.

Eight o'clock on the appointed morning finds us on the ramparts at the Gulaek Gate, on the north side of the city. A cold
breeze is blowing off the snowy mountains to the northeast, and we
decide to commence our novel walk toward the west. Following
the zigzag configuration of the ramparts, we find it at first somewhat rough and stony to the feet; on our right we look down into
the broad ditch, and beyond, over the sloping plain, our eyes follow the long, even rows of kanaat mounds stretching away to the
rolling foot-hills; towering skyward in the background, but eight
miles away, are the snowy masses of the Elburz Range. Forty
miles away, at our back, the conical peak of Demavend peeps,
white, spectral, and cold, above a bank of snow-clouds that are
piled motionless against its giant sides, as though walling it completely off from the lower world. On our left lies the city, a curious
conglomeration of dead mud-walls, flat-roofed houses, and poplar-
peopled gardens. A thin haze of smoke hovers immediately above
the streets, through which are visible the minarets and domes of
the mosques, the square, illumined towers of the Shah's *anderoon*, the
monster skeleton dome of the canvas theatre, beneath which the
Shah gives once a year the royal *tazzia* (representation of the tragedy
of "Hussein and Hassan"), and the tall chimney of the arsenal,
from which a column of black smoke is issuing. Away in the distance, far beyond the confines of the city, to the southward, glittering like a mirror in the morning sun, is seen the dome of the great

mosque at Shahabdullahzeen, said to be roofed with plates of pure gold.

As we pass by we can see inside the walls of the English Legation grounds; a magnificent garden of shady avenues, asphalt walks, and dark-green banks of English ivy that trail over the ground and climb half-way up the trunks of the trees. A square-turreted clock-tower and a building that resembles some old ancestral manor, imparts to "the finest piece of property in Teheran" a home-like appearance; the representative of Her Majesty's Government, separated from the outer world by a twenty-four-foot brick wall, might well imagine himself within an hour's ride of London.

Beyond the third gate, the character of the soil changes from the stone-strewn gravel of the northern side, to red stoneless earth, and both inside and outside the ramparts fields of winter wheat and hardy vegetables form a refreshing relief from the barren character of the surface generally. The Ispahan gate, on the southern side, appears the busiest and most important entrance to the city; by this gate enter the caravans from Bushire, bringing English goods, from Bagdad, Ispahan, Yezd, and all the cities of the southern provinces. Numbers of caravans are camped in the vicinity of the gate, completing their arrangements for entering the city or departing for some distant commercial centre; many of the waiting camels are kneeling beneath their heavy loads and quietly feeding. They are kneeling in small, compact circles, a dozen camels in a circle with their heads facing inward. In the centre is placed a pile of chopped straw; as each camel ducks his head and takes a mouthful, and then elevates his head again while munching it with great gusto, wearing meanwhile an expression of intense satisfaction mingled with timidity, as though he thinks the enjoyment too good to last long, they look as cosey and fussy as a gathering of Puritanical grand-dames drinking tea and gossiping over the latest news.

Within a mile of the Ispahan gate are two other gates, and between them is an area devoted entirely to the brick-making industry. Here among the clay-pits and abandoned kilns we obtain a momentary glimpse of a jackal, drinking from a ditch. He slinks off out of sight among the caves and ruins, as though conscious of acting an ungenerous part in seeking his living in a city already full of gaunt, half-starved pariahs, who pass their lives in wandering listlessly and hungrily about for stray morsels of offal. Several of these pariahs have been so unfortunate as to get down into the rampart ditch ; we can see the places where they have repeatedly made frantic rushes for liberty up the almost per-pendicular escarp, only to fall helplessly back to the bottom of their roofless dungeon, where they will gradually starve to death. The natives down in this part of the city greet us with curious looks ; they are wondering at the sight of two Ferenghis prome-nading the ramparts, far away from the European quarter ; we can hear them making remarks to that effect, and calling one another's attention. The sun gets warm, although it is January, as we pass the Doshan Tepe and the Meshed gates, remarking as we go past that the Shah's summer palace on the hill to the east compares favorably in whiteness with the snow on the neighboring moun-tains. As we again reach the Gulaek gate and descend from the ramparts at the place we started, the clock in the English Lega-tion tower strikes twelve.

" How many miles do you call it ? " asks my companion.

"Just about twelve miles," I reply ; " what do you make it ? "

" That's about it," he agrees ; " twelve miles round, and eleven gates. We have walked or climbed over the archway of eight of the gates ; and at the other three we had to climb off the ramparts and on again."

As far as can be learned, this is the first time any Ferenghi has walked clear around the ramparts of Teheran. It is nothing worth boasting about ; only a little tramp of a dozen miles, and there is little of anything new to be seen. All around the out-side is the level plain, verdureless, except an occasional cultivated field, and the orchards of the tributary villages scattered here and there.

In certain quarters of Teheran one happens across a few re-maining families of *guebres*, or fire-worshippers ; remnant represen-tatives of the ancient Parsee religion, whose devotees bestowed their strange devotional offerings upon the fires whose devouring flames they constantly fed, and never allowed to be extinguished. These people are interesting as having kept their heads above the over-whelming flood of Mohammedanism that swept over their country, and clung to their ancient belief through thick and thin—or, at all events, to have steadfastly refused to embrace any other. Little evidence of their religion remains in Persia at the present day, except their " towers of silence " and the ruins of their old fire-temples. These latter were built chiefly of soft adobe bricks, and after the lapse of centuries, are nothing more than shapeless reminders of the past. A few miles southeast of Teheran, in a desolate, unfrequented spot, is the *guebre* " tower of silence, " where they dispose of their dead. On top of the tower is a kind of balcony with an open grated floor ; on this the naked corpses are placed until the carrion crows and the vultures pick the skeleton perfectly clean ; the dry bones are then cast into a common receptacle in the tower. The *guebre* communities of Persia are too impecunious or too indifferent to keep up the ever-burning-fires nowadays ; the fires of Zoroaster, which in olden and more prosperous times were fed with fuel night and day, are now extinguished forever, and the scattering survivors of this ancient form of worship form a unique item in the sum total of the population of Persia.

The head-quarters—if they can be said to have any head-quarters —of the Persian *guebres* are at Yezd, a city that is but little known to Europeans, and which is all but isolated from the remainder of the country by the great central desert. One great result of this geographical isolation is to be observed to-day, in the fact that the *guebres* of Yezd held their own against the unsparing sword of Islam better than they did in more accessible quarters ; consequently they are found in greater numbers there now than in other Persian cities. Curiously enough, the chief occupation—one might say the sole occupation—of the *guebres* throughout Persia, is taking care of the suburban gardens and premises of wealthy people. For this purpose I am told *guebre* families are in such demand, that if they were sufficiently numerous to go around, there would be scarcely a piece of valuable garden property in all Persia without a family of *guebres* in charge of it. They are said to be far more honest and trustworthy than the Persians, who, as Shiite Mohammedans, consider themselves the holiest people on earth ; or the Armenians, who hug the flattering unction of being Christians and not Mohammedans to their souls, and expect all Christendom to regard them benignly on that account. It is doubtless owing to this invaluable trait of their character, that the *guebres* have naturally drifted to their level of guardians over the private property of their wealthy neighbors.

The costume of the *guebre* female consists of Turkish trousers

Ayoob Khan and his Attendant.

with very loose, baggy legs, the material of which is usually calico print, and a mantle of similar material is wrapped about the head and body. Unlike her Mohammedan neighbor, she makes no pretence of concealing her features ; her face is usually a picture of pleasantness and good-nature rather than strikingly handsome or passively beautiful, as is the face of the Persian or Armenian belle.

The costume of the men differs but little from the ordinary costume of the lower-class Persians. Like all the people in these Mohammedan countries, who realize the weakness of their position as a small body among a fanatical population, the Teheran *guebres* have long been accustomed to consider themselves as under the protecting shadow of the English Legation ; whenever they meet a " *Sahib* " on the street, they seem to expect a nod of recognition.

Among the people who awaken special interest in Europeans here, may be mentioned Ayoob Khan, and his little retinue of attendants, who may be seen on the streets almost any day. Ayoob Khan is in exile here at Teheran in accordance with some mutual arrangement between the English and Persian governments. On almost any afternoon, about four o'clock, he may be met with riding a fine, large chestnut stallion, accompanied by another Afghan on an iron gray. I have never seen them riding faster than a walk, and they are almost always accompanied by four foot-runners, also Afghans, two of whom walk behind their chieftain and two before. These runners carry stout staves with which to warn off mendicants, and with a view to making it uncomfortable for any irrepressible Persian rowdy who should offer any insults. Both Ayoob Khan and his attendants retain their national costume, the main distinguishing features being a huge turban with about two feet of the broad band left dangling down behind ; besides this, they wear white cotton pantalettes even in mid-winter. They wear European shoes and overcoats, as though they had profited by their intercourse with Anglo-Indians to the extent of at least shoes and coat. The foot-runners have their legs below the knee bound tightly with strips of dark felt. Judging from outward appearances, Ayoob Khan wears his exile lightly, for his rotund countenance looks pleasant always, and I have never yet met him when he was not chatting gayly with his companion.

Of the interesting scenes and characters to be seen every day on the streets of Teheran, their name is legion. The peregrinating *tchai*-venders, who, with their little cabinet of tea and sugar in one hand, and samovar with live charcoals in the other, wander about the city picking up stray customers, for whom they are prepared to make a glass of hot tea at one minute's notice ; the scores of weird-looking mendicants and dervishes with their highly fantastic costumes, assailing you with " huk, yah huk ! " the barbers shaving the heads of their customers on the public streets—shaving their pates clean, save little tufts to enable Mohammed to pull them up to Paradise ; and many others the description and enumeration of which would, of themselves, fill a good-sized volume.

www.ingramcontent.com/pod-product-compliance
Lightning Source LLC
Chambersburg PA
CBHW021907211225
37087CB00073B/1090